LONDON MATHEMATICAL SOCIETY LECTURE NOTE SERIES

The titles below are available from booksellers, or from Cambridge University Press at www.cambridge.org/mathematics

400 Dense sphere packings: a blueprint for formal proofs, T. HALES
401 A double Hall algebra approach to affine quantum Schur–Weyl theory, B. DENG, J. DU & Q. FU
402 Mathematical aspects of fluid mechanics, J.C. ROBINSON, J.L. RODRIGO & W. SADOWSKI (eds)
403 Foundations of computational mathematics, Budapest 2011, F. CUCKER, T. KRICK, A. PINKUS & A. SZANTO (eds)
404 Operator methods for boundary value problems, S. HASSI, H.S.V. DE SNOO & F.H. SZAFRANIEC (eds)
405 Torsors, étale homotopy and applications to rational points, A.N. SKOROBOGATOV (ed)
406 Appalachian set theory, J. CUMMINGS & E. SCHIMMERLING (eds)
407 The maximal subgroups of the low-dimensional finite classical groups, J.N. BRAY, D.F. HOLT & C.M. RONEY-DOUGAL
408 Complexity science: the Warwick master's course, R. BALL, V. KOLOKOLTSOV & R.S. MACKAY (eds)
409 Surveys in combinatorics 2013, S.R. BLACKBURN, S. GERKE & M. WILDON (eds)
410 Representation theory and harmonic analysis of wreath products of finite groups, T. CECCHERINI-SILBERSTEIN, F. SCARABOTTI & F. TOLLI
411 Moduli spaces, L. BRAMBILA-PAZ, O. GARCÍA-PRADA, P. NEWSTEAD & R.P. THOMAS (eds)
412 Automorphisms and equivalence relations in topological dynamics, D.B. ELLIS & R. ELLIS
413 Optimal transportation, Y. OLLIVIER, H. PAJOT & C. VILLANI (eds)
414 Automorphic forms and Galois representations I, F. DIAMOND, P.L. KASSAEI & M. KIM (eds)
415 Automorphic forms and Galois representations II, F. DIAMOND, P.L. KASSAEI & M. KIM (eds)
416 Reversibility in dynamics and group theory, A.G. O'FARRELL & I. SHORT
417 Recent advances in algebraic geometry, C.D. HACON, M. MUSTAŢĂ & M. POPA (eds)
418 The Bloch–Kato conjecture for the Riemann zeta function, J. COATES, A. RAGHURAM, A. SAIKIA & R. SUJATHA (eds)
419 The Cauchy problem for non-Lipschitz semi-linear parabolic partial differential equations, J.C. MEYER & D.J. NEEDHAM
420 Arithmetic and geometry, L. DIEULEFAIT *et al* (eds)
421 O-minimality and Diophantine geometry, G.O. JONES & A.J. WILKIE (eds)
422 Groups St Andrews 2013, C.M. CAMPBELL *et al* (eds)
423 Inequalities for graph eigenvalues, Z. STANIĆ
424 Surveys in combinatorics 2015, A. CZUMAJ *et al* (eds)
425 Geometry, topology and dynamics in negative curvature, C.S. ARAVINDA, F.T. FARRELL & J.-F. LAFONT (eds)
426 Lectures on the theory of water waves, T. BRIDGES, M. GROVES & D. NICHOLLS (eds)
427 Recent advances in Hodge theory, M. KERR & G. PEARLSTEIN (eds)
428 Geometry in a Fréchet context, C.T.J. DODSON, G. GALANIS & E. VASSILIOU
429 Sheaves and functions modulo p, L. TAELMAN
430 Recent progress in the theory of the Euler and Navier–Stokes equations, J.C. ROBINSON, J.L. RODRIGO, W. SADOWSKI & A. VIDAL-LÓPEZ (eds)
431 Harmonic and subharmonic function theory on the real hyperbolic ball, M. STOLL
432 Topics in graph automorphisms and reconstruction (2nd Edition), J. LAURI & R. SCAPELLATO
433 Regular and irregular holonomic D-modules, M. KASHIWARA & P. SCHAPIRA
434 Analytic semigroups and semilinear initial boundary value problems (2nd Edition), K. TAIRA
435 Graded rings and graded Grothendieck groups, R. HAZRAT
436 Groups, graphs and random walks, T. CECCHERINI-SILBERSTEIN, M. SALVATORI & E. SAVA-HUSS (eds)
437 Dynamics and analytic number theory, D. BADZIAHIN, A. GORODNIK & N. PEYERIMHOFF (eds)
438 Random walks and heat kernels on graphs, M.T. BARLOW
439 Evolution equations, K. AMMARI & S. GERBI (eds)
440 Surveys in combinatorics 2017, A. CLAESSON *et al* (eds)
441 Polynomials and the mod 2 Steenrod algebra I, G. WALKER & R.M.W. WOOD
442 Polynomials and the mod 2 Steenrod algebra II, G. WALKER & R.M.W. WOOD
443 Asymptotic analysis in general relativity, T. DAUDÉ, D. HÄFNER & J.-P. NICOLAS (eds)
444 Geometric and cohomological group theory, P.H. KROPHOLLER, I.J. LEARY, C. MARTÍNEZ-PÉREZ & B.E.A. NUCINKIS (eds)
445 Introduction to hidden semi-Markov models, J. VAN DER HOEK & R.J. ELLIOTT
446 Advances in two-dimensional homotopy and combinatorial group theory, W. METZLER & S. ROSEBROCK (eds)

447 New directions in locally compact groups, P.-E. CAPRACE & N. MONOD (eds)

448 Synthetic differential topology, M.C. BUNGE, F. GAGO & A.M. SAN LUIS

449 Permutation groups and cartesian decompositions, C.E. PRAEGER & C. SCHNEIDER

450 Partial differential equations arising from physics and geometry, M. BEN AYED *et al* (eds)

451 Topological methods in group theory, N. BROADDUS, M. DAVIS, J.-F. LAFONT & I. ORTIZ (eds)

452 Partial differential equations in fluid mechanics, C.L. FEFFERMAN, J.C. ROBINSON & J.L. RODRIGO (eds)

453 Stochastic stability of differential equations in abstract spaces, K. LIU

454 Beyond hyperbolicity, M. HAGEN, R. WEBB & H. WILTON (eds)

455 Groups St Andrews 2017 in Birmingham, C.M. CAMPBELL *et al* (eds)

456 Surveys in combinatorics 2019, A. LO, R. MYCROFT, G. PERARNAU & A. TREGLOWN (eds)

457 Shimura varieties, T. HAINES & M. HARRIS (eds)

458 Integrable systems and algebraic geometry I, R. DONAGI & T. SHASKA (eds)

459 Integrable systems and algebraic geometry II, R. DONAGI & T. SHASKA (eds)

460 Wigner-type theorems for Hilbert Grassmannians, M. PANKOV

461 Analysis and geometry on graphs and manifolds, M. KELLER, D. LENZ & R.K. WOJCIECHOWSKI

462 Zeta and L-functions of varieties and motives, B. KAHN

463 Differential geometry in the large, O. DEARRICOTT *et al* (eds)

464 Lectures on orthogonal polynomials and special functions, H.S. COHL & M.E.H. ISMAIL (eds)

465 Constrained Willmore surfaces, Á.C. QUINTINO

466 Invariance of modules under automorphisms of their envelopes and covers, A.K. SRIVASTAVA, A. TUGANBAEV & P.A. GUIL ASENSIO

467 The genesis of the Langlands program, J. MUELLER & F. SHAHIDI

468 (Co)end calculus, F. LOREGIAN

469 Computational cryptography, J.W. BOS & M. STAM (eds)

470 Surveys in combinatorics 2021, K.K. DABROWSKI *et al* (eds)

471 Matrix analysis and entrywise positivity preservers, A. KHARE

472 Facets of algebraic geometry I, P. ALUFFI *et al* (eds)

473 Facets of algebraic geometry II, P. ALUFFI *et al* (eds)

474 Equivariant topology and derived algebra, S. BALCHIN, D. BARNES, M. KĘDZIOREK & M. SZYMIK (eds)

475 Effective results and methods for Diophantine equations over finitely generated domains, J.-H. EVERTSE & K. GYŐRY

476 An indefinite excursion in operator theory, A. GHEONDEA

477 Elliptic regularity theory by approximation methods, E.A. PIMENTEL

478 Recent developments in algebraic geometry, H. ABBAN, G. BROWN, A. KASPRZYK & S. MORI (eds)

479 Bounded cohomology and simplicial volume, C. CAMPAGNOLO, F. FOURNIER-FACIO, N. HEUER & M. MORASCHINI (eds)

480 Stacks Project Expository Collection (SPEC), P. BELMANS, W. HO & A.J. DE JONG (eds)

481 Surveys in combinatorics 2022, A. NIXON & S. PRENDIVILLE (eds)

482 The logical approach to automatic sequences, J. SHALLIT

483 Rectifiability: a survey, P. MATTILA

484 Discrete quantum walks on graphs and digraphs, C. GODSIL & H. ZHAN

485 The Calabi problem for Fano threefolds, C. ARAUJO *et al*

486 Modern trends in algebra and representation theory, D. JORDAN, N. MAZZA & S. SCHROLL (eds)

487 Algebraic combinatorics and the Monster group, A.A. IVANOV (ed)

488 Maurer–Cartan methods in deformation theory, V. DOTSENKO, S. SHADRIN & B. VALLETTE

489 Higher dimensional algebraic geometry, C. HACON & C. XU (eds)

490 C^∞-algebraic geometry with corners, K. FRANCIS-STAITE & D. JOYCE

491 Groups and graphs, designs and dynamics, R.A. BAILEY, P.J. CAMERON & Y. WU (eds)

492 Homotopy theory of enriched Mackey functors, N. JOHNSON & D. YAU

493 Surveys in combinatorics 2024, F. FISCHER & R. JOHNSON (eds)

494 K-theory and representation theory, R. PLYMEN & M.H. ŞENGÜN (eds)

495 Polygraphs: from rewriting to higher categories, D. ARA *et al*

496 Groups St Andrews 2022 in Newcastle, C.M. CAMPBELL *et al* (eds)

497 Proof complexity generators, J. KRAJÍČEK

498 Polynomial functors, N. NIU & D.I. SPIVAK

499 Moduli, motives and bundles, P.L. DEL ÁNGEL R., F. NEUMANN & A.H.W. SCHMITT (eds)

500 The Toda lattice and universality for the computation of the eigenvalues of a random matrix, P. DEIFT, G. DUBACH, C. TOMEI & T. TROGDON

501 Topics in number theory, J. MORALES

502 The geometry of jet bundles, D. J. SAUNDERS

503 Multifunctorial equivariant algebraic K-theory, D. YAU

504 Spectra of signed graphs, Z. STANIĆ

London Mathematical Society Lecture Note Series: 504

Spectra of Signed Graphs

ZORAN STANIĆ
University of Belgrade, Serbia

CAMBRIDGE
UNIVERSITY PRESS

CAMBRIDGE
UNIVERSITY PRESS

Shaftesbury Road, Cambridge CB2 8EA, United Kingdom

One Liberty Plaza, 20th Floor, New York, NY 10006, USA

477 Williamstown Road, Port Melbourne, VIC 3207, Australia

314–321, 3rd Floor, Plot 3, Splendor Forum, Jasola District Centre, New Delhi – 110025, India

Cambridge University Press is part of Cambridge University Press & Assessment, a department of the University of Cambridge.

We share the University's mission to contribute to society through the pursuit of education, learning and research at the highest international levels of excellence.

www.cambridge.org
Information on this title: www.cambridge.org/9781009853347
DOI: 10.1017/9781009853316

When citing this work, please include a reference to the DOI 10.1017/9781009853316

First published 2026

A catalogue record for this publication is available from the British Library

A Cataloging-in-Publication data record for this book is available from the Library of Congress

ISBN 978-1-009-85334-7 Paperback

Contents

	Preface	*page* vii
1	**Introduction**	1
	1.1 Fundamental Concepts	4
	1.2 Enumeration	6
	1.3 Matrices	7
	1.4 Switching and Eigenvector Components	15
	1.5 Relations Between Laplacian and Net Laplacian Eigenvalues	15
	1.6 Spectral Criteria for Balance	19
	Problems	20
2	**Eigenvalues**	22
	2.1 Main Eigenvalues	23
	2.2 Determinant	28
	2.3 Operations and Resulting Spectra	29
	2.4 Symmetric Spectrum	37
	2.5 Eigenvalue Bounds	42
	2.6 Hoffman Program	57
	2.7 Spectral Distances and Related Functions	58
	2.8 Hoffman Polynomial	61
	Problems	62
3	**Strong Regularity**	64
	3.1 Disconnected and Bipartite SRSGs	65
	3.2 Relations with Association Schemes	69
	3.3 A Classification of Inhomogeneous SRSGs	76
	3.4 Net Regularity, Strong Regularity and Walk Regularity	82
	3.5 Net-Regular SRSGs with Degree at Most Four	88
	Problems	93

4 **Star Complements** 95
4.1 Eigenvalue Multiplicity 100
4.2 Decompositions of Signed Graphs with Two Eigenvalues 102
Problems 108

5 **Signed Line Graphs** 110
5.1 Definitions of a Signed Line Graph 110
5.2 Balance of Signed Line Graphs 114
5.3 Exceptional Signed Graphs 116
5.4 Signed Graph Foundations 118
5.5 Eigenbases for -2 in Signed Line Graphs 122
Problems 127

6 **Signed Graphs with a Small Number of Eigenvalues** 130
6.1 Two Eigenvalues 130
6.2 Regular with Three Eigenvalues 139
6.3 Common and Computational Results 142
6.4 Two or Three Net Laplacian Eigenvalues 145
Problems 148

7 **Integrality and Cospectrality** 150
7.1 Integral Signed Graphs 150
7.2 Cospectral Signed Graphs 156
Problems 167

8 **Polynomial Reconstruction** 169
8.1 Disconnected Signed Graphs 172
8.2 Signed Graphs with Pendant Vertices 174
8.3 Further Results 177
Problems 181

9 **Related Structures** 183
9.1 Linear Ternary Codes 183
9.2 Line Systems 189
9.3 Oriented Graphs 191
Problems 204

10 **Generalizations** 207
10.1 Graphs of Hermitian Matrices 208
10.2 Gain Graphs 209
10.3 Mixed Graphs and Extended Mixed Graphs 211
Problems 215
References 217
Index 230

Preface

This monograph is primarily intended for mathematicians working in the spectral theory of signed graphs. It also contains material likely to be of interest to graduate students entering this area, as well as researchers from related fields. While the focus is predominantly theoretical, the scope of the book reveals interactions with other disciplines, making it relevant to computer scientists, chemists, physicists, biologists, electrical engineers, and others whose work involves graph-based modelling.

A *signed graph* is a graph whose edges carry positive or negative signs. Formally, a simple undirected graph is not recognized as a specific signed graph. However, informally, and particularly within spectral theory, a graph is often interpreted as a signed graph where all edge signs are positive – a convention adopted throughout this book. From this perspective, it is common to say that the concept of a signed graph generalizes or extends that of ordinary (unsigned) graphs.

The study of signed graphs traces its roots to the early 1950s, with precursors appearing even decades earlier in ideas related to distinguishing edges in ways now aligned with signed graphs. Over the past 70 years, the associated matrix theory, eigenvalues, eigenspaces and broader spectral framework have developed in parallel with the underlying combinatorial structures. In particular, the last two decades have witnessed an exceptional acceleration in the development of the spectral theory of signed graphs. The field has matured into a rich and sophisticated discipline, offering a wide range of challenging and fruitful research directions. The present moment therefore appears especially suitable for a monograph that revisits, synthesizes and reinterprets a curated selection of the most significant concepts and results.

I should note that this selection is deliberately narrow, for two reasons. First, while I have been deeply involved in the field, I do not claim exhaustive awareness of all developments. Second, the primary aim of the book is to present the most advanced ideas and the most compelling results in a manner accessible to the broadest possible audience over an extended period. In this regard, the book should not be considered as a systematic treatment of spectra of signed graphs, but rather as a unifying collection of material on that subject. Accordingly, the structure and spirit of the book are inspired by several foundational monographs on the spectra of (unsigned) graphs, notably [73] and [75], as well as the more compact and differently structured treatments in [19] and [45].

The book develops the foundational concepts of signed graphs in full detail, with the goal of making the material as self-contained as possible in this regard. However, it assumes a foundational understanding of the basic spectral theory of unsigned graphs. Standard results – such as the Perron–Frobenius theory, spectral properties of various graph classes (including, but not limited to, regular and bipartite graphs), the theory of regular two-graphs and the relationship between spectral moments and walks – are assumed to be known and are not reproduced in detail. This approach reflects the belief that, in the modern era, well-established and frequently referenced mathematical ideas are readily accessible. In addition, signed graphs constitute a refinement rather than a starting point; therefore, they are not the primary subject to be studied first, and it is natural to expect that readers of this book already possess a solid foundational knowledge.

The principal emphasis of the book is placed on the *eigenvalues* and the corresponding *eigenspaces* of the *adjacency matrix*, the *Laplacian matrix* and the *net Laplacian matrix* associated with finite simple undirected signed graphs. In addition, we examine signed graphs that allow parallel edges and explore several meaningful generalizations of the concept. While many techniques and patterns will be familiar to readers well versed in spectral graph theory, the transition to signed graphs often reveals surprising differences and unexpected phenomena. These differences make the spectral theory of signed graphs not only more nuanced, but also more stimulating and ripe with open questions.

The terminology and notation employed in this work are primarily adopted from [73, 75, 209, 222], and remain broadly consistent with the established conventions in related literature.

Approximately 70 of the references listed in the bibliography correspond to works I have authored or co-authored. Results derived from these contributions are, as a rule, presented with complete proofs. A substantial number of results of other authors have also been incorporated, with selected proofs provided to ensure clarity and self-containment.

Outline of the Book

Chapter 1 establishes the terminology and notation, introduces the fundamental concepts of signed graphs that distinguish them from ordinary graphs, and defines the associated matrices. Basic properties of their characteristic polynomials and eigenvalues are also presented.

Chapter 2 delves deeper into the analysis of eigenvalues by introducing several central notions such as *main eigenvalues*, the *Hoffman program* and the *spectral distance*.

The next eight chapters may be viewed as extensions, generalizations or counterparts of established results in the domain of unsigned graphs. Remarkably, many of these results undergo substantial transformation when transferred to the setting of signed graphs. In numerous instances they diverge from standard concepts, thereby revealing novel perspectives and opening up new avenues for research.

Chapter 3 is entirely devoted to the *strong regularity* of signed graphs, a generalization of the corresponding notion in the unsigned case. These signed graphs typically exhibit a high degree of symmetry, which often results in large automorphism groups.

In Chapter 4, the technique of *star complements*, along with the associated concept of *star sets*, is extended to the setting of signed graphs. This method facilitates the construction of large graphs from smaller subgraphs, rendering it a valuable tool for the classification, construction and analysis of signed graphs exhibiting prescribed spectral properties.

Chapter 5 explores the concept of *signed line graphs*, their structural properties and their relations to other signed graphs with spectra bounded below by -2.

Chapter 6 focuses on signed graphs with a *small number of distinct eigenvalues*. The exposition flows seamlessly between general characterization theorems and detailed classifications of particular families.

Integral and *cospectral* signed graphs are studied in Chapter 7, with particular attention paid to their algebraic and combinatorial aspects. The reader will encounter standard constructions, as well as recent developments specific to signed graphs, which elevate the theory to an exceptional level of depth and richness.

Chapter 8 addresses the *reconstruction of the characteristic polynomial* of a signed graph from the collection of characteristic polynomials of its vertex-deleted subgraphs. While this problem remains open in the unsigned case (and is closely tied to the classic Ulam conjecture), signed graphs admit a simple counterexample. Nevertheless, outside of this case, the problem leads to a range of interesting open questions supported by various particular developments.

Chapter 9 explores connections between signed graphs and other mathematical structures, including *linear ternary codes*, *systems of lines in Euclidean spaces* and *oriented graphs*. Special emphasis is placed on the spectral theory of oriented graphs via their skew adjacency matrices, which creates a strong link between the two domains and facilitates the transfer of problems from oriented to signed graphs.

Finally, Chapter 10 discusses natural generalizations of signed graphs, focusing on *graphs of Hermitian matrices*, *gain graphs*, *mixed graphs* and *extended mixed graphs*.

The chapters are intended to be read in the order presented, as each builds upon concepts and results developed in the preceding ones. Each chapter includes theoretical results alongside commentary, discussions, illustrative examples and open problems. Chapters conclude with a set of exercises, which include both original problems and a small number of results whose proofs have been omitted from the main text.

1

Introduction

Signed graphs have been extensively studied over the last 70 years. Some of the pioneering papers are Harary's [113], published in 1953, which considers a mathematical formalization of a problem in social psychology, and König's [138] (from 1936), which addressed structures now recognized as signed graphs, albeit without explicitly defining them through edge labellings by the two-element group. According to [260], this group-theoretic labelling is fundamental to the formal theory of signed graphs.

Over the decades, signed graphs have found applications across numerous fields including physics, chemistry, biology and engineering, as well as in various branches of computer science such as coding theory, complex networks, clustering, data mining and virus prevention. More recently, they have gained prominence in control theory and the study of social networks. A broad overview of these applications can be found in Zaslavsky's dynamic bibliography survey [258].

The spectral theory of signed graphs has been studied in the literature for more than 60 years, with particularly rapid advancements occurring over the past two decades. The term 'spectrum' in this context refers to the spectrum of a prescribed square matrix associated with a signed graph.

We proceed with a formal definition followed by some terminology and notation. A *signed graph* $\dot{G}$ is a graph in which each edge is declared positive or negative. More precisely, $\dot{G}$ is a pair (G, σ) where $G = (V, E)$ is an ordinary finite simple unoriented graph, called the *underlying graph*, and $\sigma\colon E \to \{1, -1\}$ is the *sign function* or the *signature*. The vertex set $V(\dot{G})$ coincides with the set $V = V(G)$. Needless to add, the edges mapped to 1 are *positive*, and those mapped to -1 are *negative*. Accordingly, we write $E(\dot{G}) = E^{+}(\dot{G}) \sqcup E^{-}(\dot{G})$, where the subsets of the edge set contain positive and negative edges, respectively. The sign group $\{1, -1\}$ can also be written $\{+, -\}$, and these notations will be treated as equivalent. The *order* (resp. *size*) of a signed graph, denoted by n (resp. m), is the number of its vertices (resp. edges). We interpret an unsigned graph as a signed graph with $E^-(\dot{G}) = \emptyset$, that is, a signed graph with an all-positive signature.

Remark 1.1 Throughout the book, the term graph will refer to a finite simple undirected graph, unless stated otherwise. A signed graph is always explicitly identified with the prefix 'signed'. To avoid ambiguity, the terms *ordinary graph* and *unsigned graph* will be used interchangeably for a graph.

We adopt a *dot notation* for signed graphs: symbols with a dot (e.g., $\dot{G}$, $\dot{H}$, $\dot{K}_n$, $\dot{C}_n$, etc.) are reserved exclusively for signed graphs, while the same symbols without a dot (G, H, K_n, C_n, etc.) denote the corresponding underlying unsigned graphs. This notation proves especially useful when interpreting an ordinary graph as a signed graph, as the presence or absence of the dot clearly indicates the intended structure. An exception to the dot notation occurs in the case of signed graphs derived from association schemes, considered in Section 3.2, where the standard notation of that domain is used instead.

It is worth noting that different notational conventions appear in the literature; for further details, see Zaslavsky's survey [258].

Many familiar notions from graph theory extend directly to signed graphs. For example, the *degree* d_v of a vertex v in $\dot{G}$ is simply its degree in G; the *neighbourhood* $N(v)$ of v is the set of vertices adjacent to v; $\dot{G}$ is *regular* or *bipartite* if the same holds for G. In particular, for vertex degree three (resp. four) we deal with *cubic* (resp. *quartic*) regular signed graphs. A connected signed graph is called *k-cyclic* if its order and size satisfy $n = m - k + 1$. Trees, unicyclic, bicyclic and tricyclic signed graphs are obtained for $k = 0$, $k = 1$, $k = 2$ and $k = 3$, respectively. The definition of a walk in a signed graph does not deviate from the same definition in the case of graphs. Thus, a *walk* is a sequence of alternate vertices and edges such that consecutive vertices are incident with the corresponding edge. The standard notation for basic graph families extends naturally to the signed setting. For example, if K_n, P_n, C_n and $K_{s,t}$ denote a complete graph, path, cycle and complete bipartite graph respectively, then the corresponding signed graphs are denoted by $\dot{K}_n$, $\dot{P}_n$, $\dot{C}_n$ and $\dot{K}_{s,t}$, where a particular signature has been introduced. As will be demonstrated, any signature on P_n produces, from a spectral viewpoint, essentially the same signed graph.

There are notions exclusive to signed graphs. For example, the *positive degree* d_v^+ is the number of positive edges incident with v, and the *positive neighbourhood* $N^+(v)$ is the set of vertices joined to v by a positive edge. The *negative degree* d_v^- and the *negative neighbourhood* $N^-(v)$ are defined analogously. The *net degree* of v is defined as $d_v^{\pm} = d_v^+ - d_v^-$. A signed graph is *net-regular*

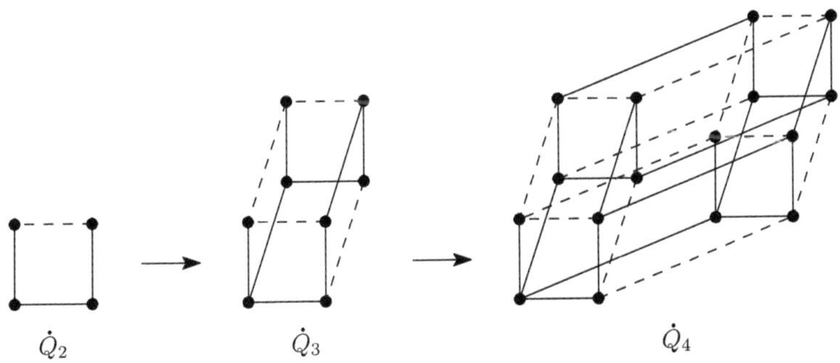

Figure 1.1 The r-dimensional signed cubes with negative quadrangles for $r \leq 4$. Here and in other figures, negative edges are dashed.

if the net degree considered as a function on the vertex set is a constant. In this case, the common net degree is denoted by ϱ.

The previous notation is used interchangeably with the one that emphasizes the signed graph; for example, $d_{\dot{G}}(v)$ is the degree of v in $\dot{G}$, and similarly in other cases.

If the vertices u and v are adjacent (i.e., joined by either a positive or a negative edge), then we write $u \sim v$. In particular, $u \overset{+}{\sim} v$ (resp. $u \overset{-}{\sim} v$) designates the existence of a positive (resp. negative) edge between them. A walk in a signed graph is *positive* if the number of its negative edges (counted with their multiplicity if there are repeated edges) is even; otherwise, it is *negative*. Accordingly, a cycle $\dot{C}$ is *positive* if the product of its edge signs $\sigma(\dot{C})$ is 1; otherwise, it is *negative*. The signed graph obtained from $\dot{G}$ by deleting a subgraph $\dot{H}$ is denoted by $\dot{G} - \dot{H}$; in particular, $\dot{G} - u$ denotes the signed graph obtained by deleting a single vertex u.

At this stage, it is necessary to introduce a family of signed graphs, presented here primarily as examples containing negative cycles.

Example 1.2 The *r-dimensional cube* is the bipartite r-regular graph of order 2^r with the vertex set $\{0,1\}^r$ (all possible binary r-tuples) in which two vertices are adjacent if they differ in exactly one coordinate. Accordingly, an *r-dimensional signed cube* is a signed graph whose underlying graph is the r-dimensional cube. For $r \geq 2$, we introduce *r-dimensional signed cubes with negative quadrangles* $\dot{Q}_r$, that is, those in which every four-vertex cycle is negative.

These cubes are constructed iteratively as follows (see Figure 1.1). The signed cube $\dot{Q}_r$ is formed by taking two copies of $\dot{Q}_{r-1}$ and connecting corresponding vertices u from each copy with an edge. This connecting edge is assigned a positive sign if u belongs to the first colour class of $\dot{Q}_{r-1}$, and a negative sign if u belongs to the second colour class, where an arbitrary colour class of $\dot{Q}_{r-1}$ is designated as the first.

The *negation* $-\dot{G}$ is obtained by reversing the sign of every edge of $\dot{G}$. We say that $\dot{G}$ is *homogeneous* if all edges have the same sign or its edge set is empty; otherwise, it is *inhomogeneous*. Accordingly, every graph G and its negation $-G$ are homogeneous signed graphs.

1.1 Fundamental Concepts

There are two fundamental concepts in the theory of signed graphs: balance and switching equivalence.

A signed graph or any of its subgraphs is called *balanced* if every cycle within it, whenever such cycles exist, is positive; otherwise, it is termed *unbalanced*. Clearly, every unsigned graph, viewed as a signed graph with all positive edges, is balanced. In the case of an unbalanced signed graph, the notion of frustration arises as a quantitative measure of the deviation of $\dot{G}$ from its underlying unsigned graph G. The *frustration index* ℓ (resp. *frustration number* f) of $\dot{G}$ is the minimum number of edges (resp. vertices) whose removal results in a balanced signed graph. A signed graph $\dot{G}$ such that $-\dot{G}$ is balanced is called *antibalanced*. Acyclic signed graphs are simultaneously balanced and antibalanced.

Theorem 1.3 (Harary balance theorem [113]**)** *A signed graph is balanced if and only if its vertex set can be divided into two disjoint subsets (where one of them is allowed to be empty) in such a way that every negative edge is located between these sets.*

If $\dot{G}$ is connected then, up to interchanging, the sets of Theorem 1.3 are uniquely determined. We mention in passing another balance criterion, also obtained by Harary [113] – in fact a consequence of the previous theorem. A signed graph is balanced if and only if, for any two vertices, every connecting path has the same sign.

We now consider the other fundamental concept. For $U \subseteq V(\dot{G})$, let $\dot{G}^U$ be a signed graph obtained by reversing the sign of every edge between U and $V(\dot{G}) \setminus U$. We say that $\dot{G}$ and $\dot{G}^U$ are *switching equivalent*, denoted by $\dot{G} \simeq \dot{G}^U$. In the context of signed graphs, isomorphism is typically considered in conjunction with switching equivalence as follows: $\dot{G}$ and $\dot{H}$ are *switching isomorphic*, denoted by $\dot{G} \cong \dot{H}$, if one is isomorphic to a signed graph that is switching equivalent to the other.

If U consists of a single vertex, say u, then we are dealing with a switching at u. Signed graphs are switching equivalent if and only if one of them is obtained by a series of switchings at vertices of the other.

Switching equivalence is an equivalence relation on signatures of a signed graph, with the equivalence class called a *switching equivalence class*. Similarly, switching isomorphism is an equivalence relation on all signed graphs, and the corresponding class is called a *switching isomorphism class*.

In many contexts, switching isomorphism is not distinguished from switching equivalence; indeed, switching isomorphic signed graphs are often identified. In particular, when we assert that a signed graph is unique with respect to a certain property, we mean that it is unique up to switching isomorphism.

According to the previous definitions, a signed graph is balanced if and only if it switches to its underlying graph [260], which is another balance criterion. Switching-equivalent signed graphs share the set of negative cycles (and then they also share the set of positive cycles).

It is not difficult to see that a signed graph $\dot{G}$ switches to its negation if and only if $\dot{G}$ is bipartite. Indeed, if $\dot{G}$ contains an odd cycle then this particular cycle is positive in exactly one of $\dot{G}$ and $-\dot{G}$, which means that they are not switching equivalent. On the other hand, a switching does not alter the sign of even cycles, which leads to the desired conclusion.

If we involve the switching isomorphism, then the situation is more complicated. For example, a signed graph consisting of two triangles (one positive, the other negative) that share exactly one vertex does not switch into its negation, but is switching isomorphic to it.

Example 1.4 For a given graph G, one can obtain numerous signed graphs by assigning different signatures. However, many of these will be switching equivalent. For instance, every signature on a tree T is switching equivalent to T itself, implying that, up to switching, there is essentially a single signed tree regardless of the signature. In the case of a unicyclic signed graph, the unique cycle may be either positive or negative, yielding exactly two distinct signed graphs arising from a fixed unicyclic graph. One may apply the same reasoning to discuss bicyclic graphs.

Naturally, the situation becomes more intricate as the cyclomatic number increases. In [264], it is shown that the Petersen graph gives rise to exactly six distinct signed graphs, which are illustrated in Figure 1.2. The first corresponds to the underlying unsigned graph, while the last is its negation.

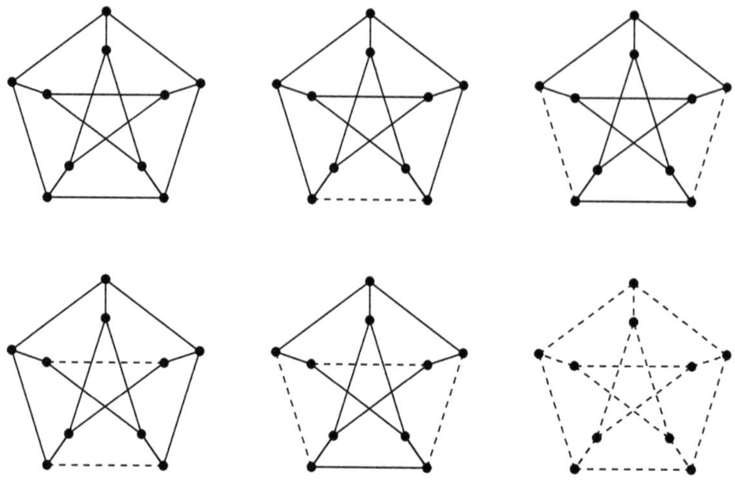

Figure 1.2 Representatives of the six switching classes of the Petersen graph. Reproduced from [108], licensed under CC BY 4.0.

1.2 Enumeration

We present several theoretical results alongside numerical data to offer insight into the subject under consideration.

A *labelled (signed) graph* is a (signed) graph with each vertex labelled differently, so they are distinguished from one another. It can be seen that the number of (not necessarily connected) labelled unsigned graphs with n vertices is $2^{\binom{n}{2}}$. In the following theorem we consider labelled signed graphs that share the same underlying graph.

Theorem 1.5 ([220]) *If $\mathscr{S}(G)$ denotes the set of all labelled signed graphs that are not pairwise switching equivalent and share a fixed connected k-cyclic underlying graph G, then $|\mathscr{S}(G)| = 2^k$.*

The proof is based on an induction argument. If $e \in E(\dot{G})$ is contained in at least one cycle, then $\dot{G} - e$ is connected and, by hypothesis, $\mathscr{S}(G-e)$ counts 2^{k-1} labelled signed graphs. Then, by adding e with a positive and then with a negative sign, we get $|\mathscr{S}(G)| = 2^k$. Signed graphs of $\mathscr{S}(G)$ are characterized in the next theorem.

Theorem 1.6 ([220]) *Let T be a spanning tree of a k-cyclic graph G, and let $e_1, e_2, \ldots, e_k$ denote the edges in G not contained in T. Then the 2^k labelled signed graphs in $\mathscr{S}(G)$ are obtained by assigning all possible combinations of signatures to the edges $e_1, e_2, \ldots, e_k$.*

Table 1.1 *The number $T(n)$ of connected switching non-isomorphic signed graphs with $n \leq 8$ vertices.*

n	3	4	5	6	7	8
$T(n)$	3	12	79	1123	42 065	4 880 753
% of graphs	66.67	50.00	26.58	9.97	2.03	0.23

We now present the results of a computer search for connected switching non-isomorphic signed graphs on at most eight vertices. These results are reported in [220], and representatives of the switching isomorphism classes are available online at [224]. For each underlying graph G, Theorem 1.6 enumerates all signed graphs derived from it, and the classification is completed by identifying those that are switching isomorphic.

The proportion of distinct underlying graphs relative to the total number decreases markedly.

1.3 Matrices

We introduce several standard and some less conventional matrices associated with signed graphs. In the case of square matrices, we also consider their characteristic polynomials, eigenvalues and corresponding eigenspaces. For any such matrix, the *spectral radius* is defined as the largest absolute value among its eigenvalues. Throughout the book, the same term is used for other square matrices, even when they are not associated with signed graphs.

1.3.1 Adjacency Matrix

The *adjacency matrix* $A_{\dot{G}}$ of a signed graph $\dot{G}$ is the $n \times n$ $(1,0,-1)$-matrix obtained from the adjacency matrix of its underlying graph by reversing the signs of all entries corresponding to negative edges. The *characteristic polynomial* $\Phi_{\dot{G}}$ of $\dot{G}$ is defined as the characteristic polynomial of its adjacency matrix, that is, $\Phi_{\dot{G}}(x) = \det(xI - A_{\dot{G}})$. The roots of $\Phi_{\dot{G}}$ are the *eigenvalues* of $\dot{G}$, which collectively form the *spectrum* of $\dot{G}$. The corresponding eigenvectors are likewise referred to as the *eigenvectors* of $\dot{G}$. Since $A_{\dot{G}}$ is symmetric, all its eigenvalues are real. We denote them by $\lambda_1, \lambda_2, \ldots, \lambda_n$, indexed in non-increasing order. The largest eigenvalue λ_1 is called the *index* of $\dot{G}$. Observe that the spectral radius may differ from the index. The symbols m_λ and $m_{\dot{G}}(\lambda)$ are used interchangeably to denote the multiplicity of λ.

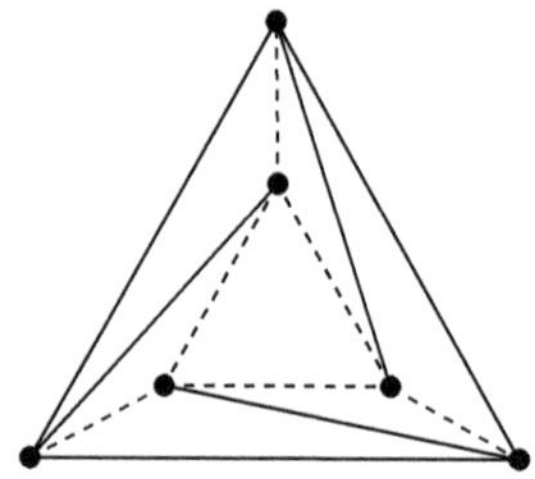

Figure 1.3 A signed graph with spectrum $[2^3, (-2)^3]$. Reproduced from [233].

Figure 1.3 illustrates a signed graph with exactly two distinct eigenvalues, 2 and -2. Its spectrum reveals that, in contrast to unsigned graphs, the adjacency matrix of a signed graph may be reducible. In particular, the index need not be a simple eigenvalue.

Remark 1.7 Throughout the book, the statement '$\dot{G}$ has k eigenvalues' is understood to mean that $\dot{G}$ possesses exactly k distinct eigenvalues. The spectrum of every square matrix associated with a signed graph is expressed in exponential notation, where the exponent indicates the multiplicity of the corresponding eigenvalue.

Clearly, for $k \in \mathbb{N}$, the (i, j) entry of the kth power $A^k_{\dot{G}}$ counts the difference between the number of positive and the number of negative walks that start at i, terminate at j and traverse k edges. Since $A_{\dot{G}}$ is a Hermitian matrix, the eigenvalues of a signed graph are interlaced by the eigenvalues of any of its subgraph, see [73, Theorem 0.10]. More precisely, the following theorem holds.

Theorem 1.8 (Cauchy interlacing theorem) *If $\lambda'_1 \geq \lambda'_2 \geq \cdots \geq \lambda'_{n'}$ are the eigenvalues of an induced subgraph $\dot{H}$ of a signed graph $\dot{G}$, then $\lambda_i \geq \lambda'_i \geq \lambda_{n-n'+i}$, $1 \leq i \leq n'$, where $\lambda_1 \geq \lambda_2 \geq \cdots \geq \lambda_n$ are the eigenvalues of $\dot{G}$.*

Switching-equivalent signed graphs share the spectrum. Indeed, if the vertex labelling is transferred from the underlying graph common to $\dot{G}$ and $\dot{H}$, then $\dot{G}$ and $\dot{H}$ are switching equivalent if and only if there is a diagonal matrix S, called the *switching matrix*, with ± 1 on the diagonal such that $A_{\dot{H}} = S^{-1} A_{\dot{G}} S$. According to the definition of switching isomorphism, switching-isomorphic signed graphs also have the same spectrum.

The spectrum gives information on whether an unsigned graph is bipartite; namely, a graph is bipartite if and only if its spectrum is symmetric about the origin [75, 209]. On the contrary, the spectrum of a signed graph does not provide the same information, see again Figure 1.3.

We proceed by computing the coefficients of the characteristic polynomial of a signed graph. The *elementary figures* are the complete signed graph with two vertices $\dot{K}_2$ and the signed cycle with n vertices $\dot{C}_n$; a *basic figure* is the disjoint union of elementary figures.

Theorem 1.9 (Sachs formula [34]) *Let*

$$\Phi_{\dot{G}}(x) = x^n + a_1 x^{n-1} + \cdots + a_n$$

be the characteristic polynomial of a signed graph $\dot{G}$. Then

$$a_i = \sum_{B \in \mathscr{B}_i} (-1)^{p(B)} 2^{|c(B)|} \sigma(B),$$

where $\mathscr{B}_i$ is the set of basic figures with i vertices in $\dot{G}$, $p(B)$ is the number of components of B, $c(B)$ the set of cycles in B and $\sigma(B) = \prod_{\dot{C} \in B} \sigma(\dot{C})$.

A *k-matching* in a graph is a collection of k vertex-disjoint edges. The *matching polynomial* of a graph G is defined as

$$\varpi_G(x) = \sum_{k=0}^{m(G)} (-1)^k m_k(G) x^{n-2k},$$

where $m(G)$ is the size of the largest matching and $m_k(G)$ is the number of k-matchings in G. We can express $\Phi_{\dot{G}}$ in terms of the matching polynomials.

Theorem 1.10 ([34, 96]) *Let $\mathscr{R}$ be the set of two-regular signed graphs of $\dot{G} = (G, \sigma)$. Then*

$$\Phi_{\dot{G}}(x) = \varpi_G(x) + \sum_{\dot{R} \in \mathscr{R}} \sigma(\dot{R})(-2)^{p(\dot{R})} \varpi_{G-R}(x),$$

where $p(\dot{R})$ and $\sigma(\dot{R})$ respectively denote the number of cycles and the product of the signs of $\dot{R}$.

It is assumed that $\varpi_{G-R}(x) = 1$ if $G - R$ has no vertices. Since a basic figure consists of disjoint matchings and cycles, the previous result is just a reformulation of the Sachs formula.

We quote some reduction procedures for computing the characteristic polynomial. Using the Sachs formula, or some more general results (see, for example, [31, 184]), we obtain the following Schwenk-like formula:

$$\Phi_{\dot{G}}(x) = x\Phi_{\dot{G}-u}(x) - \sum_{v \sim u} \Phi_{\dot{G}-u-v}(x) - 2 \sum_{\dot{C} \in \mathscr{C}_u} \sigma(\dot{C}) \Phi_{\dot{G}-\dot{C}}(x), \tag{1.1}$$

where, as in [195], $\dot{\mathscr{C}}_u$ denotes the set of signed cycles passing through u. If u is a *pendant vertex* (i.e., of degree 1) and v its neighbour, then (1.1) reduces to

$$\Phi_{\dot{G}}(x) = x\Phi_{\dot{G}-u}(x) - \Phi_{\dot{G}-u-v}(x). \tag{1.2}$$

Example 1.11 Following [195], we compute the characteristic polynomial of a signed cycle $\dot{C}_n$. By using (1.1), we obtain

$$\Phi_{C_n}(x) = x\Phi_{P_{n-1}}(x) - 2\Phi_{P_{n-2}}(x) - 2,$$
$$\Phi_{\dot{C}_n}(x) = x\Phi_{P_{n-1}}(x) - 2\Phi_{P_{n-2}}(x) + 2,$$

which yields $\Phi_{\dot{C}_n}(x) = \Phi_{C_n}(x) + 4$. Thus, the eigenvalues of $\dot{C}_n$ coincide with the points for which Φ_{C_n} takes -4. The characteristic polynomial of C_n is known (see [73, p. 73]) and can be written as

$$\Phi_{C_n}(x) = 2\cos\left(n\arccos\frac{x}{2}\right) - 2. \tag{1.3}$$

Observe that $\Phi'_{C_n}(x) = n\Phi_{P_{n-1}}(x)$. The eigenvalues of P_{n-1} are $2\cos(\pi j/n)$ for $1 \leq j \leq n-1)$ – see again [73, p. 73] or use an induction argument based on (1.2). Substituting these eigenvalues into (1.3), we arrive at the equality $\Phi_{C_n}(2\cos(\pi j/n)) = -4$ for j odd, which means that these numbers belong to the spectrum of $\dot{C}_n$. Since the eigenvalues of P_{n-1} interlace those of $\dot{C}_n$, we conclude that each of the numbers mentioned must be a double eigenvalue (of multiplicity 2) of $\dot{C}_n$. For n odd, an additional point for which Φ_{C_n} takes -4 is -2. Therefore, we have the following:

- For n even, the spectrum of $\dot{C}_n$ consists of the numbers $2\cos(\pi j/n)$, $j \in \{1, 3, \ldots, n-1\}$, all with multiplicity 2.
- For n odd, -2 is an additional eigenvalue.

Note that when n is odd, the eigenvalues of $\dot{C}_n$ correspond precisely to the negation of the eigenvalues of C_n.

1.3.2 Incidence Matrix

One way to define the vertex–edge incidence matrix of a signed graph is to extend the classical definition from the unsigned domain. Specifically, the incidence matrix of $\dot{G}$ is an $n \times m$ matrix whose (v, e) entry equals $\sigma(e)$ if the edge e is incident with the vertex v, and 0 otherwise. This definition is intuitive and uniquely determines $\dot{G}$ from the matrix.

However, it is often more advantageous to adopt an alternative definition introduced by Zaslavsky [263], the significance of which becomes apparent in the study of signed line graphs, as discussed in Chapter 5.

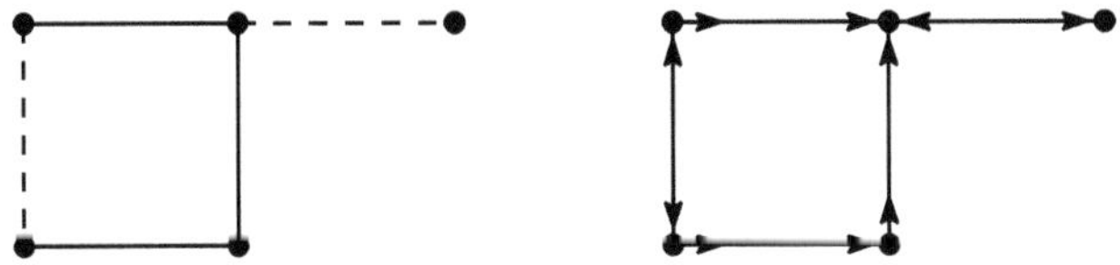

$\dot{G}$ $\dot{G}_\eta$

Figure 1.4 A signed graph $\dot{G}$ and a bi-directed graph $\dot{G}_\eta$.

For a signed graph $\dot{G}$, we introduce the vertex–edge *orientation*

$$\eta : V(\dot{G}) \times E(\dot{G}) \to \{1, 0, -1\}$$

formed by obeying the following rules:

(O1) $\eta(u, vw) = 0$ if $u \notin \{v, w\}$;
(O2) $\eta(u, uv) = 1$ or $\eta(u, uv) = -1$;
(O3) $\eta(u, uv)\eta(v, uv) = -\sigma(uv)$.

The *incidence matrix* $B_\eta = (\eta_{ve})$ is a vertex-edge incidence matrix derived from $\dot{G}$ such that its (v, e) entry is equal to $\eta(v, e)$. However, this matrix is not uniquely determined by $\dot{G}$ alone. Namely, one can randomly choose an entry $\eta(u, uv)$ to be either $+1$ or -1, but the entry $\eta(v, uv)$ is then determined by $\sigma(uv)$. Accordingly, η is called an *orientation of* $\dot{G}$ (and a *bi-orientation of* G). From a signed graph $\dot{G}$ we obtain many bi-directed graphs $\dot{G}_\eta$, but each of them leads back to $\dot{G}$. An example is illustrated in Figure 1.4.

1.3.3 Laplacian Matrix

Let $D_{\dot{G}}$ denote the diagonal matrix of vertex degrees. The *Laplacian matrix* of the signed graph $\dot{G}$ is then defined as $L_{\dot{G}} = D_{\dot{G}} - A_{\dot{G}}$. Notably, it is straightforward to verify that the Laplacian matrix can be expressed as the product of the incidence matrix B_η with its transpose, independent of the choice of orientation η; that is, $L_{\dot{G}} = B_\eta B_\eta^{\intercal}$. The *Laplacian characteristic polynomial* $\Psi_{\dot{G}}$ of $\dot{G}$ is defined as the characteristic polynomial of $L_{\dot{G}}$. Analogous to the adjacency matrix, the *Laplacian eigenvalues*, the *Laplacian spectrum* and the *Laplacian eigenvectors* of $\dot{G}$ correspond to the matrix $L_{\dot{G}}$. Since $L_{\dot{G}}$ is positive semidefinite, all of its eigenvalues are non-negative. We denote these eigenvalues by $\mu_1, \mu_2, \ldots, \mu_n$, indexed in non-increasing order.

The Laplacian spectrum of the signed graph depicted in Figure 1.3 is given by $[6^3, 2^3]$, which aligns with the fact that for this graph $L_{\dot{G}} = 4I - A_{\dot{G}}$. In general, the Laplacian eigenvalues of a regular signed graph $\dot{G}$ are obtained by adding the common vertex degree to the eigenvalues of the adjacency matrix of $-\dot{G}$. The Laplacian matrix of a signed graph may be reducible, but we already knew this since the same holds for the Laplacian matrix of a graph.

If μ and $\mathbf{x} = (x_1, x_2, \ldots, x_n)^{\intercal}$ are a Laplacian eigenvalue and an associated Laplacian eigenvector, then we have

$$(d_v - \mu)x_v = \sum_{uv \in E(\dot{G})} \sigma(uv)x_u, \tag{1.4}$$

the so-called *eigenvalue equation* for μ.

Here is a counterpart to Theorem 1.8.

Theorem 1.12 (Laplacian interlacing) *Let $\dot{H}$ be a signed graph obtained by deleting k edges of a signed graph $\dot{G}$. Then $\mu_i \geq \mu_i' \geq \mu_{i+k}$, $1 \leq i \leq n-k$, where $\mu_1 \geq \mu_2 \geq \cdots \geq \mu_n$ and $\mu_1' \geq \mu_2' \geq \cdots \geq \mu_n'$ are the Laplacian eigenvalues of $\dot{G}$ and $\dot{H}$, respectively.*

The proof is obtained by applying the eigenvalue interlacing to $B_\eta^{\intercal} B_\eta$ and taking into account that this matrix shares the non-zero eigenvalues with the Laplacian $L_{\dot{G}} = B_\eta B_\eta^{\intercal}$.

As before, switching-equivalent signed graphs have the same Laplacian spectrum. Moreover, the matrix transformation is realized by the same (switching) matrix S.

If $E^-(\dot{G}) = \emptyset$, then $L_{\dot{G}}$ coincides with the standard Laplacian matrix of G. Conversely, if $E^+(\dot{G}) = \emptyset$, then $L_{\dot{G}}$ reduces to the well-known signless Laplacian matrix of G. In this sense, the spectral theory of signed graphs based on the Laplacian matrix both encompasses and generalizes the spectral theories of graphs founded on the Laplacian and the signless Laplacian.

The Laplacian spectrum encodes information regarding the connectivity of an unsigned graph: specifically, the number of connected components equals the multiplicity of the zero eigenvalue [75, 209]. Furthermore, if a graph is known to be connected, the signless Laplacian spectrum indicates whether it is bipartite; similarly, if a graph is bipartite, the signless Laplacian spectrum reveals information about its connectivity. This information is closely related to the existence and multiplicity of the zero eigenvalue [75, 209]. However, the Laplacian spectrum of a signed graph does not convey any of these connectivity or bipartiteness properties.

To obtain a counterpart to the Sachs formula, we need the following notions. A *TU-subgraph* of a signed graph $\dot{G}$ is a subgraph whose components are trees or unbalanced unicyclic graphs. If $\dot{H}$ is a TU-subgraph, then $\dot{H} \cong \dot{T}_1 \cup \dot{T}_2 \cup \cdots \cup \dot{T}_r \cup \dot{U}_1 \cup \dot{U}_2 \cup \cdots \cup \dot{U}_s$, where the T_i are trees and the U_i are unbalanced unicyclic graphs. The weight of $\dot{H}$ is defined as $w(\dot{H}) = 4^s \prod_{i=1}^{r} |V(\dot{T}_i)|$. In the particular case of graphs, the formula for the coefficients of the Laplacian characteristic polynomial based on TU-subgraphs was obtained by Dedò [83]. This formula is a special case of the Kel'mans–Chaiken type formula for weighted digraphs [61]. Its signed variant is proved in [34].

Theorem 1.13 (Dedò formula) *Let*

$$\Psi_{\dot{G}}(x) = x^n + b_1 x^{n-1} + \cdots + b_{n-1}x + b_n$$

be the Laplacian characteristic polynomial of a signed graph $\dot{G}$*. Then*

$$b_i = (-1)^i \sum_{\dot{H} \in \mathscr{H}_i} w(\dot{H}),$$

where $\mathscr{H}_i$ *denotes the set of* TU*-subgraphs of* $\dot{G}$ *containing i edges.*

One of the most significant results for the Laplacian matrix of graphs is the matrix-tree theorem stating that for every connected graph the number of spanning trees (also known as the complexity) equals $\det(L_G^i)$, where L_G^i is the principal submatrix of L_G obtained by deleting the ith row and column. It follows that $b_{n-1} = \sum_{i=1}^{n} |L_G^i|$, and thus the complexity is b_{n-1}/n. In this spirit, the Dedò formula for signed graphs can be interpreted as a variant of the same theorem. Another variant is given below. A *pseudoforest* is a signed graph for which every component is a tree or a unicyclic signed graph.

Theorem 1.14 (Matrix-tree theorem [263]**)** *The determinant* $\det(L_{\dot{G}})$ *is the sum, over all pseudoforests F with n edges and with no positive cycles, of* $4^{c(F)}$*, where* $c(F)$ *is the number of components of F.*

The summands in Theorem 1.14 correspond to the coefficients in the Dedò formula. A slightly different formulation can be found in [141].

1.3.4 Net Laplacian Matrix

Let $D_{\dot{G}}^{\pm}$ denote the diagonal matrix of vertex net degrees. The *net Laplacian matrix* of the signed graph $\dot{G}$ is defined as $N_{\dot{G}} = D_{\dot{G}}^{\pm} - A_{\dot{G}}$. In a manner analogous to the preceding subsections, the eigenvalues, spectrum and eigenvectors of $N_{\dot{G}}$ are referred to as the *net Laplacian eigenvalues*, the *net Laplacian spectrum* and the *net Laplacian eigenvectors* of $\dot{G}$, respectively. We denote the net Laplacian eigenvalues of $\dot{G}$ by $\nu_1, \nu_2, \ldots, \nu_n$, ordered non-increasingly.

In general, the net Laplacian spectrum may contain both positive and negative values, with zero always included as an eigenvalue associated with the all-ones eigenvector $\mathbf{j}$. Moreover, if $\dot{G}$ is disconnected, the multiplicity of the zero eigenvalue is at least equal to the number of connected components.

For example, the net Laplacian spectrum of the signed graph depicted in Figure 1.3 is given by $[\sqrt{12}^2, 0^2, (-\sqrt{12})^2]$. If $\dot{G}$ is net-regular with common net degree ϱ, then its net Laplacian eigenvalues are $-\lambda_1+\varrho, -\lambda_2+\varrho, \ldots, -\lambda_n+\varrho$, where $\lambda_1, \lambda_2, \ldots, \lambda_n$ are the eigenvalues of $A_{\dot{G}}$. Furthermore, if $\dot{G}$ is taken to be net-regular and regular, with all vertices having the same negative degree d^-, then the net Laplacian eigenvalues satisfy $\mu_1 - 2d^-, \mu_2 - 2d^-, \ldots, \mu_n - 2d^-$, where $\mu_1, \mu_2, \ldots, \mu_n$ denote the Laplacian eigenvalues of $\dot{G}$.

The *eigenvalue equation* of a net Laplacian eigenvalue ν and the corresponding eigenvector $\mathbf{x} = (x_1, x_2, \ldots, x_n)^\intercal$ is slightly different from the one for a Laplacian eigenvalue:

$$(d_v^{\pm} - \nu)x_v = \sum_{uv \in E(\dot{G})} \sigma(uv)x_u. \tag{1.5}$$

In contrast to the spectra of the preceding two matrices, the net Laplacian spectrum is not invariant under switching equivalence.

Take the $n \times m$ matrix such that its (v, e) entry is 1 if the edge e is incident with the vertex v and $\sigma(e) = 1$, i (the imaginary unit) if the edge e is incident with the vertex v and $\sigma(e) = -1$, and 0 otherwise. Now, let B be obtained from the previous matrix by multiplying one of two non-zero entries of each column by -1. It is straightforward to verify that $N_{\dot{G}} = BB^\intercal$ [148], and thus, in this context, the matrix B can be regarded as an analogue of the incidence matrix B_η.

We conclude this subsection with the following result, which can be viewed as a counterpart to Theorem 1.14. A *positive* (resp. *negative*) spanning tree in $\dot{G}$ is defined as a spanning tree containing an even (resp. odd) number of negative edges.

Theorem 1.15 ([148]) *Let $N^i_{\dot{G}}$ be a principal submatrix of $N_{\dot{G}}$ obtained by deleting the ith row and column. Then $\det(N^i_{\dot{G}})$ is the difference between the number of positive spanning trees and the number of negative spanning trees in $\dot{G}$.*

In Section 1.5, we will further discuss the net Laplacian eigenvalues, particularly by transferring certain results from the unsigned case and elaborating on their interplay with the Laplacian eigenvalues. These connections provide additional insight into how structural properties of signed graphs influence their spectral characteristics.

1.4 Switching and Eigenvector Components

We have already established that switching does not alter the spectrum or the Laplacian spectrum of a signed graph. Here, we analyse the effect of the switching operation on the eigenspaces of both the adjacency matrix and the Laplacian matrix.

Let D be a diagonal matrix with diagonal entries d_i. It is known that in the product DA (or AD), all entries in the ith row (resp. jth column) are multiplied by d_i (resp. d_j). In particular, if each $d_i = \pm 1$, then the entries in the ith row (or jth column) change sign precisely when d_i (or d_j) equals -1.

Let $\dot{G}' = \dot{G}^U$, where U is a subset of the vertex set of $\dot{G}$. The switching induced by U produces an edge sign switching on the cut $[U, \dot{G} - U]$, and such a switching is represented by the switching matrix $S = \operatorname{diag}(d_1, d_2, \ldots, d_n)$, where $d_i = 1$ if $v_i \in U$ and $d_i = -1$ otherwise. It has already been observed that $S^{-1}A_{\dot{G}}S = A_{\dot{G}'}$ and $S^{-1}L_{\dot{G}}S = L_{\dot{G}'}$. Hence, we immediately deduce the following result.

Theorem 1.16 *Let* $\mathbf{x}$ *be a* λ*-eigenvector (resp.* μ*-eigenvector) of* $A_{\dot{G}}$ *(resp.* $L_{\dot{G}}$*). If* $\dot{G}' = \dot{G}^U$ *is obtained from the switching realized by the switching matrix* S*, then* $S\mathbf{x}$ *is a* λ*-eigenvector (resp.* μ*-eigenvector) of* $A_{\dot{G}'}$ $(L_{\dot{G}'})$*.*

1.5 Relations Between Laplacian and Net Laplacian Eigenvalues

We establish several relationships between the eigenvalues of the Laplacian matrix $L_{\dot{G}}$ and those of the net Laplacian matrix $N_{\dot{G}}$ for an arbitrary signed graph $\dot{G}$. In particular, we investigate which classical results pertaining to the Laplacian spectrum of unsigned graphs extend to the Laplacian and net Laplacian matrices of signed graphs.

We start with the following result.

Theorem 1.17 ([228]) *Let* $\dot{G}$ *be a signed graph of order* n*.*

(i) If $\dot{H}$ *is obtained from* $\dot{G}$ *by either*

(a) reversing the sign of a negative edge,

(b) adding a positive edge or

(c) removing a negative edge,

then $\nu_i(\dot{G}) \leq \nu_i(\dot{H})$ *for* $1 \leq i \leq n$*. The inequality is strict for at least one* i*.*

(ii) $\nu_i(\dot{G}) \leq \mu_i(\dot{G})$ *for* $1 \leq i \leq n$. *If* $\dot{G}$ *is unbalanced, then the inequality is strict for at least one* i. *If every vertex of* $\dot{G}$ *is incident with at least one negative edge, the inequality is strict for all* i.

We proceed with the following principle proposed for the Laplacian of an unsigned graph by Merris [157]. One will recognize the equality cases of Theorem 1.17(i).

Theorem 1.18 (Edge principle [218, 157]) *Let* ξ *be either a Laplacian or a net Laplacian eigenvalue associated with an eigenvector* $\mathbf{x} = (x_1, x_2, \ldots, x_n)^\mathsf{T}$ *of a signed graph* $\dot{G}$. *If, for some vertices* u *and* v *we have* $x_u = x_v$, *then* ξ *is an eigenvalue associated with the same eigenvector of a signed graph obtained by*

- *adding or removing a positive edge between* u *and* v *if* ξ *is a Laplacian eigenvalue, or*
- *adding or removing (either a positive or a negative) edge between* u *and* v *if* ξ *is a net Laplacian eigenvalue,*

where 'adding or removing' depends on whether or not uv *is an edge of* $\dot{G}$.

The result follows by considering the eigenvalue equations (1.4) and (1.5) for $\dot{G}$ and the signed graph obtained.

Remark 1.19 The edge principle holds for an eigenvalue of the adjacency matrix whenever the corresponding entries x_u and x_v are zero.

It turns out that the following result from [157] applies solely to the net Laplacian spectrum. For signed graphs $\dot{G}_1$ and $\dot{G}_2$, the *positive join* (resp. *negative join*) $\dot{G}_1 \nabla^+ \dot{G}_2$ (resp. $\dot{G}_1 \nabla^- \dot{G}_2$) is the signed graph obtained by adding all possible positive (resp. negative) edges between the vertices of $\dot{G}_1$ and those of $\dot{G}_2$.

Theorem 1.20 ([218]) *Let* $\dot{G}_1$ *and* $\dot{G}_2$ *be signed graphs with net Laplacian eigenvalues* $\nu_1(\dot{G}_1), \nu_2(\dot{G}_1), \ldots, \nu_{n_1}(\dot{G}_1) = 0$ *and* $\nu_1(\dot{G}_2), \nu_2(\dot{G}_2), \ldots, \nu_{n_2}(\dot{G}_2) = 0$, *respectively, and let* $*$ *be a fixed element of* $\{+, -\}$. *The net Laplacian eigenvalues of* $\dot{G}_1 \nabla^* \dot{G}_2$ *are* $*(n_1 + n_2)$, $\nu_1(\dot{G}_1) * n_2$, $\nu_2(\dot{G}_1) * n_2$, $\ldots$, $\nu_{n_1-1}(\dot{G}_1) * n_2$, $\nu_1(\dot{G}_2) * n_1$, $\nu_2(\dot{G}_2) * n_1$, $\ldots$, $\nu_{n_2-1}(\dot{G}_2) * n_1$ *and* 0.

Let $\mathbf{x}$ *be an eigenvector of* $\dot{G}_1$, *orthogonal to* $\mathbf{j}$, *associated with the eigenvalue* ν. *Then its extension to the vertices of* $\dot{G}_1 \nabla^* \dot{G}_2$ *defined by assigning zero to each vertex of* $\dot{G}_2$ *is an eigenvector corresponding to the eigenvalue* $\nu * n_2$. *A similar statement holds for eigenvectors of* $\dot{G}_2$. *Moreover, the eigenvalue* $*(n_1 + n_2)$ *corresponds to the eigenvector that takes the value* $-n_2$ *on each vertex of* $\dot{G}_1$ *and* n_1 *on each vertex of* $\dot{G}_2$.

Proof For $\mathbf{y} = (\mathbf{x}^\intercal, \mathbf{0}^\intercal)^\intercal$, where $\mathbf{x}$ is an eigenvector as described in the theorem, we have

$$N_{\dot{G}_1 \nabla^* \dot{G}_2}\mathbf{y} = \begin{pmatrix} (N_{\dot{G}_1} * n_2 I)\mathbf{x} \\ \mathbf{0} \end{pmatrix} = (\nu * n_2)\mathbf{y}.$$

Similarly, for $\mathbf{y} = (\underbrace{-n_2, -n_2, \dots, -n_2}_{n_1}, \underbrace{n_1, n_1, \dots, n_1}_{n_2})^\intercal$, we have

$$\begin{aligned} N_{\dot{G}_1 \nabla^* \dot{G}_2}\mathbf{y} &= \begin{pmatrix} (N_{\dot{G}_1} * n_2 I)(-n_2\mathbf{j}_{\mathbf{n_1}}) * n_1(-n_2\mathbf{j}_{\mathbf{n_1}}) \\ (N_{\dot{G}_2} * n_1 I)(n_1\mathbf{j}_{\mathbf{n_2}}) * n_2(n_1\mathbf{j}_{\mathbf{n_2}}) \end{pmatrix} \\ &= *(n_1+n_2)\begin{pmatrix} -n_2\mathbf{j}_{\mathbf{n_1}} \\ n_1\mathbf{j}_{\mathbf{n_2}} \end{pmatrix} = *(n_1+n_2)\mathbf{y}, \end{aligned}$$

since $N_{\dot{G}_i}\mathbf{j}_{\mathbf{n_i}} = \mathbf{0}$ for $i \in \{1,2\}$. □

Henceforth, the symbols ∇^+ and ∇ will be used interchangeably.

We now turn to the following relation between the smallest Laplacian eigenvalue and the property of being balanced.

Theorem 1.21 ([25]) *For a signed graph $\dot{G}$ of order n, $\mu_n(\dot{G}) \leq f(\dot{G}) \leq \ell(\dot{G})$, where $f(\dot{G})$ and $\ell(\dot{G})$ denote the frustration number and the frustration index of $\dot{G}$.*

The proof is an extension of the proof of [84, Theorem 2.1] to the case of signed graphs. Due to this result, the smallest Laplacian eigenvalue is referred to as the *algebraic frustration*. In the next theorem, the algebraic frustration μ_n is used to bound the frustration index $\ell(\dot{G})$.

Theorem 1.22 ([150]) *For a signed graph $\dot{G}$ of order n, maximum degree Δ, least Laplacian eigenvalue μ_n and frustration index $\ell(\dot{G})$, we have*

$$\frac{n}{4}\mu_n \leq \ell(\dot{G}) \leq \frac{n}{\sqrt{2}}\sqrt{\mu_n(2\Delta - \mu_n)}.$$

In [87], Fiedler gave a notable formula for the second smallest Laplacian eigenvalue of a graph. Here are its counterparts for signed graphs.

Theorem 1.23 *Let $\dot{G}$ be a connected signed graph and $\mathbf{x} = (x_1, x_2, \dots, x_n)^\intercal$ a real vector. Then:*

(i) (cf. [87]) *For the least Laplacian eigenvalue,*

$$\mu_n(\dot{G}) = 2n \min_{\mathbf{x} \neq \mathbf{0}} \frac{\sum_{ij \in E^+(\dot{G})}(x_i - x_j)^2 + \sum_{ij \in E^-(\dot{G})}(x_i + x_j)^2}{\sum_{i=1}^n x_i^2}; \tag{1.6}$$

(ii) ([228]) *For the largest net Laplacian eigenvalue* ν *associated with a non-constant eigenvector,*

$$\nu_1(\dot{G}) = 2n \max_{\mathbf{x}\neq\mathbf{0},\mathbf{x}\cdot\mathbf{j}=0} \frac{\sum_{ij\in E^+(\dot{G})}(x_i-x_j)^2 - \sum_{ij\in E^-(\dot{G})}(x_i-x_j)^2}{\sum_{i,j\in V(\dot{G})}(x_i-x_j)^2}. \quad (1.7)$$

Proof We demonstrate the proof of (ii), since (i) is proved analogously. The Rayleigh principle gives

$$\nu_1(\dot{G}) = \max_{\mathbf{x}\neq\mathbf{0},\mathbf{x}\cdot\mathbf{j}=0} \frac{\mathbf{x}^\intercal N_{\dot{G}}\mathbf{x}}{\mathbf{x}^\intercal\mathbf{x}}.$$

We compute

$$\begin{aligned}\mathbf{x}^\intercal N_{\dot{G}}\mathbf{x} &= \sum_{i=1}^{n} d_i^{\pm}x_i^2 - 2\sum_{ij\in E(\dot{G})}\sigma(ij)x_ix_j\\ &= \sum_{ij\in E^+(\dot{G})}(x_i^2+x_j^2) - \sum_{ij\in E^-(\dot{G})}(x_i^2+x_j^2)\\ &\quad -2\Big(\sum_{ij\in E^+(\dot{G})}x_ix_j - \sum_{ij\in E^-(\dot{G})}x_ix_j\Big)\\ &= \sum_{ij\in E^+(\dot{G})}(x_i-x_j)^2 - \sum_{ij\in E^-(\dot{G})}(x_i-x_j)^2\end{aligned}$$

and

$$\mathbf{x}^\intercal\mathbf{x} = \sum_{i=1}^{n}x_i^2 = \frac{1}{n}\Big(\frac{1}{2}\sum_{i,j\in V(\dot{G})}(x_i-x_j)^2 + \Big(\sum_{i=1}^{n}x_i\Big)^2\Big) = \frac{1}{2n}\sum_{i,j\in V(\dot{G})}(x_i-x_j)^2,$$

where the former equality follows by the Lagrange identity and the latter by the orthogonality of $\mathbf{x}$ and $\mathbf{j}$. The desired result follows from the previous computation. □

For $\mu_n = 0$, equality (1.6) reduces to the classical Fiedler equality for the second smallest Laplacian eigenvalue, with the denominator as in (1.7). Clearly, if $\nu(\dot{G}) > 0$, then the assumption of Theorem 1.23(ii), imposing that the associated eigenvector is non-constant, is automatically satisfied. Note, however, that the largest net Laplacian eigenvalue may be zero, corresponding to a constant eigenvector.

We conclude this section with the following result concerning particular spanning subgraphs of signed graphs.

Theorem 1.24 *Let* $\dot{G}_1, \dot{G}_2, \ldots, \dot{G}_k$ *be edge-disjoint spanning subgraphs of a signed graph* $\dot{G}$*, such that every edge of* $\dot{G}$ *belongs to some of them. Then:*

(i) ([212]) $\mu_n(\dot{G}) \geq \sum_{i=1}^{k} \mu_n(\dot{G}_k)$;
(ii) ([218]) $\nu_1(\dot{G}) \leq \sum_{i=1}^{k} \nu_1(\dot{G}_k)$.

The proof is based on the Rayleigh principle. For (i), it follows from

$$\begin{aligned}\mu_n(\dot{G}) &= \min_{\mathbf{x}\in\mathbb{R}^n,\,||\mathbf{x}||=1} \mathbf{x}^\mathsf{T} L_{\dot{G}}\mathbf{x} = \min_{\mathbf{x}\in\mathbb{R}^n,\,||\mathbf{x}||=1} \mathbf{x}^\mathsf{T}\Big(\sum_{i=1}^{k} L_{\dot{G}_i}\Big)\mathbf{x}\\ &= \min_{\mathbf{x}\in\mathbb{R}^n,\,||\mathbf{x}||=1} \sum_{i=1}^{k} \mathbf{x}^\mathsf{T} L_{\dot{G}_i}\mathbf{x}\\ &\geq \sum_{i=1}^{k} \min_{\mathbf{x}_i\in\mathbb{R}^n,\,||\mathbf{x}_i||=1} \mathbf{x}_i^\mathsf{T} L_{\dot{G}_i}\mathbf{x}_i = \sum_{i=1}^{k} \mu_n(\dot{G}_i),\end{aligned}$$

By the previous theorem, if $\dot{H}$ is obtained by removing an arbitrary set of edges from $\dot{G}$, then $\mu_n(\dot{H}) \leq \mu_n(\dot{G})$, which also follows by the interlacing argument.

One can easily obtain results that are analogous to those of the last two theorems, but concern μ_1 and ν_n.

1.6 Spectral Criteria for Balance

We now present several spectral criteria for determining whether a signed graph is balanced. Since the spectrum of a signed graph can be computed in polynomial time, these criteria offer a significant computational advantage over the previously discussed structural approaches. Moreover, spectral methods provide a unified framework that can be applied across different classes of signed graphs, allowing for rapid detection of balance without requiring exhaustive examination of cycles or edge signs.

The first result is a classical theorem of Acharya, originally formulated in the context of weighted graphs.

Theorem 1.25 (Acharya's spectral criterion [3]) *A connected signed graph $\dot{G} = (G, \sigma)$ is balanced if and only if the spectra of $\dot{G}$ and G coincide.*

One direction is straightforward: if a signed graph is balanced, then it is switching equivalent to its underlying (unsigned) graph, and thus shares the same spectrum. For a concise proof of the converse implication, we refer the reader to [110].

The second criterion relies solely on the index λ_1 of $\dot{G}$ and provides a refinement of the previous result.

Theorem 1.26 ([210]) *A connected signed graph $\dot{G} = (G, \sigma)$ is balanced if and only if the indices of $\dot{G}$ and G coincide.*

Proof Again, one direction (necessity) is trivial. To restate the proof of the opposite one, we first notice that by the Rayleigh principle, we have the following chain of equalities and inequalities:

$$\lambda_1(\dot{G}) = 2\left|\sum_{i \overset{+}{\sim} j} x_i x_j - \sum_{i \overset{-}{\sim} j} x_i x_j\right| \leq 2\left(\sum_{i \overset{+}{\sim} j} |x_i x_j| + \sum_{i \overset{-}{\sim} j} |x_i x_j|\right) \leq \lambda_1(G), \quad (1.8)$$

where $\mathbf{x} = (x_1, x_2, \ldots, x_n)^\intercal$ is a unit eigenvector with $\lambda_1(\dot{G})$.

Since $\lambda_1(\dot{G}) = \lambda_1(G)$, all the inequalities in (1.8) are equalities. This means that $|\mathbf{x}|$ is an eigenvector with $\lambda_1(G)$, and so $x_i \neq 0$ for all i (by the Perron–Frobenius theorem). By switching with respect to the set of vertices that correspond to negative entries of $\mathbf{x}$, we arrive at a signed graph, say $\dot{H}$, such that $|\mathbf{x}|$ is associated with $\lambda_1(\dot{H})$ as well.

Since $\lambda_1(\dot{H}) = \lambda_1(G)$, it follows (by (1.8), with $\dot{H}$ and $|\mathbf{x}|$ in the roles of $\dot{G}$ and $\mathbf{x}$) that $\dot{H}$ does not contain negative edges, that is, $\dot{G}$ switches to a graph, necessarily its underlying graph G. □

In [62, 110, 182], this result is referred to as *Stanić's spectral criterion.* Another criterion, based on the least Laplacian eigenvalue, was established by Zaslavsky.

Theorem 1.27 (Zaslavsky's spectral criterion [260]) *A connected signed graph $\dot{G}$ of order n is balanced if and only if $\mu_n = 0$.*

Since the Laplacian eigenvalues of a signed graph are non-negative, and since zero is a Laplacian eigenvalue of every (unsigned) graph, it follows immediately that $\mu_n(\dot{G}) \geq \mu_n(G) = 0$. The preceding theorem gives additional information, stating that equality holds if and only if $\dot{G}$ is switching equivalent to G.

Problems

1.1 Prove that the frustration index is equal to the minimum number of edges whose negation results in a balanced signed graph.

1.2 What is the maximum possible value of the frustration index (resp. number) in a signed graph of size m and order n?

1.3 Calculate the frustration index and the frustration number of the r-dimensional signed cube with negative quadrangles.

1.4 Show that Theorem 1.3 is equivalent to the criterion stating that a signed graph is balanced if and only if it can be switched to its underlying graph.

1.5 Provide a combinatorial interpretation of the coefficients a_1, a_2, a_3 of the characteristic polynomial of the adjacency matrix of a signed graph. (Use the Sachs formula.)

1.6 Find a condensed form of the eigenvalues of signed cycles as given in Example 1.11.

1.7 Show that $B_\eta B_\eta^\intercal$ yields the same (Laplacian) matrix regardless of the orientation chosen.

1.8 Prove Theorems 1.12 and 1.14.

1.9 Follow the proof of Theorem 1.26 to establish Theorem 1.27.

2

Eigenvalues

Eigenvalues are a central topic throughout the entire book. The purpose of this chapter is to present a collection of results that do not rely on the more advanced concepts introduced in later chapters.

We begin with a general framework in which the matrix under consideration is assumed to be square, real and symmetric. For ease of reading, it may be helpful to keep in mind that such matrices typically arise as standard matrix representations associated with signed graphs. In the following sections, particular emphasis will be placed on the adjacency matrix.

For an $n \times n$ real symmetric matrix M with distinct eigenvalues $\lambda_1, \lambda_2, \dots, \lambda_t$, let P_i represent the orthogonal projection of $\mathbb{R}^n$ onto the eigenspace $\mathscr{E}(\lambda_i)$ with respect to the canonical basis $\{\mathbf{e}_1, \mathbf{e}_2, \dots, \mathbf{e}_n\}$. Then the *spectral decomposition* of M is given by

$$M = \sum_{i=1}^{t} \lambda_i P_i.$$

If $\{\mathbf{x}_1, \mathbf{x}_2, \dots, \mathbf{x}_d\}$ is an orthonormal basis of $\mathscr{E}(\lambda_i)$, then $P_i = \sum_{i=1}^{d} \mathbf{x}_i \mathbf{x}_i^\intercal$, and therefore the matrices $P_1, P_2, \dots, P_t$ are symmetric, idempotent and mutually orthogonal (in the sense that $P_i P_j = O$, the zero matrix, for $i \neq j$). Accordingly, for any polynomial p over $\mathbb{R}$, we have

$$p(M) = \sum_{i=1}^{t} p(\lambda_i) P_i.$$

Using the previous formula, one can derive the following expression for the characteristic polynomial of a signed graph in terms of the characteristic polynomial of its vertex-deleted subgraph.

Theorem 2.1 ([172]) *Let $\dot{G}$ be obtained from a signed graph $\dot{H}$ of order n by adding a new vertex whose neighbourhood in $\dot{H}$ is determined by the charac-*

teristic $(1,0,-1)$*-vector* $\mathbf{r}$. *The characteristic polynomial of* $\dot{G}$ *is given by*

$$\Phi_{\dot{G}}(x) = \Phi_{\dot{H}}(x)\left(x - \sum_{i=1}^{t} \frac{||Q_i\mathbf{r}||^2}{x-\lambda_i}\right),$$

where $\lambda_1, \lambda_2, \ldots, \lambda_t$ *are the distinct eigenvalues of* $\dot{H}$ *and* $Q_1, Q_2, \ldots, Q_t$ *are the matrices of the orthogonal projections of* $\mathbb{R}^n$ *onto the eigenspaces of* $\dot{H}$ *with respect to the canonical basis.*

Proof Using Schur's complement formula together with

$$\operatorname{adj}(xI - A_{\dot{H}}) = \det(xI - A_{\dot{H}})(xI - A_{\dot{H}})^{-1},$$

we obtain

$$\Phi_{\dot{G}}(x) = \det\begin{pmatrix} x & -\mathbf{r}^\intercal \\ -\mathbf{r} & xI - A_{\dot{H}} \end{pmatrix} = x\Phi_{\dot{H}}(x) - \mathbf{r}^\intercal \operatorname{adj}(xI - A_{\dot{H}})\mathbf{r}$$
$$= \Phi_{\dot{H}}(x)\left(x - \mathbf{r}^\intercal (xI - A_{\dot{H}})^{-1}\mathbf{r}\right).$$

Since $(xI - A_{\dot{H}})^{-1}$ has spectral decomposition $\sum_{i=1}^{t} Q_i/(x-\mu_i)$, we have

$$\Phi_{\dot{G}}(x) = \Phi_{\dot{H}}(x)\left(x - \mathbf{r}^\intercal\left(\sum_{i=1}^{t} \frac{1}{x-\mu_i} Q_i\right)\mathbf{r}\right).$$

The desired equality follows from

$$\mathbf{r}^\intercal Q_i \mathbf{r} = \mathbf{r}^\intercal Q_i Q_i \mathbf{r} = \mathbf{r}^\intercal Q_i^\intercal Q_i \mathbf{r} = (Q_i\mathbf{r})^\intercal Q_i \mathbf{r} = ||Q_i\mathbf{r}||^2,$$

and the proof is complete. □

The unsigned version of the previous result is has been recognized in [75, Theorem 2.2.8]. For the matrix M, we also have

$$\operatorname{sum}(M^k) = \mathbf{j}^\intercal M^k \mathbf{j} = \sum_{i=1}^{t} \lambda_i^k ||P_i\mathbf{j}||^2.$$

where $\operatorname{sum}(M)$ is the sum of the entries of M.

2.1 Main Eigenvalues

We continue to denote by M a real symmetric $n \times n$ matrix. The numbers $\beta_i = ||P_i\mathbf{j}||/\sqrt{n}$ are called the *main angles* of the matrix M. More precisely, they represent the cosines of the angles between the eigenspaces $\mathscr{E}(\lambda_i)$ and the principal direction $\mathbf{j}$. An eigenvalue of M is said to be *main* if its corresponding main angle is non-zero, that is, if there exists an associated eigenvector that is not orthogonal to $\mathbf{j}$. The main eigenvalues of M are denoted by $\mu_1, \mu_2, \ldots, \mu_s$.

In the spirit of [75, Proposition 2.1.3], we perform the following generalization. Let A be a real symmetric $n_1 \times n_1$ matrix with spectral decomposition $A = \mu_1 P_1 + \mu_2 P_2 + \cdots + \mu_a P_a$, and let B be a real symmetric $n_2 \times n_2$ matrix with spectral decomposition $B = \lambda_1 Q_1 + \lambda_1 Q_2 + \cdots + \lambda_b Q_b$. We write $A^* = J - A$, $B^* = J - B$,

$$q_1(x) = n_1 \sum_{i=1}^{a} \frac{\beta_i^2}{x - \mu_i}, \qquad q_2(x) = n_2 \sum_{j=1}^{b} \frac{\gamma_j^2}{x - \lambda_j},$$

where $\beta_i = ||P_i\mathbf{j}||/\sqrt{n_1}$ and $\gamma_j = ||Q_j\mathbf{j}||/\sqrt{n_2}$. Thus, β_i and λ_j are the main angles of μ_i and λ_j, respectively. For the characteristic polynomial of A^* we have

$$\begin{aligned}\Phi_{A^*}(x) &= \det(xI + A - J)\\ &= \det(xI + A) - \mathbf{j}^\intercal \mathrm{adj}(\mathrm{xI + A})\mathbf{j}\\ &= \det(xI + A) - \mathbf{j}^\intercal \det(xI + A)(xI + A)^{-1}\mathbf{j}\\ &= \det(xI + A)\left(1 - \sum_{i=1}^{a} \frac{||P_i\mathbf{j}||^2}{x + \mu_i}\right).\end{aligned}$$

Similarly,

$$\Phi_{B^*}(x) = \det(xI + B)\left(1 - \sum_{j=1}^{b} \frac{||Q_j\mathbf{j}||^2}{x + \lambda_j}\right).$$

Hence,

$$\begin{aligned}\Phi_{A^*}(x) &= (-1)^{n_1} P_A(-x)(1 + q_1(-x)),\\ \Phi_{B^*}(x) &= (-1)^{n_1} P_B(-x)(1 + q_2(-x)).\end{aligned}$$

Note that the matrix $\begin{pmatrix} A & O \\ O & B \end{pmatrix}^*$ has characteristic polynomial

$$(-1)^{n_1+n_2}\Phi_A(-x)\Phi_B(-x)(1 + q_1(-x) + q_2(-x)),$$

which can be rewritten as

$$\begin{aligned}(-1)^{n_1+n_2}(-\Phi_A(-x)\Phi_B(-x) + \Phi_B(-x)(-1)^{n_1}\Phi_{A^*}(x)\\ + \Phi_A(-x)(-1)^{n_2}\Phi_{B^*}(x)).\end{aligned}$$

Now suppose that A, B are the adjacency matrices of signed graphs $\dot{G}, \dot{H}$, and let $C = \begin{pmatrix} A^* & O \\ O & B^* \end{pmatrix}$. The adjacency matrix of the positive join $\dot{G}\nabla^+\dot{H}$ is C^*. Replacing A with A^* and B with B^* in the previous equality, we obtain

$$(-1)^{n_1+n_2}\Phi_{C^*}(x) = -\Phi_{A^*}(-x)\Phi_{B^*}(-x) + \Phi_{B^*}(-x)(-1)^{n_1}\Phi_A(x)$$

$$
\begin{aligned}
&+\Phi_{A^*}(-x)(-1)^{n_2}\Phi_B(x)\\
&=-(-1)^{n_1}\Phi_A(x)(1+q_1(x))(-1)^{n_2}\Phi_B(x)(1+q_2(x))\\
&\quad+(-1)^{n_1+n_2}\Phi_B(x)(1+q_2(x))\Phi_A(x)\\
&\quad+(-1)^{n_1+n_2}\Phi_A(x)(1+q_1(x))\Phi_B(x)\\
&=(-1)^{n_1+n_2}\Phi_A(x)\Phi_B(x)(-(1+q_1(x))(1+q_2(x))\\
&\qquad\qquad+1+q_1(x)+1+q_2(x)).
\end{aligned}
$$

Hence, we arrive at the following result.

Theorem 2.2 ([173]) *Let $\dot{G}$ be a signed graph of order n_1 with distinct eigenvalues $\mu_1,\mu_2,\dots,\mu_a$ and corresponding main angles $\beta_1,\beta_2,\dots,\beta_a$, and $\dot{H}$ be a signed graph of order n_2 with distinct eigenvalues $\lambda_1,\lambda_2,\dots,\lambda_b$ and corresponding main angles $\gamma_1,\gamma_2,\dots,\gamma_b$. Then*

$$\Phi_{\dot{G}\nabla^+\dot{H}}(x)=\Phi_{\dot{G}}(x)\Phi_{\dot{H}}(x)\left(1-n_1n_2\sum_{i=1}^{a}\sum_{j=1}^{b}\frac{\beta_i^2\gamma_j^2}{(x-\mu_i)(x-\gamma_j)}\right).$$

The following result is a direct consequence of the previous theorem; we recall that a *cone* is a positive join of a signed graph with a single vertex.

Corollary 2.3 ([173]) *The cone over a signed graph $\dot{H}$ of order n has characteristic polynomial*

$$\Phi_{K_1\nabla^+\dot{H}}(x)=\Phi_{\dot{H}}(x)\left(x-\sum_{i=1}^{a}\frac{n\beta_i^2}{x-\mu_i}\right),$$

where $\mu_1,\mu_2,\dots,\mu_a$ are the distinct eigenvalues of $\dot{H}$ and $\beta_1,\beta_2,\dots,\beta_a$ are the corresponding main angles.

If the main eigenvalues of M are $\mu_1,\mu_2,\dots,\mu_s$, we set

$$\mu(x)=\sum_{i=1}^{s}(x-\mu_i). \tag{2.1}$$

Let $\{\Delta_1,\Delta_2,\dots,\Delta_k\}$ be a partition of $\{1,2,\dots,n\}$ that determines a blocking $M=(M_{ij})$ such that each block M_{ij} has a constant row sum, say f_{ij}. Observe that such a blocking exists for any M, as we can always take $k=n$. The corresponding partition is called an *equitable partition* and the $k\times k$ matrix (f_{ij}) is denoted by F. We quote some basic results drawn from [213]. They can also be derived from the results concerning the main eigenvalues of (the adjacency matrix) of a graph, which can be found in [179]; see also [75].

Theorem 2.4 ([213]) *With the previous notation, let M be an $n \times n$ real symmetric matrix with s main eigenvalues and, for $1 \leq i \leq s$, let the main eigenvalues and the corresponding main angles be denoted by μ_i and β_i, respectively. The following statements hold true.*

(i) $\beta_i = \sqrt{\mathrm{sum}(P_i)/n}$ *for* $1 \leq i \leq s$.

(ii) $s \neq 0$, $\sum_{i=1}^{s} \beta_i^2 = 1$ *and*

$$\mathrm{sum}(M^k) = n \sum_{i=1}^{s} \mu_i^k \beta_i^2. \tag{2.2}$$

(iii) *For a polynomial p over $\mathbb{R}$, $p(M)\mathbf{j} = \mathbf{0}$ if and only if the polynomial μ (defined in* (2.1)*) divides p. In particular, $\mu(M)\mathbf{j} = \mathbf{0}$.*

(iv) *The largest k such that the vectors $\mathbf{j}, M\mathbf{j}, \dots, M^k\mathbf{j}$ are linearly independent is equal to $s-1$.*

(v) *The polynomial μ divides Φ_F.*

(vi) *The spectrum of F is contained in the spectrum of M (taking into account the repetition of eigenvalues).*

(vii) *$s = 1$ holds if and only if $\mathbf{j}$ is an eigenvector of M. The unique main eigenvalue is associated with $\mathbf{j}$.*

(viii) *$s = 2$ holds if and only if $\mathbf{j}$ is not an eigenvector of M, and $(M^2 - aM + bI)\mathbf{j} = \mathbf{0}$ holds for some $a, b \in \mathbb{R}$. The main eigenvalues μ_1, μ_2 are determined by $\mu_1 + \mu_2 = a$, $\mu_1\mu_2 = b$.*

The matrix F of Theorem 2.4(vi) is called the *divisor* (or the *front divisor*) of M.

We proceed with the following result concerning signed graphs. In particular, we count walks in terms of the main eigenvalues. For a signed graph $\dot{G}$ of order n, we set $N_k = \sum_{i=1}^{n} w_k(i)$. Here, $w_k(i)$ is the difference between the number of positive walks and the number of negative walks of length k starting at vertex i, and analogously N_k is the difference between the number of positive walks and the number of negative walks of length k in $\dot{G}$.

Lemma 2.5 ([213]) *Let $\dot{G}$ be a signed graph with n vertices and main eigenvalues $\mu_1, \mu_2, \dots, \mu_s$. The following statements hold:*

(i) $N_k = n\sum_{i=1}^{s} \mu_i^k \beta_i^2$.

(ii) *An eigenvalue λ of $\dot{G}$ is main if and only if $-\lambda$ is a main eigenvalue of $-\dot{G}$. Moreover, they share the same main angle.*

(iii) *If λ is a non-main eigenvalue of $\dot{G}$, then there exists a switching equivalent signed graph in which λ is main.*

Proof As $N_k = \text{sum}(A^k_{\dot{G}})$, the result of (i) follows from (2.2). The result of (ii) follows since λ (of $\dot{G}$) and $-\lambda$ of $(-\dot{G})$ share the same eigenspace.

For (iii), if $A_{\dot{G}}\mathbf{x} = \lambda\mathbf{x}$ for $\mathbf{x} = (x_1, x_2, \ldots, x_n)^\intercal$, and $S = (s_{ij})$ is the diagonal matrix with

$$s_{ii} = \begin{cases} -1 & \text{if } x_i < 0, \\ 1 & \text{if } x_i \geq 0, \end{cases}$$

Theorem 1.16 ensures that $S\mathbf{x}$ is an eigenvector associated with λ in $S^{-1}A_{\dot{G}}S$. Moreover, the entries of $S\mathbf{x}$ are non-negative, and we are done. □

In the particular case of graphs, Lemma 2.5(i) reduces to the well-known result expressing the number of walks of length k in terms of main eigenvalues and main angles [179].

Remark 2.6 We can deduce from the proof of item (iii) of the previous theorem that for every eigenvalue (main or non-main) λ, there exists a switching-equivalent signed graph for which the eigenspace of λ contains a non-negative eigenvector.

Example 2.7 By Theorem 2.4, any matrix associated with a signed graph has exactly one main eigenvalue if and only if $\mathbf{j}$ is its eigenvector. Consequently, a signed graph $\dot{G}$ has exactly one main eigenvalue if and only if $\dot{G}$ is net-regular (with the net degree as the main eigenvalue). Similarly, $\dot{G}$ has exactly one main Laplacian eigenvalue if and only if $\dot{G}^-$, induced by negative edges, is regular (with $2d_i^-$ as the main eigenvalue). Finally, zero is the unique main eigenvalue of the net Laplacian matrix of every signed graph.

Motivated by Lemma 2.5(iii), which asserts that even if an eigenvalue is not main, there exists a switching of $\dot{G}$ for which it becomes main, we highlight the following conjecture proposed by Akbari et al.

Conjecture 2.8 ([8]) *Let G be a graph distinct from K_2 and $K_4 - e$ (obtained from K_4 by deleting an edge). There exists a switching in which all eigenvalues are main.*

It is not difficult to verify that the excluded graphs cannot be switched into a signed graph possessing the desired property. At present, no additional counterexamples are known. On the other hand, the conjecture has been confirmed for all graphs with at most nine vertices [8], as well as for particular connected graphs without repeated eigenvalues, namely those with at most ten vertices and regular graphs with a comparatively small order [15]. Furthermore, the conjecture holds for several special classes of graphs, including Cayley graphs, distance-regular graphs, vertex-transitive and edge-transitive graphs, double stars and paths [8].

In [15], Conjecture 2.8 is also considered in the context of Laplacian eigenvalues. It is shown to hold for all graphs with at most seven vertices, while exactly 22 connected graphs with eight vertices provide exceptions. This motivates the following problem, which, in the particular case of the adjacency matrix, constitutes a weaker version of the preceding conjecture.

Problem 2.9 ([15]) For a given graph G, determine all switching-equivalent signed graphs $\dot{G}$ such that all the eigenvalues of $A_{\dot{G}}$ (or $L_{\dot{G}}$) are main.

If G is a cograph without repeated eigenvalues, then [15] provides an iterative solution for the previous problem and the complete solution in the particular case when G is a threshold graph. We recall that a *cograph* is a P_4-free graph (i.e., it does not contain the path P_4 as an induced subgraph) and a *threshold graph* is a $\{2K_2, P_4, C_4\}$-free graph.

2.2 Determinant

The *determinant* of a signed graph is the determinant of its adjacency matrix, meaning $\det(\dot{G}) = (-1)^n a_n = \prod_{i=1}^{n} \lambda_i$, that is, $\det(\dot{G})$ equals the product of the eigenvalues taken with multiplicities. Another way to compute the determinant is based on the basic figures considered in Theorem 1.9. In what follows, we use Theorem 2.1 to demonstrate a computation in terms of distinct eigenvalues of a fixed vertex-deleted subgraph. The result is used in Section 8.2.

Theorem 2.10 ([10]) *Let $\dot{G}$ be a signed graph of order n, and $\dot{H}$ a subgraph obtained by deleting a vertex whose neighbourhood in $\dot{H}$ is determined by the characteristic $(1,0,-1)$-vector $\mathbf{r}$. If $\lambda_1, \lambda_2, \ldots, \lambda_k$ are distinct eigenvalues of $\dot{H}$ and $m_1, m_2, \ldots, m_k$ are their multiplicities, then the determinant of $\dot{G}$ is*

$$\det(\dot{G}) = -\sum_{i=1}^{k} \alpha_i \lambda_i^{m_i - 1} \prod_{j \neq i} \lambda_j^{m_j},$$

where $\alpha_i = \sum_{j=1}^{m_i} (\mathbf{r}^\intercal \mathbf{x}_j^i)^2$ and $\mathbf{x}_1^i, \mathbf{x}_2^i, \ldots, \mathbf{x}_{m_i}^i$ are mutually orthogonal unit vectors that make an eigenbasis of λ_i.

Proof The orthogonal projection of $\mathbb{R}^{n-1}$ onto the eigenspace of λ_i is realized by the matrix $Q_i = \sum_{j=1}^{m_i} \mathbf{x}_j^i \mathbf{x}_j^{i\intercal}$ (see the beginning of this chapter). We have

$$||Q_i\mathbf{r}||^2 = \Big(\Big(\sum_{j=1}^{m_i} \mathbf{x}_j^i \mathbf{x}_j^{i\intercal}\Big)\mathbf{r}\Big)^\intercal \Big(\sum_{j=1}^{m_i} \mathbf{x}_j^i \mathbf{x}_j^{i\intercal}\Big)\mathbf{r} = \mathbf{r}^\intercal \Big(\sum_{j=1}^{m_i} \mathbf{x}_j^i \mathbf{x}_j^{i\intercal}\Big)^\intercal \Big(\sum_{j=1}^{m_i} \mathbf{x}_j^i \mathbf{x}_j^{i\intercal}\Big)\mathbf{r}$$

$$=\mathbf{r}^{\mathsf{T}}\Big(\sum_{j=1}^{m_i}\mathbf{x}^i_j\mathbf{x}^{i\,\mathsf{T}}_j\Big)\Big(\sum_{j=1}^{m_i}\mathbf{x}^i_j\mathbf{x}^{i\,\mathsf{T}}_j\Big)\mathbf{r}=\mathbf{r}^{\mathsf{T}}\Big(\sum_{j=1}^{m_i}\mathbf{x}^i_j\mathbf{x}^{i\,\mathsf{T}}_j\mathbf{x}^i_j\mathbf{x}^{i\,\mathsf{T}}_j\Big)\mathbf{r}$$

$$=\mathbf{r}^{\mathsf{T}}\Big(\sum_{j=1}^{m_i}\mathbf{x}^i_j\mathbf{x}^{i\,\mathsf{T}}_j\Big)\mathbf{r}=\sum_{j=1}^{m_i}(\mathbf{r}^{\mathsf{T}}\mathbf{x}^i_j)^2=\alpha_i,$$

where the fourth equality follows from the mutual orthogonality of distinct eigenvectors, while the fifth equality holds because the eigenvectors are normalized to unit length. If $\phi_{\dot{H}}$ denotes the minimal polynomial of $\dot{H}$, then Theorem 2.1 leads to

$$\Phi_{\dot{G}}(x)=\Phi_{\dot{H}}(x)\Big(x-\sum_{i=1}^{k}\frac{\alpha_i}{x-\lambda_i}\Big)=x\Phi_{\dot{H}}(x)-\frac{\Phi_{\dot{H}}(x)}{\phi_{\dot{H}}(x)}\sum_{i=1}^{k}\alpha_i\prod_{j\neq i}(x-\lambda_j).$$

In particular, this means that the constant term a_n of $\Phi_{\dot{G}}(x)$ is equal to the constant term of $-(\Phi_{\dot{H}}(x)/\phi_{\dot{H}}(x))\sum_{i=1}^{k}\alpha_i\prod_{j\neq i}(x-\lambda_j)$, and thus

$$\begin{aligned}a_n&=-\frac{(-1)^{n-1}\prod_{i=1}^{k}\lambda_i^{m_i}}{(-1)^k\prod_{i=1}^{k}\lambda_i}\sum_{i=1}^{k}\alpha_i(-1)^{k-1}\prod_{j\neq i}\lambda_j\\&=(-1)^{n-1}\prod_{i=1}^{k}\lambda_i^{m_i-1}\sum_{i=1}^{k}\alpha_i\prod_{j\neq i}\lambda_j=(-1)^{n-1}\sum_{i=1}^{k}\alpha_i\lambda_i^{m_i-1}\prod_{j\neq i}\lambda_j^{m_j}.\end{aligned}$$

The desired result follows from $\det(\dot{G})=(-1)^n a_n$. □

Consequently, if $\dot{H}$ has no repeated eigenvalues, then

$$\det(\dot{G})=-\sum_{i=1}^{n-1}\alpha_i\prod_{j\neq i}\lambda_j. \tag{2.3}$$

Also, if 0 is an eigenvalue of $\dot{H}$, then

$$\det(\dot{G})=\begin{cases}-(\mathbf{r}^{\mathsf{T}}\mathbf{x})^2\prod_{\lambda_j\neq 0}\lambda_j^{m_j} & \text{if 0 is simple,}\\ 0 & \text{otherwise,}\end{cases} \tag{2.4}$$

where $\mathbf{x}$ is a unit eigenvector associated with 0.

2.3 Operations and Resulting Spectra

As shown in Section 2.1, the spectrum of the join, either positive or negative, of two signed graphs can be expressed in terms of their main eigenvalues. In this section, we examine several operations on signed graphs and analyze the resulting spectra.

Our focus is on the complementary signed graph introduced by Caveleri,

Donno and Spessato [58], a family of operations commonly known as a non-complete extended p-sum (NEPS) of signed graphs, the corona product and few other products. In the remainder of the book one will frequently meet spectral properties of signed graphs obtained by other operations (say, signed line graphs). More results that belong to this topic but are not covered in the book can be found in [50, 51] (lexicographic product), [170, 192] (corona-like products) and [241, 242] (more sophisticated products).

2.3.1 A Complement

At present, there is no commonly accepted definition of the complement of a signed graph. However, there are several attempts used for particular purposes. Needless to add, the requirement that the underlying graph of a signed graph $\dot{G}$ is the complement of the underlying graph of the complement of $\dot{G}$ features as a natural one. The approaches differ in signature on the complementary signed graph. Here will explain one of them.

In [58], the authors define pseudo-potentials on a signed graph $\dot{G}$, which coincide with usual potential functions in the balanced case. On this basis they have defined the signature on the complementary signed graph in such a way that the complete signed graph obtained by combining $\dot{G}$ and its complement is stable under switching equivalence.

Accordingly, for a connected signed graph $\dot{G} = (G, \sigma)$, let T be a fixed spanning tree of G and v a fixed vertex of $V(G)$. The *pseudo-potential* $U_{T,v,\sigma} \colon V \to \{1, -1\}$ is the function defined for each $w \in V$ as

$$U_{T,v,\sigma}(w) = \sigma(\gamma_{v,w}),$$

where $\gamma_{v,w}$ is any walk contained in T from v to w. Observe that $U_{T,v,\sigma}$ does not depend on the choice of the walk, and $U_{T,v,\sigma}(v) = 1$. If $\dot{G}$ has k connected components, then a pseudo-potential function is the union of k pseudo-potentials, one for each component. In what follows, $U_{T,v,\sigma}(w)$ is simultaneously used for the diagonal matrix whose (w,w) entry coincides with $U_{T,v,\sigma}(w)$. Here are some basic properties.

Lemma 2.11 ([58]) *Let (G, σ) be a signed graph and T a spanning tree of G.*

(i) If G is connected, then for any $v, v' \in V$, either $U_{T,v,\sigma} = U_{T,v',\sigma}$ or $U_{T,v,\sigma} = -U_{T,v',\sigma}$.
(ii) If (G, σ) is balanced, then $U_{T,v,\sigma}$ is independent of T.
(iii) If σ and σ' correspond to switching-equivalent signed graphs $\dot{G}$ and $\dot{H}$, then there exists a diagonal matrix S with entries in $\{1, -1\}$ such that

$$A_{\dot{H}} = S^{-1} A_{\dot{G}} S, \quad U_{T,v,\sigma'} = S U_{T,v,\sigma} = U_{T,v,\sigma} S.$$

We now define the signature on the complementary signed graph $\dot{G}^c$ of $\dot{G} = (G, \sigma)$. Fix a spanning tree T of G, a vertex $v \in V(G)$, and let $U_{T,v,\sigma}$ be the corresponding pseudo-potential. Then the signature on the complement $\dot{G}^c$ is defined by

$$\sigma_T^c(xy) = U_{T,v,\sigma}(x) U_{T,v,\sigma}(y).$$

From the previous definition and Lemma 2.11, we deduce the following:

- If $\overline{G}$ is the complement of G, $U_{T,v,\sigma}$ is a potential for $(\overline{G}, \sigma_T^c)$, so $\dot{G}^c$ is balanced.
- If $\dot{G}$ is connected, σ_T^c is independent of v.
- If $\dot{G}$ is balanced, σ_T^c is independent of T.
- The adjacency matrix of $\dot{G}^c = (\overline{G}, \sigma_T^c)$ is $A_{\dot{G}^c} = U_{T,v,\sigma} A_{\overline{G}} U_{T,v,\sigma}$.

We proceed with the complete signed graph obtained by combining $\dot{G}$ and $\dot{G}^c$. Let $\dot{G} = (G, \sigma)$ be a signed graph and K_n the complete graph, each on n vertices. The *extended signature* $\overline{\sigma}_T$ on K_n is defined as

$$\overline{\sigma}_T(xy) = \begin{cases} \sigma(xy) & \text{if } xy \in E(G), \\ \sigma_T^c(xy) & \text{otherwise.} \end{cases}$$

Stability under the switching operation is ensured by the following result.

Theorem 2.12 ([58]) *Let G be a graph and let T be a spanning tree of G. Consider a switching class $[\sigma]$ of signatures on G. Then $[\overline{\sigma}_T]$ does not depend on the choice of the fixed vertex of G (or list of vertices, one for each connected component, if G is disconnected), nor on the choice of the representative for $[\sigma]$.*

The proof is based on the basic properties listed in and below Lemma 2.11.

A well-known property of the spectrum of an ordinary regular graph G is that the spectrum of its complement is determined by the spectrum of G. Specifically, if G is an r-regular graph with eigenvalues $r = \lambda_1 \geq \lambda_2 \geq \cdots \geq \lambda_n$, then its complement $\overline{G}$ is $(n-r-1)$-regular with eigenvalues

$$\overline{\lambda}_1 = n - \lambda_1 - 1 = n - r - 1,$$
$$\overline{\lambda}_j = -\lambda_j - 1, \quad 2 \leq j \leq n;$$

see [222, Theorem 2.2.1].

In contrast, no general analogue of this result holds for the complement of a signed graph. Nevertheless, it has been shown in [58] that, under certain additional regularity conditions, some analogous spectral relationships can be established.

2.3.2 NEPS Products

The NEPS of a signed graph is the generalization of the homonym operation on a simple graph [91]. NEPS was defined by Cvetković to generalize the usual graph products as the Cartesian product or the strong product. It is generalized to signed graphs as follows. Let $\dot{G}_1, \dot{G}_2, \ldots, \dot{G}_p$ be p signed graphs, where the underlying graphs are $G_i = (V_i, E_i)$, and let $\beta = (b_1, b_2, \ldots, b_p) \in \{0,1\}^p$.

By $\mathrm{NEPS}(\dot{G}_1, \dot{G}_2, \ldots, \dot{G}_p; \beta)$ we denote the signed graph such that:

- the vertex set is given by $V_1 \times V_2 \times \ldots \times V_p$;
- the edge set is $E = \{(u_1, u_2, \ldots, u_p)(v_1, v_2, \ldots, v_p) \colon u_i = v_i \text{ if } b_i = 0,\ u_iv_i \in E_i \text{ if } b_i = 1\}$;
- the sign is given by $\sigma((u_1, u_2, \ldots, u_p)(v_1, v_2, \ldots, v_p)) = \prod_{i\colon b_i=1} \sigma(u_iv_i)$.

In the general setting, we have a set

$$\mathscr{B} = \{\beta_1, \beta_2, \ldots, \beta_r \colon \beta_i \in \{0,1\}^p \setminus \{(0,0,\ldots,0)\}\}$$

and define

$$\mathrm{NEPS}(\dot{G}_1, \dot{G}_2, \ldots, \dot{G}_p; \mathscr{B}) \cong \bigcup_{\beta \in \mathscr{B}} \mathrm{NEPS}(\dot{G}_1, \dot{G}_2, \ldots, \dot{G}_p; \beta).$$

The following theorem establishes the spectrum of $\mathrm{NEPS}(\dot{G}_1, \dot{G}_2, \ldots, \dot{G}_p; \mathscr{B})$.

Theorem 2.13 ([91]) *Let $\dot{G} \cong \mathrm{NEPS}(\dot{G}_1, \dot{G}_2, \ldots, \dot{G}_p; \mathscr{B})$. The eigenvalues of $\dot{G}$ are*

$$\lambda_{j_1 j_2 \ldots, j_p} = \sum_{\beta \in \mathscr{B}} \lambda_{1 j_1}^{b_1} \lambda_{2 j_2}^{b_2} \cdots \lambda_{p j_p}^{b_p},$$

where $\beta = (b_1, b_2, \ldots, b_p)$ and λ_{ij} is the jth eigenvalue of $\dot{G}_i$.

Suitable choices of $\mathscr{B}$ give some standard products. If $\mathscr{B} = \{(0,1),(1,0)\}$, then we get the signed version of the standard Cartesian product. For $\mathscr{B} = \{(1,1)\}$, we get the tensor product. Here are more details.

The set of vertices of the *Cartesian product* $\dot{G}_1 \,\square\, \dot{G}_2$ and the *tensor product* $\dot{G}_1 \times \dot{G}_2$ is the Cartesian product of the sets of vertices of $\dot{G}_1$ and $\dot{G}_2$.

For $\dot{G}_1 \,\square\, \dot{G}_2$, the vertices (u_i, v_j) and (u_k, v_l) are connected by a positive (resp. negative) edge if and only if $u_i = u_k$ and v_j, v_l are joined by a positive (resp. negative) edge or u_i, u_k are joined by a positive (resp. negative) edge and $v_j = v_l$. Observe that $\dot{G}_1 \,\square\, \dot{G}_2$ consists of n_2 copies of $\dot{G}_1$ and additional edges which occur between a vertex in the jth copy and the same vertex in the lth copy precisely if v_j and v_l are adjacent in $\dot{G}_2$; the sign of such an edge is transferred from $\dot{G}_2$. Clearly, $\dot{G}_1 \,\square\, \dot{G}_2$ is connected if and only if $\dot{G}_1$ and $\dot{G}_2$ are connected.

For $\dot{G}_1 \times \dot{G}_2$, the vertices (u_i, v_j) and (u_k, v_l) are joined by an edge if and only if $u_i \sim u_k$ and $v_j \sim v_l$. The sign equals $\sigma(u_iu_k)\sigma(v_jv_l)$.

According to Theorem 2.13, the spectrum of $\dot{G}_1 \,\square\, \dot{G}_2$ is comprised of all possible sums of the eigenvalues of $\dot{G}_1$ and $\dot{G}_2$. Similarly, all possible products of the eigenvalues of the same signed graphs form the spectrum of $\dot{G}_1 \times \dot{G}_2$.

A particular tensor product $\mathrm{bd}(\dot{G}) \cong \dot{G} \times K_2$ is called the *bipartite double* (also known as the bipartite cover or 2-lift) of $\dot{G}$. The bipartite double is connected if and only if $\dot{G}$ is non-bipartite; otherwise, it consists of two disjoint copies of $\dot{G}$. The eigenvalues of $\mathrm{bd}(\dot{G})$ are $\pm\lambda_1, \pm\lambda_2, \ldots, \pm\lambda_n$, where $\lambda_1, \lambda_2, \ldots, \lambda_n$ are the eigenvalues of $\dot{G}$.

2.3.3 Corona Product

Let $\dot{G}_1$ and $\dot{G}_2$ be two signed graphs with n_1 and n_2 vertices, respectively. The *positive corona* (resp. *negative corona*), denoted by $\dot{G}_1 \circ^+ \dot{G}_2$ (resp. $\dot{G}_1 \circ^- \dot{G}_2$), is the signed graph obtained by taking one copy of $\dot{G}_1$ and n_1 copies of $\dot{G}_2$, and joining the ith vertex of $\dot{G}_1$ with positive (resp. negative) edges to every vertex in the ith copy of $\dot{G}_2$.

The previous definition generalizes the definition of the corona of ordinary graphs. We demonstrate a technique to compute the spectrum of the net Laplacian matrix of $\dot{G}_1 \circ^* \dot{G}_2$, where $*$ stands for $+$ or $-$. The following notion is needed: for an $n \times n$ matrix M, the *coronal* $\zeta_M(t)$ is the entry sum of the matrix $(tI_n - M)^{-1}$. It is established that, if every row sum of M is equal to a fixed real number k, then $\zeta_M(t) = n/(t-k)$ [153].

Theorem 2.14 ([192]) *Let $\dot{G} = \dot{G}_1 \circ^* \dot{G}_2$ (where $* \in \{+,-\}$) be a corona of signed graphs $\dot{G}_1$ and $\dot{G}_2$ having n_1 and n_2 vertices, respectively. Let the net Laplacian eigenvalues of these signed graphs be $\nu_1(\dot{G}_1), \nu_2(\dot{G}_1), \ldots, \nu_{n_1}(\dot{G}_1)$ and $\nu_1(\dot{G}_2), \nu_2(\dot{G}_2), \ldots, \nu_{n_2}(\dot{G}_2)$, where $\nu_k(\dot{G}_2) = 0$ for some k $(1 \leq k \leq n_2)$. Then the net Laplacian spectrum of $\dot{G}_1 \circ^* \dot{G}_2$ is comprised of:*

(i) $\nu_j(\dot{G}_2) * 1$ with multiplicity n_1, for every $\nu_j(\dot{G}_2)$ $(j \neq k)$;

(ii) $\left[(\nu_i(\dot{G}_1) * (n_2+1)) \pm \sqrt{(\nu_i(\dot{G}_1) * (n_2+1))^2 *^{\mathrm{c}} 4\nu_i(\dot{G}_1)}\right]/2$ for every $\nu_i(\dot{G}_1)$, where $*^{\mathrm{c}}$ is a sign opposite to $*$.

Proof If the vertex sets of the two constituents are $V_{\dot{G}_1} = \{v_1, v_2, \ldots, v_{n_1}\}$ and $V_{\dot{G}_2} = \{u_1, u_2, \ldots, u_{n_2}\}$, respectively, let $\{u_1^j, u_2^j, \ldots, u_{n_2}^j\}$, $1 \leq j \leq n_1$, denote the vertex set of the jth copy of $\dot{G}_2$, with the convention that u_i^j is the copy of u_i for each i. We fix $V_i = \{u_i^1, u_i^2, \ldots, u_i^{n_1}\}$, $1 \leq i \leq n_2$. Then $V_{\dot{G}_1} \cup V_1 \cup V_2 \cup \cdots \cup V_{n_2}$ is a partition of the vertex set $V_{\dot{G}}$. With this partition, the corresponding net

Laplacian matrix is given by

$$N_{\dot{G}} = \begin{pmatrix} N_{\dot{G}_1} * n_2 I_{n_1} & *^{\mathrm{c}}(\mathbf{j}_{n_2}^{\mathsf{T}} \otimes I_{n_1}) \\ *^{\mathrm{c}}\left(\mathbf{j}_{n_2}^{\mathsf{T}} \otimes I_{n_1}\right)^{\mathsf{T}} & (N_{\dot{G}_2} * I_{n_2}) \otimes I_{n_1} \end{pmatrix}.$$

By employing Schur's complement formula and simple transformations on the Kronecker product, we compute the characteristic polynomial

$$\begin{aligned}
\Phi_{N_{\dot{G}}}(t) &= \det\left(tI_{n_1(n_2+1)} - N_{\dot{G}}\right) \\
&= \det\begin{pmatrix} tI_{n_1} - (N_{\dot{G}_1} * n_2 I_{n_1}) & *(\mathbf{j}_{n_2}^{\mathsf{T}} \otimes I_{n_1}) \\ *\left(\mathbf{j}_{n_2}^{\mathsf{T}} \otimes I_{n_1}\right)^{\mathsf{T}} & tI_{n_1 n_2} - \left(N_{\dot{G}_2} * I_{n_2}\right) \otimes I_{n_1} \end{pmatrix} \\
&= \det\begin{pmatrix} tI_{n_1} - (N_{\dot{G}_1} * n_2 I_{n_1}) & *(\mathbf{j}_{n_2}^{\mathsf{T}} \otimes I_{n_1}) \\ *\left(\mathbf{j}_{n_2}^{\mathsf{T}} \otimes I_{n_1}\right)^{\mathsf{T}} & \left(tI_{n_2} - \left(N_{\dot{G}_2} * I_{n_2}\right)\right) \otimes I_{n_1} \end{pmatrix} \\
&= \det\left(\left(tI_{n_2} - \left(N_{\dot{G}_2} * I_{n_2}\right)\right) \otimes I_{n_1}\right) \times \det\Big(tI_{n_1} - (N_{\dot{G}_1} * n_2 I_{n_1}) \\
&\quad - (\mathbf{j}_{n_2}^{\mathsf{T}} \otimes I_{n_1})\left(\left(tI_{n_2} - (N_{\dot{G}_2} * I_{n_2})\right) \otimes I_{n_1}\right)^{-1} (\mathbf{j}_{n_2} \otimes I_{n_1})\Big) \\
&= \left(\det\left((t *^{\mathrm{c}} 1)I_{n_2} - N_{\dot{G}_2}\right)\right)^{n_1} \times \det\Big(tI_{n_1} - (N_{\dot{G}_1} * n_2 I_{n_1}) \\
&\quad - \left(\mathbf{j}_{n_2}^{\mathsf{T}}\left((t *^{\mathrm{c}} 1)I_{n_2} - N_{\dot{G}_2}\right)^{-1} \mathbf{j}_{n_2} \otimes I_{n_1}\right)\Big) \\
&= \left(\det\left((t *^{\mathrm{c}} 1)I_{n_2} - N_{\dot{G}_2}\right)\right)^{n_1} \det\left(\left(t - \zeta_{N_{\dot{G}_2}}(t *^{\mathrm{c}} 1) *^{\mathrm{c}} n_2\right) I_{n_1} - N_{\dot{G}_1}\right) \\
&= \left(\Phi_{N_{\dot{G}_2}}(t *^{\mathrm{c}} 1)\right)^{n_1} \Phi_{N_{\dot{G}_1}}\left(t *^{\mathrm{c}} n_2 - \zeta_{N_{\dot{G}_2}}(t *^{\mathrm{c}} 1)\right).
\end{aligned}$$

Since the entry sum of every row of the net Laplacian matrix is zero, for the net Laplacian coronal we have

$$\zeta_{N_{\dot{G}_2}}(t) = \frac{n_2}{t}.$$

Clearly, $t = 0$ is a pole of $\zeta_{N_{\dot{G}_2}}(t)$ of order 1. Thus, taking into account the above expression for the characteristic polynomial, we obtain that $\nu_j(\dot{G}_2) * 1$ is a net Laplacian eigenvalue of $N_{\dot{G}}$ of multiplicity n_1 for $j \neq k$. The remaining $2n_1$ net Laplacian eigenvalues are obtained by solving $t *^{\mathrm{c}} n_2 - n_2/(t *^{\mathrm{c}} 1) - \nu_i(\dot{G}_1) = 0$ for every i. □

Spectra of many corona-like products of signed graphs are computed in [170, 192].

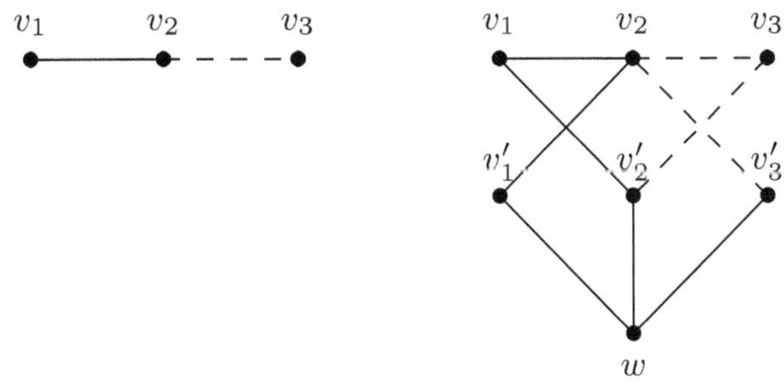

Figure 2.1 A signed path and its Mycielskian.

2.3.4 Signed Mycielskian

The Mycielskian of a graph was introduced by Mycielski in 1955 [161] as a transformation of a graph into a triangle-free graph with an arbitrary chromatic number. Accordingly, for $V = \{v_1, v_2, \ldots, v_n\}$ and $E = \{e_1, e_2, \ldots, e_n\}$, the vertex set of the Mycielskian $M(G)$ is the disjoint union of V, $V' = \{v'_1, v'_2, \ldots, v'_n\}$ and an isolated vertex w. Simultaneously, its edge set is $E \cup \{v'_i v_j \colon v_i v_j \in E\} \cup \{v'_i w : v'_i \in V'\}$.

Recently, Mathew and Germina [151] have introduced the concept of a Mycielskian for signed graphs. The *signed Mycielskian* of $\dot{G} = (G, \sigma)$ is the signed graph $M(\dot{G}) = (M(G), \sigma_M)$, where $M(G)$ is the Mycielskian of the underlying graph G, and the signature σ_M is defined as $\sigma_M(v_i v_j) = \sigma_M(v'_i v_j) = \sigma(v_i v_j)$ and $\sigma_M(v'_i w) = 1$. An example of the signed Mycielskian is illustrated in Figure 2.1.

Now, we compute the characteristic polynomial of $M(\dot{G})$ when $\dot{G}$ is an arbitrary signed graph.

Theorem 2.15 ([242]) *Let $\dot{G}$ be a signed graph with n vertices and eigenvalues $\lambda_1, \lambda_2, \ldots, \lambda_n$. The characteristic polynomial of the signed Mycielskian $M(\dot{G})$ is*

$$\phi_{M(\dot{G})}(x) = (x - \zeta_B(x)) \prod_{i=1}^{n} \left(x^2 - \lambda_i x - \lambda_i^2\right),$$

where $B = A_{\dot{G}}(xI_n - A_{\dot{G}})^{-1} A_{\dot{G}}$ and ζ_B is the coronal of B.

Proof With a consistent vertex labelling, the adjacency matrix of $M(\dot{G})$ is given by

$$A_{M(\dot{G})} = \begin{pmatrix} A_{\dot{G}} & A_{\dot{G}} & \mathbf{0}_n \\ A_{\dot{G}} & O_{n\times n} & \mathbf{j}_n \\ \mathbf{0}_n^\intercal & \mathbf{j}_n^\intercal & 0 \end{pmatrix}.$$

Therefore, for the corresponding characteristic polynomial, we find

$$\Phi_{M(\dot{G})}(x) = \det\left(xI_{2n+1} - A_{M(\dot{G})}\right) = \det\begin{pmatrix} xI_n - A_{\dot{G}} & -A_{\dot{G}} & \mathbf{0}_n \\ -A_{\dot{G}} & xI_n & -\mathbf{j}_n \\ \mathbf{0}_n^\intercal & -\mathbf{j}_n^\intercal & x \end{pmatrix}.$$

By the application of the row operation $R_2 + (\mathbf{j}_n/x)R_3 \to R_2$, we obtain

$$\Phi_{M(\dot{G})}(x) = \det\begin{pmatrix} xI_n - A_{\dot{G}} & -A_{\dot{G}} & \mathbf{0}_n \\ -A_{\dot{G}} & xI_n - (J_{n\times n}/x) & \mathbf{0}_n \\ \mathbf{0}_n^\intercal & -\mathbf{j}_n^\intercal & x \end{pmatrix} = x\det(N), \quad (2.5)$$

where $\det(M)$ is computed by Schur's complement formula as

$$\begin{aligned}
\det(N) &= \det\begin{pmatrix} xI_n - A_{\dot{G}} & -A_{\dot{G}} \\ -A_{\dot{G}} & xI_n - (J_{n\times n}/x) \end{pmatrix} \\
&= \det\left(xI_n - A_{\dot{G}}\right)\det\left(xI_n - (J_{n\times n}/x) - A_{\dot{G}}(xI_n - A_{\dot{G}})^{-1}A_{\dot{G}}\right) \\
&= \left(1 - \frac{\zeta_B(x)}{x}\right)\det\left(xI_n - A_{\dot{G}}\right)\det\left(xI_n - B\right).
\end{aligned}$$

We have replaced $B = A_{\dot{G}}\left(xI_n - A_{\dot{G}}\right)^{-1}A_{\dot{G}}$, and the latter identity follows from the general identity $\det(xI_n - A - \alpha J_{n\times n}) = (1 - \alpha\zeta_A(x))\det(xI_n - A)$.

Thus,

$$\begin{aligned}
\det(N) &= \left(1 - \frac{\zeta_B(x)}{x}\right)\prod_{i=1}^{n}(x - \lambda_i)\prod_{i=1}^{n}\left(x - \lambda_i(x - \lambda_i)^{-1}\lambda_i\right) \\
&= \left(1 - \frac{\zeta_B(x)}{x}\right)\prod_{i=1}^{n}(x - \lambda_i)\prod_{i=1}^{n}\left(x - \frac{\lambda_i^2}{x - \lambda_i}\right) \\
&= \frac{x - \zeta_B(x)}{x}\prod_{i=1}^{n}(x - \lambda_i)\prod_{i=1}^{n}\left(\frac{x^2 - \lambda_i x - \lambda_i{}^2}{x - \lambda_i}\right) \\
&= \frac{x - \zeta_B(x)}{x}\prod_{i=1}^{n}(x^2 - \lambda_i x - \lambda_i{}^2). \quad (2.6)
\end{aligned}$$

The desired result follows from (2.5) and (2.6). □

The following result is a consequence when $\dot{G}$ is net-regular.

Corollary 2.16 ([242]) *Under the assumptions of Theorem 2.15, let $\dot{G}$ be net-regular with net degree ϱ and $\lambda_k(\dot{G}) = \varrho$ for some fixed k $(1 \leq k \leq n)$. Then the spectrum of $M(\dot{G})$ consists of*

(i) $\frac{1}{2}(1 \pm \sqrt{5})\lambda_i$ *for* $i \neq k$*, and*
(ii) the roots of $x^3 - \varrho x^2 - (\varrho^2 + n)x + n\varrho = 0$.

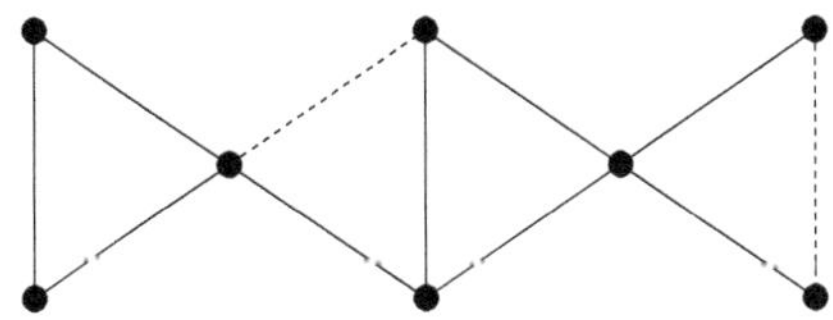

Figure 2.2 A sign-symmetric signed graph.

Proof Every row sum of $B = A_{\dot{G}}(xI_n - A_{\dot{G}})^{-1}A_{\dot{G}}$ is $s^2/(x-s)$, and thus

$$x - \zeta_B(x) = x - \frac{n}{x - \varrho^2/(x-\varrho)} = x - \frac{n(x-\varrho)}{x(x-\varrho)-\varrho^2}$$
$$= \frac{x^3 - \varrho x^2 - (\varrho^2 + n)x + n\varrho}{x^2 - \varrho x - \varrho^2},$$

and the result follows from Theorem 2.15. □

For the Laplacian and net Laplacian spectrum of particular Mycielskian products, we refer to [242].

2.4 Symmetric Spectrum

In this section we discuss signed graphs whose spectrum (of the adjacency matrix) is symmetric with respect to the origin, that is, if λ is an eigenvalue of $\dot{G}$ then $-\lambda$ is also an eigenvalue of $\dot{G}$ with the same multiplicity. We mentioned in Subsection 1.3.1 that a graph is bipartite if and only if its spectrum is symmetric. Moreover, a connected graph is bipartite if and only if the negation of the least eigenvalue equals the index.

With signed graphs these results no longer hold. If $\dot{G}$ is switching isomorphic to its negation $-\dot{G}$, then $\dot{G}$ is called *sign-symmetric*. We also have the following straightforward observation.

Theorem 2.17 *Let $\dot{G}$ be a sign-symmetric signed graph, that is, $\dot{G} \cong -\dot{G}$. Then the spectrum of $\dot{G}$ coincides with the spectrum of $-\dot{G}$, and it is symmetric with respect to the origin.*

Bipartite signed graphs are naturally sign-symmetric.

In general, sign-symmetric signed graphs can have non-bipartite underlying graphs, as shown in Figure 2.2. In such a case, $\dot{G}$ and $-\dot{G}$ are not switching equivalent, but only switching isomorphic.

For a signed graph, having a symmetric spectrum seems not to lead to significant combinatorial restrictions. However, there is a limitation on the triangles.

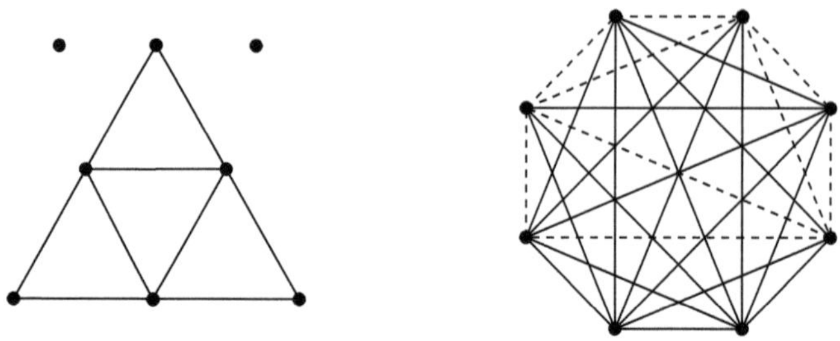

Figure 2.3 The graph G and the corresponding signed complete graph $\dot{S}$. Reproduced from [30], licensed under CC BY 4.0.

Corollary 2.18 *If a signed graph has a symmetric spectrum, then the number of positive triangles equals the number of negative ones. In particular, the total number of triangles cannot be odd.*

Proof A signed graph with symmetric spectrum has a characteristic polynomial whose odd-indexed coefficients are null. In view of the Sachs formula, Theorem 1.9, for $i = 3$, a_3 equals the difference between the positive and negative triangles, and $a_3 = 0$. Hence, such a signed graph has an equal number of positive and negative triangles. □

A natural question arises as to whether the converse of Theorem 2.17 holds. If it does not, then there must exist signed graphs whose spectra are symmetric, yet are not switching isomorphic to their own negation, that is, signed graphs with symmetric spectrum that are not sign-symmetric. For brevity, we refer to such graphs as *SSNSS signed graphs*.

Examples of signed graphs for which the converse of Theorem 2.17 fails can be found in the context of the Seidel spectral theory of graphs. Recall that the Seidel matrix of a (simple and unsigned) graph G is $S_G = J - I - 2A_G$. In this formulation, adjacent vertex pairs correspond to entries of -1, and non-adjacent pairs to entries of 1. Consequently, the Seidel matrix of an unsigned graph can be interpreted as the adjacency matrix of a complete signed graph with a particular signature. More broadly, Seidel spectral theory may be regarded as a sub-theory within the spectral theory of signed graphs. In this setting, signature switching corresponds precisely to the Seidel switching, the latter having historically inspired the former.

Figure 2.3 corresponds to the aforementioned example for Theorem 2.17. It depicts the graph G and the corresponding signed graph $\dot{S}$, that is, a signed graph whose adjacency matrix coincides with the Seidel matrix of G. It is readily verified that the spectrum of $\dot{S}$ is symmetric, and that the complement of G yields $-\dot{S}$, which is not switching isomorphic to $\dot{S}$. Hence, $\dot{S}$ is an SSNSS signed graph.

In [30], collecting some open problems for spectral signed graph theory, the question of whether there are some non-complete SSNSS signed graphs has attracted the attention of some scholars [9, 93, 106, 206]. We mention here some interesting results from these papers.

The first one is based on the Cartesian product defined in Subsection 2.3.2. We start with the following lemma.

Lemma 2.19 ([206]) *If the spectra of signed graphs $\dot{G}_1$ and $\dot{G}_2$ are symmetric, then the spectrum of $\dot{G}_1 \,\square\, \dot{G}_2$ is also symmetric.*

Proof Assume that the spectrum of $\dot{G}_1 \,\square\, \dot{G}_2$ is not symmetric. Then there is a non-zero eigenvalue of the form $\lambda_i + \mu_j$ (for fixed eigenvalues λ_i of $\dot{G}_1$ and μ_j of $\dot{G}_2$) such that its negation $-\lambda_i - \mu_j$ is not an eigenvalue of the Cartesian product. But since the spectra of $\dot{G}_1$ and $\dot{G}_2$ are symmetric, $-\lambda_i$ is an eigenvalue of $\dot{G}_1$ and $-\mu_j$ is an eigenvalue of $\dot{G}_2$, which means that their sum belongs to the spectrum of $\dot{G}_1 \,\square\, \dot{G}_2$ and contradicts the assumption. □

Here is our first construction of SSNSS signed graphs.

Theorem 2.20 ([206]) *Let $\dot{G}_1$ be a complete SSNSS signed graph and $\dot{G}_2$ be a connected signed graph with a symmetric spectrum, such that the underlying graph G_1 is not an induced subgraph of the underlying graph G_2. Then $\dot{G}_1 \,\square\, \dot{G}_2$ is a connected SSNSS signed graph.*

Proof We have that $\dot{G}_1 \,\square\, \dot{G}_2$ is connected (since $\dot{G}_1$ and $\dot{G}_2$ are connected). Next, from Lemma 2.19, we immediately get that the spectrum of $\dot{G}_1 \,\square\, \dot{G}_2$ is symmetric, and so it remains to show that $\dot{G}_1 \,\square\, \dot{G}_2$ is not switching isomorphic to its negation.

As an intermediate step, we establish that if n_1 and n_2 denote the orders of graphs G_1 and G_2, respectively, then a maximal clique (i.e., a maximal complete induced subgraph) in the Cartesian product $G_1 \,\square\, G_2$ has n_1 vertices, and there are exactly n_2 such cliques, each isomorphic to one of the n_2 disjoint copies of G_1. Assume that K is either a larger clique or a clique with n_1 vertices but distinct from the previous ones. In either case, K contains a pair of vertices $(u_i, v_j), (u_k, v_l)$ where i, k are fixed distinct integers from $\{1, \ldots, n_1\}$; otherwise, K would be one of the n_2 existing cliques. If so, then we must have $j = l$; otherwise, the mentioned vertices would not be adjacent. Moreover, K cannot contain a vertex whose second coordinate differs from v_j, since such a vertex is not adjacent to both $(u_i, v_j), (u_k, v_j)$. In other words, K is an induced

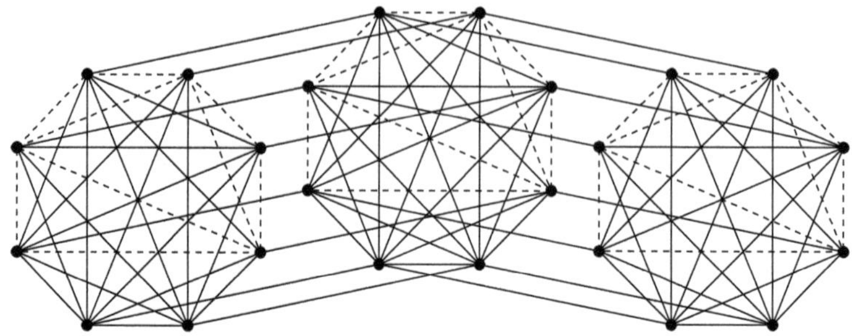

Figure 2.4 The connected non-complete SSNSS signed graph $\dot{S}\square P_3$. Reproduced from [206], licensed under CC BY 4.0.

subgraph of G_2. Since G_2 does not contain G_1, we conclude that K has fewer than n_1 vertices, which contradicts the initial assumption.

Assume now that $\dot{G}_1 \square \dot{G}_2$ is switching isomorphic to its negation. By the previous part of the proof, a corresponding switching isomorphism must map $\dot{G}_1$ onto a copy of $-\dot{G}_1$ in $-\dot{G}_1 \square \dot{G}_2$, but this is impossible since $\dot{G}_1$ is not switching isomorphic to its negation. □

It follows by definition that the Cartesian product is complete if and only if one of $\dot{G}_1, \dot{G}_2$ consists of a single vertex, while the other is complete. Therefore, apart from this simple situation, the previous theorem necessarily produces non-complete SSNSS signed graphs. We proceed with an example.

Example 2.21 The signed graph $\dot{S}\square P_3$ is illustrated in Figure 2.4. It is a connected non-complete SSNSS signed graph, by Theorem 2.20. In this example, the second signed graph of the product (so, P_3) is bipartite, but non-bipartite ones with symmetric spectrum can also be taken into account. Such a signed graph can be obtained, say, by taking two triangles, one positive and the other negative, and inserting a path between a pair of their vertices.

The next theorem establishes an infinite family of connected non-complete SSNSS signed graphs.

Theorem 2.22 *For $i \geq 0$, the signed graph $\dot{G}_i$ obtained by the iterative procedure*

$$\begin{cases} \dot{G}_0 \cong \dot{S}, \\ \dot{G}_{i+1} \cong \dot{G}_i \square K_2 \end{cases}$$

is a connected SSNSS signed graph. It is non-complete for $i \geq 1$.

Proof $\dot{G}_i$ is connected since it is the Cartesian product of connected signed graphs.

We prove that $\dot{G}_i$ is an SSNSS signed graph by induction. The base case is immediate as $\dot{S}$ is an SSNSS signed graph. Assume that the claim holds for $\dot{G}_i$ and consider $\dot{G}_{i+1} \cong \dot{G}_i \square K_2$. By the definition of the Cartesian product, every complete induced subgraph of $\dot{G}_{i+1}$ with eight vertices is isomorphic to $\dot{S}$, and every complete induced subgraph of $-\dot{G}_{i+1}$ with eight vertices is isomorphic to $-\dot{S}$. Therefore, a switching isomorphism between $\dot{G}_{i+1}$ and $-\dot{G}_{i+1}$ necessarily maps $\dot{S}$ onto $-\dot{S}$, which is impossible as they are not switching isomorphic.

Evidently, for $i \geq 1$, $\dot{G}_i$ is non-complete by the argument mentioned below Theorem 2.20. □

The signed graph $\dot{G}_1$ of Theorem 2.22 is obtained by deleting the third copy of $\dot{S}$ from Figure 2.4.

In what follows, we consider the special case of the corona product introduced in Subsection 2.3.3, where $\dot{G}_2 \cong K_1$. It begins by computing the spectrum of such a corona.

Lemma 2.23 ([206]) *If $\lambda_1, \lambda_2, \ldots, \lambda_n$ are the eigenvalues of $\dot{G}$, then all the eigenvalues of $\dot{G} \circ K_1$ are determined by the equations $x^2 - \lambda_i x - 1 = 0$, for $1 \leq i \leq n$.*

The proof is based on the Schur's complement formula. If the spectrum of $\dot{G}$ is symmetric, so is the spectrum of $\dot{G} \circ K_1$. Indeed, from Lemma 2.23, if λ is an eigenvalue of $\dot{G}$, then $\frac{1}{2}(\lambda \pm \sqrt{4+\lambda^2})$ are the eigenvalues of $\dot{G} \circ K_1$ and they are symmetric to the eigenvalues $\frac{1}{2}(-\lambda \mp \sqrt{4+\lambda^2})$ that arise from the eigenvalue $-\lambda$ of $\dot{G}$.

We are ready to present another iterative construction.

Theorem 2.24 *For $i \geq 0$, the signed graph $\dot{G}_i$ obtained by the iterative procedure*

$$\begin{cases} \dot{G}_0 \cong \dot{S}, \\ \dot{G}_{i+1} \cong \dot{G}_i \circ K_1, \end{cases}$$

is a connected SSNSS signed graph. It is non-complete for $i \geq 1$.

Proof The graph G_i is connected and non-complete for $i \geq 1$ by the definition of corona, while its spectrum is symmetric by the previous discussion. Observe that, for every i, G_i contains exactly one complete subgraph with eight vertices (this is $\dot{S}$) and therefore is not isomorphic to its negation by the argument exploited in the proof of Theorem 2.22. □

The signed graph $\dot{G}_1$ of Theorem 2.24 is illustrated in Figure 2.5. The following intuitive construction was proposed by Ghorbani et al.

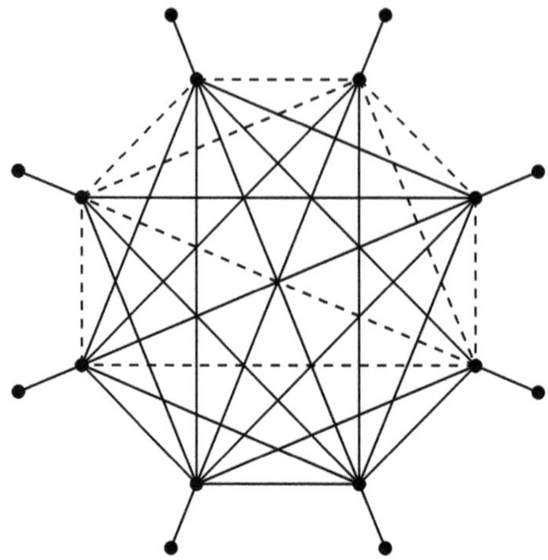

Figure 2.5 The connected non-complete SSNSS signed graph $\dot{S} \circ K_1$. Reproduced from [206], licensed under CC BY 4.0.

Theorem 2.25 ([93]) *Let B and C be symmetric matrices of order k with entries in* $\{1,0,-1\}$. *If B has a zero diagonal, then the signed graph associated with the adjacency matrix*

$$\begin{pmatrix} B & C \\ C & -B \end{pmatrix}$$

is sign-symmetric.

It is evident that B represents the adjacency matrix of a signed graph. In the specific case where C is likewise an adjacency matrix, the preceding theorem provides a construction that combines the two signed graphs into a new one exhibiting the desired spectral characteristic. Some additional infinite families of SSNSS signed graphs are established in the same reference.

2.5 Eigenvalue Bounds

Theorem 1.26 implies that $\lambda_1(\dot{G}) \leq \lambda_1(G) \leq n-1$, and taking the negation yields $\lambda_n(\dot{G}) \geq -n+1$. Therefore, the spectrum of any signed graph $\dot{G}$ is contained in the closed interval $[-n+1, n-1]$.

It is established in [121] that $\mu_1(\dot{G}) \leq \mu_1(-G)$, with equality if and only if $\dot{G}$ is switching equivalent to $-G$, which is also an easy exercise. Since $\mu_1(-G) \leq 2(n-1)$, with equality if and only if $-G$ switches to $-K_n$, we conclude that the entire Laplacian spectrum of a signed graph lies within the closed interval $[0, 2(n-1)]$.

Finally, Theorem 1.17(ii) implies that $\nu_1(\dot{G}) \leq \mu_1(G) \leq n$, where the upper bound is attained if and only if $\dot{G}$ is a positive join [157]. Thus, the net Laplacian spectrum of a signed graph lies in $[-n, n]$.

The most investigated bounds for the eigenvalues of a graph concern the largest eigenvalue and the least eigenvalue. In the framework of signed graphs there is no significant difference in the study of λ_1 and λ_n since every upper (resp. lower) bound for λ_1 of $\dot{G}$ is simultaneously the negation of the lower (resp. upper) bound for $-\lambda_n$ of $-\dot{G}$. We have exactly the same situation for ν_1 and ν_n (since $N_{\dot{G}} = -N_{-\dot{G}}$). Therefore, we will restrict ourselves to the bounds for λ_1, μ_1, μ_n and ν_1. We remind the reader that some bounds for particular Laplacian or net Laplacian eigenvalues are already met in Section 1.5.

2.5.1 Upper Bounds for λ_1, μ_1 and ν_1

As discussed previously, any upper bound for $\lambda_1(G)$ (resp. $\mu_1(-G)$ or $\mu_1(G)$) also serves as an upper bound for $\lambda_1(\dot{G})$ (resp. $\mu_1(\dot{G})$ or $\nu_1(\dot{G})$). A selection of such bounds is summarized in Table 2.1. These inequalities are adapted from [209], with the focus limited to bounds expressed in terms of vertex degrees and related invariants. In particular, the symbol m_i denotes the *average 2-degree*, that is, the average degree of the vertices in the neighbourhood of vertex i. All the other notation used in the table is introduced in the opening section.

Equality cases are discussed in [209]. One may recall that the second and the third bounds are known as the *Stanley bound* and the *Schwenk bound*, respectively. It is worth noting that the possibility of obtaining sharper bounds is to be expected, as the current ones are direct transfers from the spectral framework of ordinary graphs.

The first bound of Table 2.1 yields

$$\lambda_1(\dot{G}) \leq \lambda_1(G) \leq \max\{\sqrt{d_i d_j} \colon 1 \leq i, j \leq n\}.$$

The last bound gives

$$\nu_1(\dot{G}) \leq \mu_1(G) \leq \max\{d_i + d_j \colon i \sim j\}$$

(the *Anderson–Morley bound*) and

$$\nu_1(\dot{G}) \leq \mu_1(G) \leq \max\{d_i + m_i \colon 1 \leq i \leq n\}$$

(the *Merris bound*). The previous three bounds can be found in [120]. The latter two remain valid for $\mu_1(\dot{G})$ (see [209, p. 207]).

Table 2.1 *Upper bounds for $\lambda_1(\dot{G})$, $\mu_1(\dot{G})$ and $\nu_1(\dot{G})$ transferred from the domain of graphs.*

Eigenvalue	Upper bound	Remark
λ_1	$\max\{\sqrt{d_im_i}\colon 1\leq i\leq n\}$	Finer bound in [167]
λ_1	$\frac{1}{2}(\sqrt{8m+1}-1)$	Finer bound in [118, 164]
λ_1	$\sqrt{2m-n+1}$	$\dot{G}$ connected; same as before
λ_1	$\max\left\{\sqrt{\frac{1}{2}(d_i^2+d_im_i)}\colon 1\leq i\leq n\right\}$	See [85]
ν_1,μ_1	$\max\{\sqrt{2d_i(d_i+m_i)}\colon 1\leq i\leq n\}$	For ν_1, finer bound in [82]
ν_1,μ_1	$\max\left\{\frac{1}{2}(d_i+\sqrt{d_i^2+8d_im_i})\colon 1\leq i\leq n\right\}$	Finer bound in [103]
ν_1	$\max\{d_i+d_j-\lvert N(i)\cap N(j)\rvert\colon i\sim j\}$	See [82]

We continue with more sophisticated upper bounds. All of them are proved by standard eigenvector techniques adapted for signed graphs. In the domain of graphs, these techniques can be found in [82, 103, 116, 117]; for early works concerning upper bounds for the largest eigenvalue of Hermitian matrices, we refer to [90]. We continue to use the notation we introduced for vertex degree and related invariants.

Theorem 2.26 ([227]) *For the index λ_1 of a signed graph of order n,*

$$\lambda_1\leq\max\left\{\frac{-d_i+\sqrt{5d_i^2+4(d_im_i-4T_i^-)}}{2}\colon 1\leq i\leq n\right\},\tag{2.7}$$

where T_i^- denotes the number of negative triangles containing i.

This bound is tight, at least for balanced regular signed graphs. Since $T_i^-\geq 0$ and $d_im_i\leq\Delta^2$, we conclude that the bound in (2.7) does not exceed the value obtained for $T_i^-=0$ and $d_im_i=\Delta^2$, that is, does not exceed the maximum vertex degree Δ, which makes it always non-trivial. In the particular case of graphs, the bound in (2.7) reduces to the bound that can be expressed in the same way with the term concerning negative triangles removed, but this result is less important as we already know a finer bound: the first one from Table 2.1.

Theorem 2.27 ([205]) *For the index λ_1 of a signed graph of order n,*

$$\lambda_1^2\leq\max\{d_im_i-n_i\colon 1\leq i\leq n\},\tag{2.8}$$

where $n_i=\sum_{j\sim i}\left(\lvert N_i^{\sigma(ij)}\cap N_j\rvert-\left\lvert\lvert N_i^{\sigma(ij)}\cap N_j^{\sigma(ij)}\rvert-\lvert N_i^{\sigma(ij)}\cap N_j^{-\sigma(ij)}\rvert\right\rvert\right)$.

Proof Let $\mathbf{x}$ be an eigenvector associated with λ_1 and let x_i be an entry that is largest in absolute value. Without loss of generality, we assume that $x_i > 0$. If $w_2^+(i,k)$ (resp. $w_2^-(i,k)$) denotes the number of positive (resp. negative) walks of length 2 starting at i and terminating at k, then considering $A^2_{\dot{G}}$, we compute

$$\begin{aligned}
\lambda_1^2 x_i &= \sum_{k=1}^{n} \left(w_2^+(i,k) - w_2^-(i,k)\right) x_k \\
&= \sum_{j \overset{+}{\sim} i} \sum_{k \overset{+}{\sim} j} x_k + \sum_{j \overset{-}{\sim} i} \sum_{k \overset{-}{\sim} j} x_k - \sum_{j \overset{+}{\sim} i} \sum_{k \overset{-}{\sim} j} x_k - \sum_{j \overset{-}{\sim} i} \sum_{k \overset{+}{\sim} j} x_k.
\end{aligned}$$

Extracting the first term on the right-hand side, we obtain

$$\begin{aligned}
\sum_{j \overset{+}{\sim} i} \sum_{k \overset{+}{\sim} j} x_k &= \sum_{j \overset{+}{\sim} i} \sum_{\substack{k \overset{+}{\sim} j \\ k \overset{+}{\sim} i}} x_k + \sum_{j \overset{+}{\sim} i} \sum_{\substack{k \overset{+}{\sim} j \\ k \overset{+}{\nsim} i}} x_k \\
&= \sum_{j \overset{+}{\sim} i} |N_i^+ \cap N_j^+| x_j + \sum_{j \overset{+}{\sim} i} \sum_{\substack{k \overset{+}{\sim} j \\ k \overset{+}{\nsim} i}} x_k \\
&\leq \sum_{j \overset{+}{\sim} i} |N_i^+ \cap N_j^+| x_j + \sum_{j \overset{+}{\sim} i} (d_j^+ - |N_i^+ \cap N_j^+|) x_i.
\end{aligned}$$

In the same way,

$$\begin{aligned}
\sum_{j \overset{-}{\sim} i} \sum_{k \overset{-}{\sim} j} x_k &\leq \sum_{j \overset{-}{\sim} i} |N_i^- \cap N_j^-| x_j + \sum_{j \overset{-}{\sim} i} (d_j^- - |N_i^- \cap N_j^-|) x_i, \\
\sum_{j \overset{+}{\sim} i} \sum_{k \overset{-}{\sim} j} x_k &\geq \sum_{j \overset{+}{\sim} i} |N_i^+ \cap N_j^-| x_j - \sum_{j \overset{+}{\sim} i} (d_j^- - |N_i^+ \cap N_j^-|) x_i, \\
\sum_{j \overset{-}{\sim} i} \sum_{k \overset{+}{\sim} j} x_k &\geq \sum_{j \overset{-}{\sim} i} |N_i^- \cap N_j^+| x_j - \sum_{j \overset{-}{\sim} i} (d_j^+ - |N_i^- \cap N_j^+|) x_i.
\end{aligned}$$

Returning to the initial equality, we compute

$$\begin{aligned}
\lambda_1^2 x_i \leq & \sum_{j \overset{+}{\sim} i} \left(|N_i^+ \cap N_j^+| - |N_i^+ \cap N_j^-|\right) x_j + \sum_{j \overset{-}{\sim} i} \left(|N_i^- \cap N_j^-| - |N_i^- \cap N_j^+|\right) x_j \\
& + \sum_{j \overset{+}{\sim} i} (d_j^+ - |N_i^+ \cap N_j^+| + d_j^- - |N_i^+ \cap N_j^-|) x_i \\
& + \sum_{j \overset{-}{\sim} i} (d_j^- - |N_i^- \cap N_j^-| + d_j^+ - |N_i^- \cap N_j^+|) x_i \\
= & \sum_{j \sim i} \left(|N_i^{\sigma(ij)} \cap N_j^{\sigma(ij)}| - |N_i^{\sigma(ij)} \cap N_j^{-\sigma(ij)}|\right) x_j \\
& + \sum_{j \sim i} (d_j - |N_i^{\sigma(ij)} \cap N_j^{\sigma(ij)}| - |N_i^{\sigma(ij)} \cap N_j^{-\sigma(ij)}|) x_i
\end{aligned}$$

$$\leq \sum_{j\sim i} \big| |N_i^{\sigma(ij)} \cap N_j^{\sigma(ij)}| - |N_i^{\sigma(ij)} \cap N_j^{-\sigma(ij)}| \big| x_i + \sum_{j\sim i} \left(d_j - |N_i^{\sigma(ij)} \cap N_j|\right) x_i,$$

giving $\lambda_1^2 \leq d_i m_i - n_i$ for the vertex i. Taking the maximum over all vertices, we arrive at (2.8). □

A similar proof technique appears in [103]. Since $n_i \geq 0$, the bound in (2.8) does not exceed the maximum vertex degree. For graphs, it reduces to the first bound of Table 2.1. Bounds (2.7) and (2.8) are mutually incomparable. In an informal sense, a large number of negative triangles makes the former one finer.

The next result gives a Laplacian counterpart to Theorem 2.26.

Theorem 2.28 ([174]) *For the Laplacian index μ_1 of a signed graph $\dot{G}$ of order n,*

$$\mu_1 \leq \max\left\{\sqrt{2\left(d_i^2 + d_i m_i - 2T_i^+\right)} : 1 \leq i \leq n\right\}, \tag{2.9}$$

where T_i^+ denotes the number of positive triangles containing i.

Equality holds if and only if $\dot{G}$ is switching equivalent to a regular signed graph with all-negative signature.

For graphs, the previous upper bound reduces to the fifth bound of Table 2.1 (considered as the upper bound for the largest eigenvalue of the signless Laplacian). Comparing (2.9) and the sixth bound of Table 2.1, the former gives a better estimate whenever $T_i^+ \geq \frac{1}{8} d_i\left(3d_i - \sqrt{d_i(d_i + 8m_i)}\right)$.

We proceed with upper bounds for ν_1.

Theorem 2.29 ([175]) *For the net Laplacian index ν_1 of a signed graph of order n,*

$$\nu_1 \leq \max\left\{\sqrt{2\left((d_i^+)^2 + (d_i^-)^2 + d_i m_i^* - T_i^{\pm}\right)} : 1 \leq i \leq n\right\}, \tag{2.10}$$

$$\nu_1 \leq \max\left\{\frac{1}{2}\left(d_i^{\pm} + \sqrt{(d_i^{\pm})^2 + 8 d_i m_i^*}\right) : 1 \leq i \leq n\right\}, \tag{2.11}$$

with $m_i^* = \left(\sum_{j\sim i} d_j^*\right)/d_i$, $d_j^* = \max\{d_j^+, d_j^-\}$ and $T_i^{\pm} = \sum_{j\sim i} T_{ij}^{\sigma(ij)}$, where $T_{ij}^{\sigma(ij)}$ denotes the number of triangles containing the edge ij and whose sign is $\sigma(ij)$.

The upper bounds presented in Theorem 2.29 are, in general, not mutually comparable. Indeed, we easily get that (2.10) gives a better estimate if and only if

$$T_i^{\pm} \geq (d_i^+)^2 + (d_i^-)^2 - \frac{1}{4} d_i^{\pm}\left(d_i^{\pm} + \sqrt{(d_i^{\pm})^2 + 8 d_i m_i^*}\right). \tag{2.12}$$

If $\dot{G}$ is a complete signed graph of even order n whose negative edges form a perfect matching, then we compute that (2.10) gives $\nu_1(\dot{G}) \leq \sqrt{2}(n-1)$, while (2.11) gives $\nu_1(\dot{G}) \leq 2(n-2)$; hence, (2.10) gives a better estimate. On the other hand, if $\dot{G}$ is triangle-free signed graph such that $d_i^+ = d_i^-$ for all i then, by (2.12), (2.11) is finer than (2.10).

Theorem 2.30 ([228]) *For the net Laplacian index ν_1 of a signed graph $\dot{G}$ of order n,*

$$\nu_1 \leq \max\left\{\frac{1}{2}\left(d_i^{\pm} + \sqrt{d_i^{\pm 2} + 4(m_i^{\pm} + n_i + 2t_i)}\right) : 1 \leq i \leq n\right\}, \tag{2.13}$$

where, for a vertex i, $d_i^{\pm}$ denotes its net degree, $m_i^{\pm} = \sum_{j\sim i}|d_j^{\pm}|$, $n_i = \sum_{j\sim i}(d_j - |N_j \cap N_i|)$ and t_i denotes the difference between the number of homogeneous triangles containing i and the number of triangles whose edges incident with i are of the same sign, while the third differs in sign.

Equality holds if and only if $\dot{G}$ is bipartite regular with an all-positive signature.

The upper bounds (2.10) and (2.13) are not comparable in general. Informally, when the vertices of a signed graph have large positive degrees, small negative degrees and the graph contains many positive triangles, the former tends to yield a better estimate.

The upper bound (2.11) is never sharper than (2.13), but it has a simpler form. These two bounds coincide in the case of triangle-free signed graphs.

2.5.2 Spectral–Structural Interplay

Let $\mathbf{x} = (x_1, x_2, \ldots, x_n)^\intercal$ be a unit eigenvector associated with $\lambda_1(\dot{G})$. Due to Remark 2.6, without loss of generality we may assume that the entries of $\mathbf{x}$ are non-negative. Let $\dot{H}$ be the signed graph obtained by deleting all negative edges of $\dot{G}$, and let $A_{\dot{H}} = (a_{ij})$ for $1 \leq i, j \leq n$. We have

$$\begin{aligned}\lambda_1(\dot{G}) &= 2\Big(\sum_{i\overset{+}{\sim}j} x_ix_j - \sum_{i\overset{-}{\sim}j} x_ix_j\Big) \leq 2\sum_{i\overset{+}{\sim}j} x_ix_j = \sum_{i=1}^{n}\sum_{j=1}^{n} x_ia_{ij}x_j \\ &\leq \max_{\mathbf{y}\in\mathbb{R}^n,\ ||\mathbf{y}||=1}\Big(\sum_{i=1}^{n}\sum_{j=1}^{n} y_ia_{ij}y_j\Big) = \lambda_1(\dot{H}),\end{aligned}$$

where the first equality follows from $\lambda_1(\dot{G}) = \mathbf{x}^\intercal A_{\dot{G}}\mathbf{x}$ and the second inequality is a consequence of the Rayleigh principle. This brings us to the following result.

Theorem 2.31 ([205]) *Every signed graph $\dot{G}$ contains a balanced spanning subgraph, say $\dot{H}$, such that $\lambda_1(\dot{G}) \leq \lambda_1(\dot{H})$.*

The reader may recall a similar result related to the least eigenvalue of a graph with a bipartite graph in the role of our balanced signed graph (cf. [209, Theorem 3.1]).

Clearly the balanced signed graph $\dot{H}$ of Theorem 2.31 is obtained by deleting at least ℓ edges, where ℓ is the frustration index of $\dot{G}$. Using this along with the Stanley and Schwenk bounds of Table 2.1, we arrive at the following result estimating the index in terms of the frustration index.

Theorem 2.32 ([205]) *For a signed graph $\dot{G}$ with n vertices, m edges and the frustration index ℓ,*

$$\lambda_1(\dot{G}) \leq \frac{1}{2}\left(\sqrt{8(m-\ell)+1}-1\right).$$

If $\dot{G}$ is connected, then $\lambda_1(\dot{G}) \leq \sqrt{2(m-\ell)-n+1}$.

We proceed with a lower bound for the chromatic number. A *k-colouring* of a signed graph $\dot{G}$ is a mapping $c\colon V(\dot{G}) \to M_k$ such that $c(u) \neq \sigma(uv)c(v)$, where $M_k = \{0, \pm 1, \pm 2, \ldots, \pm(k-1)/2\}$ for k odd and $M_k = \{\pm 1, \pm 2, \ldots, \pm k/2\}$ for k even. The *chromatic number* $\chi(\dot{G})$ is the minimum number k for which $\dot{G}$ admits a k-colouring. The concept of the chromatic number of a signed graph was introduced by Máčajová et al. [147]; see also [262]. Clearly, $\chi(G)$ coincides with the standard chromatic number of a graph. Wang et al. proved the following lower bound.

Theorem 2.33 ([254]) *For the chromatic number of a signed graph $\dot{G}$ with n vertices and at least one edge,*

$$\chi(\dot{G}) \geq \max\left\{1+\frac{\lambda_1}{|\lambda_n|}, \frac{n}{n-\lambda_1}\right\}.$$

In the framework of graphs, the lower bound $\chi(G) \geq 1+\lambda_1/|\lambda_n|$ is known as the Hoffman bound [114], while $\chi(G) \geq n/(n-\lambda_1)$ is the Wilf bound [255] (also known as the Cvetković bound [69]). Evidently, Theorem 2.33 extends these classical results to signed graphs.

We proceed with the interplay between μ_n and the invariants related to the balance of a signed graph: the frustration index ℓ and the frustration number f. Theorem 1.21 gives an elegant upper bound: $0 \leq \mu_n(\dot{G}) \leq f(\dot{G}) \leq \ell(\dot{G})$, with $\ell(\dot{G}) = 0$ if and only if $\dot{G}$ is balanced. In [119], Hou offered the following range for μ_n in terms of l. We set

$$\beta(\dot{G}) = \min\left\{\frac{\ell(\dot{S})+|\partial S|}{|S|} : \emptyset \neq S \subseteq V(\dot{G})\right\}, \tag{2.14}$$

where, for $S \subseteq V(\dot{G})$, $\dot{S}$ is the subgraph induced by S and ∂S is the set of edges with one end in S and the other in $V(\dot{G}) \setminus S$.

Theorem 2.34 ([119]) *For the least Laplacian eigenvalue μ_n of a signed graph $\dot{G}$,*

$$\Delta - \sqrt{\Delta^2 - \beta(\dot{G})^2} \leq \mu_n(\dot{G}) \leq 4\beta(\dot{G}),$$

where Δ is the maximum vertex degree and $\beta(\dot{G})$ is defined by (2.14).

Similar bounds in terms of the frustration number can be found in the next theorem.

Theorem 2.35 ([12]) *For the least Laplacian eigenvalue μ_n of a signed graph $\dot{G}$ with the frustration number f,*

$$\mu_n \leq \max\left\{\frac{1}{n}\left(\frac{f}{n-f}|\partial F| + (n-f)\overline{d}_F\right), \frac{1}{n}(|\partial F| + f\overline{d}_F)\right\}, \tag{2.15}$$

where F denotes a minimal set of vertices whose removal results in a balanced signed graph and $\overline{d}_F = (1/|F|)\sum_{u \in F} d_u(\dot{G})$.

It was noted in the same reference that the values on the right-hand side of (2.15) are incomparable. Moreover, both are incomparable with f (the upper bound for μ_n of Theorem 1.21). The same reference offers a Nordhaus–Gaddum-type bound that reads

$$\mu_1 + \mu_2 \geq \frac{1}{n-f}|\partial F| + \overline{d}_F.$$

In [212] one can find some lower bounds for μ_n of particular signed graphs called unbalanced blocks. A maximal subgraph without cut-points is called a *block*; specifically, a signed graph without cut-points is also called a block. Blocks are interesting since every cycle lies entirely within a block.

Theorem 2.36 ([212]) *For the least Laplacian eigenvalue μ_n of an unbalanced block $\dot{G}$,*

$$\mu_n > \frac{4}{c_{\mathrm{u}}n},$$

where c_{u} is length of the longest negative cycle in $\dot{G}$.

Proof Let $\mathbf{x} = (x_1, x_2, \ldots, x_n)^\intercal$ be a unit eigenvector associated with $\mu_n(\dot{G})$, and let x_j be the coordinate which is largest in modulus. Since $\mathbf{x}$ is unit, we have $|x_j| \geq 1/\sqrt{n}$. Since $\dot{G}$ is a block, every vertex belongs to at least one negative cycle, which is an easy exercise mentioned in [261]. In particular, the vertex j (which corresponds to x_j) belongs to a negative cycle. Denote such a cycle by $\dot{C}$, and its vertices by $j = j_1, j_2, \ldots, j_l, j_{l+1} = j$, indexed in the natural order.

Observe that there exists a signed graph switching equivalent to $\dot{G}$ in which the edge $j_l j$ is negative, while all remaining edges of the cycle $\dot{C}$ are positive. Let the resulting (switching equivalent) signed graph be denoted again by $\dot{G}$, and let the corresponding eigenvector be denoted again by $\mathbf{x}$. By Theorem 1.16, the entries of $\mathbf{x}$ remain the same up to sign and, in particular, x_j retains the largest absolute value among them. Then, by virtue of (1.4), we obtain

$$\mu_n(\dot{G}) \geq \sum_{i=1}^{l} (x_{j_i} - \sigma(j_i j_{i+1}) x_{j_{i+1}})^2.$$

Using the Cauchy–Schwarz inequality, we get

$$\begin{aligned}\mu_n(\dot{G}) &\geq \frac{1}{l}\Big(\sum_{i=1}^{l}(x_{j_i} - \sigma(j_i j_{i+1}) x_{j_{i+1}})\Big)^2\\ &= \frac{1}{l}(x_j - x_2 + x_2 - x_3 + \cdots + x_{l-1} - x_l + x_l + x_j)^2\\ &= \frac{1}{l}(2x_j)^2 \geq \frac{4}{ln},\end{aligned}$$

which gives $\mu_n(\dot{G}) \geq 4/c_{\mathrm{u}}n$.

If the equality is attained, then $|x_j| = 1/\sqrt{n}$, which implies that every entry of $\mathbf{x}$ is either $1/\sqrt{n}$ or $-1/\sqrt{n}$. Making a switch with respect to the set of vertices which correspond to negative entries, we arrive at a signed graph, say $\dot{H}$, in which the constant eigenvector $\mathbf{y} = (1/\sqrt{n}, 1/\sqrt{n}, \ldots, 1/\sqrt{n})^{\intercal}$ is associated with $\mu_n(\dot{H}) = 4/ln$. This gives $L_{\dot{H}}\mathbf{y} = (4/ln)\mathbf{y}$, which yields $2d_i^- = 4/ln$ for $1 \leq i \leq n$ (d_i^- being the negative degree of i). Obviously, the last equality cannot hold. Therefore, $\mu(\dot{G}) > 4/c_{\mathrm{u}}n$ holds. □

The previous theorem may be regarded as a signed analogue of Mohar's lower bound for the algebraic connectivity of a graph in terms of its order and diameter [158]. Notably, Theorem 2.36 accommodates the case $c_{\mathrm{u}} = n$, corresponding to the situation where $\dot{G}$ contains a negative Hamiltonian cycle. In such a case, one obtains an alternative lower bound relating μ_n to the number of edge-disjoint negative Hamiltonian cycles in $\dot{G}$. It is evident that every *Hamiltonian signed graph* (i.e., a signed graph containing an induced cycle that visits every vertex) is necessarily a block. Although Theorem 2.37 is closely related in spirit to the preceding result, it is not a direct consequence thereof; rather, its proof relies on Example 1.11.

Theorem 2.37 ([212]) *Assume that a signed graph $\dot{G}$ contains $k \geq 1$ edge-disjoint negative Hamiltonian cycles. Then*

$$\mu_n > k\pi^2 \frac{12n^2 - \pi^2}{12n^4}.$$

The *vertex connectivity* of a signed graph is defined as the minimum number of vertices whose removal disconnects the graph. Consequently, the vertex connectivity of a block is greater than one. In this context, it is worth noting that the smallest non-zero eigenvalue of the net Laplacian matrix of a signed graph does not exceed its vertex connectivity [228].

2.5.3 Extremal Signed Graphs

Ordinary graphs that maximize the largest eigenvalue within families defined by fixed structural invariants have attracted considerable attention. Here, we highlight two classical results concerning graphs with fixed order and size. The problem of identifying graphs of fixed order and size that attain the maximal spectral radius is among the oldest in spectral graph theory; as noted in [89], this line of inquiry was initiated by Schwarz in 1965. Brualdi and Hoffman established that such extremal graphs possess adjacency matrices with a characteristic stepwise structure [47]. A complete characterization of these graphs was subsequently provided by Rowlinson.

Theorem 2.38 ([178]) *A graph of fixed order and size with maximal index is formed from a single complete graph of maximal allowed size and a single vertex adjacent to at least one vertex of the complete graph, along with the set of isolated vertices.*

This result remains valid for signed graphs with all-positive signature in the role of the desired maximizer. The case in which $\dot{G}$ is necessarily unbalanced is resolved in [251], along with slight modifications of the previous solution.

The next result gives a structure of a graph that maximizes the index within the set of connected graphs with fixed order and size.

Theorem 2.39 ([194]) *A connected graph of fixed order and size with maximal index is a threshold graph.*

There is a bipartite counterpart to Theorem 2.39 with a chain graph (a $\{C_3, C_5, 2K_2\}$-free graph, which is necessarily bipartite) in the role of the corresponding maximizer [39]. Both results remain valid for the largest Laplacian eigenvalue, and all transfer directly to the domain of signed graphs. Accordingly, at present we only have a rough structure of the corresponding maximizer.

If one insists on unbalanced signed graphs, the answer is provided by Koledin and Stanić.

Theorem 2.40 ([134]) *A connected signed graph of fixed order, size and number of negative edges that attains the maximum index contains two connected threshold subgraphs, N_1 and N_2, each consisting solely of positive edges, as well as a connected chain subgraph $\dot{D}$ composed exclusively of negative edges. Furthermore, every vertex incident with a positive edge belongs to exactly one of N_1 or N_2, while every vertex incident with a negative edge belongs to $\dot{D}$. If at least one of N_1 or N_2 is not complete, then there are no positive edges outside these subgraphs. Similarly, if $\dot{D}$ is not complete bipartite, then no negative edges exist outside $\dot{D}$.*

Further investigation is focused on complete signed graphs with a fixed number of negative edges and maximal index. For $m \leq n^2/4$ this question is resolved in [95]. Another partial result is given in [7].

It is already established that for a signed graph $\dot{G}$ of order n, $\lambda_1(\dot{G}) \leq n-1$ holds with equality if and only if $\dot{G}$ switches to the complete graph K_n. Similarly $\mu_n(\dot{G}) \geq 0$, with equality if and only if $\dot{G}$ is balanced. In this context, one might ask which connected signed graph of order n has the second largest index, then the third largest, and so on, and similarly for the least Laplacian eigenvalue. It happens that there are some results on this topic, commonly known as the ordering of (signed) graphs with respect to a fixed eigenvalue.

According to [53], among all unbalanced signed graphs of order $n \geq 3$, the one with the largest index is switching equivalent to the complete signed graph containing exactly one negative edge. Furthermore, as shown in [52], this signed graph attains the second largest index within the entire class of signed graphs on n vertices.

To present the next three signed graphs ranked by their indices, we adopt, with some abuse of notation, the approach of the latter reference. Specifically, for a signed graph of order n, we depict only the negative edges and the non-edges, assuming by default that all edges not shown are positive. By convention, negative edges are represented by solid lines, whereas non-edges are indicated by dotted lines.

Theorem 2.41 ([52]) *The first five connected signed graphs with $n \geq 5$ vertices ordered by their indices are:*

$\dot{G}_1$ $\dot{G}_2$ $\dot{G}_3$ $\dot{G}_4$ $\dot{G}_5$

We also know the first few unbalanced or complete signed graphs.

Theorem 2.42 ([52]) *The first four connected unbalanced signed graphs with $n \geq 6$ vertices ordered by their indices are:*

$\dot{U}_1$ $\dot{U}_2$ $\dot{U}_3$ $\dot{U}_4$

The first six complete signed graphs with $n \geq 10$ vertices ordered by their indices are:

$\dot{C}_1$ $\dot{C}_2$ $\dot{C}_3$ $\dot{C}_4$ $\dot{C}_5$ $\dot{C}_6$

Small signed graphs that are uncovered by the previous results are resolved in the same reference.

Connected unbalanced signed graphs on n vertices with minimal least eigenvalue μ_n are studied in [36], where it is shown that this minimum is attained by an unbalanced triangle with a hanging path.

Furthermore, signed graphs exhibiting extremal eigenvalues have been characterized within various classes. In the next subsection, we present several results along these lines.

2.5.4 Spectral Turán Problem

One of the earliest results in extremal graph theory was established by Mantel in 1907, who proved that an n-vertex graph containing no triangles has at most $n^2/4$ edges [149]. This foundational result led to the broader problem of determining the maximum number of edges in an n-vertex graph that avoids a fixed subgraph. This problem is known as the *Turán problem*, named after Pál Turán, who characterized the extremal graphs avoiding the complete graph K_r as a subgraph [239].

The spectral analogue, referred to as the *spectral Turán problem*, concerns determining the maximum spectral radius among all n-vertex graphs excluding a given subgraph [166]. This problem can be viewed as a special case of the Brualdi–Solheid problem, which seeks the maximum spectral radius within a specified class of graphs [48]; some instances were discussed in the previous subsection. Extensions of the spectral Turán problem involve characterizing the graphs that attain the maximal spectral radius.

For ordinary graphs, a relation between the classical Turán problem and its spectral counterpart follows from the simple inequality $\lambda_1 \geq 2m/n$, relying on the Rayleigh principle and stating that the largest eigenvalue is not less than the average vertex degree. Therefore, every upper bound on the spectral radius induces an upper bound on the size m. We give a brief exposition on results concerning graphs and signed graphs.

There is a well-known result obtained by Wilf [255],

$$\lambda_1 \leq \frac{\omega - 1}{\omega} n,$$

where ω denotes the clique number, that is, the size of the maximum induced clique of G. This result is deduced from the *Motzkin–Straus equality*, [159]

$$\max\Big\{ \sum_{ij \in E(G)} x_i x_j : (x_1, x_2, \ldots x_n)^\intercal \in \mathbb{R}^n, \sum_{i=1}^{n} x_i = 1, x_i \geq 0 \Big\} = \frac{1}{2}\Big(\frac{\omega - 1}{\omega}\Big),$$

and implies the result on the chromatic number given after Theorem 2.33.

One may observe that the Wilf inequality gives an upper bound for λ_1 of $K_{\omega+1}$-free graphs. This result is generalized by Nikiforov [167] to

$$\lambda_1^r \leq \frac{\omega - 1}{\omega} w_{r-1},$$

where w_r denotes the number of walks of length r. By setting $r = 2$ and using the aforementioned inequality between λ_1 and m, we arrive at

$$m \leq \frac{\omega - 1}{2\omega} n^2.$$

The graphs that attain the above inequalities are characterized in [167].

Taking into account that the clique number never exceeds the chromatic number χ, we deduce that the above inequalities remain valid if ω is replaced with χ. Once again, we recall Theorem 2.33 extending the corresponding inequality to signed graphs.

In [131], Kannan and Pragada have made a progress for signed graphs, including the *balanced clique number* and the frustration index, where the former invariant is the size of the largest balanced clique.

Theorem 2.43 ([131]) *The index λ_1 of the adjacency matrix of a signed graph satisfies*

$$\lambda_1^2 \leq 2(m - \ell)\Big(\frac{\omega_b - 1}{\omega_b}\Big),$$

where m, ℓ and ω_b are the size, the frustration index and the balanced clique number, respectively.

For signed graphs, let $\mathscr{K}_k$ be the set of unbalanced complete signed graphs of order k. The spectral radius ρ for $\dot{H}$-free signed graphs, where $\dot{H} \in \mathscr{K}_k$ and $k \in \{3,4,5\}$, are considered in [64, 244, 248]. According to [248], we have

$$\rho \leq \frac{1}{2}(\sqrt{n^2-8}+n-4)$$

for $k=3$, with equality attained for the signed graph obtained from the positive clique K_{n-2} by adding two vertices, a negative edge between them, s positive edges between one of them and vertices of K_{n-2} and $n-2-s$ edges between the other and the remaining vertices of the same clique. The authors of [64] established that $m \leq \frac{1}{2}n(n-1)-n+3$ and $\rho \leq n-2$ hold for $k=4$, and characterized the extremal signed graphs. Finally, in [244], Wang proved that for $n \geq 9$, a signed graph $\dot{G}$ contains an unbalanced five-vertex clique provided that (i) its spectral radius is at least as large as that of the signed graph obtained from K_{n-1} by adding a vertex connected to the clique vertices via three edges, two positive and one negative, and (ii) $\dot{G}$ is not isomorphic to this particular signed graph.

We now return to unsigned graphs. Concerning $K_{s,t}$-free graphs, Bollobás and Nikiforov offered the following result.

Theorem 2.44 ([43]) *Let $s \geq t \geq 2$, and let G be a $K_{s,t}$-free graph of order n and spectral radius λ_1. If $t=2$, then*

$$\lambda_1 \leq \frac{1}{2}+\sqrt{(s-1)(n-1)+\frac{1}{4}}.$$

If $t \geq 3$, then $\lambda_1 \leq (s-t+1)^{1/t}n^{1-1/t}+(t-1)n^{1-2/t}+t-2$.

It follows that if G is a $K_{s,t}$-free graph of order n, then

$$m \leq \frac{1}{2}(s-t+1)^{1/t}n^{2-1/t}+\frac{1}{2}(t-1)n^{2-2/t}+\frac{1}{2}(t-2)n,$$

giving an upper bound for the size of a $K_{s,t}$-free graph, and contributing to the so-called Zarankiewicz problem [265] asking for such a bound. The bound for $(s,t)=(2,2)$ is attained for friendship graphs (joins $K_1 \nabla((n-1)/2)K_2)$ when n is odd. For $s \geq 3$, it is attained for particular strongly regular graphs. For $(s,t)=(2,2)$, [165, 266] offer a different bound, $\lambda_1^3-\lambda_1^2-(n-1)\lambda_1+1 \leq 0$, attained for the star with $n/2-1$ additional disjoint edges. Another bound for the same parameters is offered in [165]: for $m \geq 9$, $\lambda_1^2 \leq m$ holds with equality if G is a star.

In the remainder of this subsection, our focus will be on the spectral Turán problem for (signed) graphs that avoid particular induced paths or cycles. We first deal with unsigned graphs and set

$$p_l(n) = \max\{\lambda_1(G)\colon |V(G)| = n,\ G \text{ is } P_l\text{-free}\},$$
$$c_l(n) = \max\{\lambda_1(G)\colon |V(G)| = n,\ G \text{ is } C_l\text{-free}\}.$$

For $p_l(n)$, Nikiforov [166] obtained a tight results for large n:

$$p_{2k}(n) = \frac{k-1}{2} + \sqrt{kn - \frac{3k^2+2k-1}{4}},$$

along with

$$p_{2k+1}(n) = \frac{k-1}{2} + \sqrt{kn - \frac{3k^2+2k-1}{4}} + \frac{1}{n} + O(n^{-3/2}).$$

Moreover, for every $l \geq 4$, the unique graph attaining $p_l(n)$ is known for large n.

For l odd, the function $c_l(n)$ was essentially determined in [163]: if l is odd and $n > 321l$, then

$$c_l(n) = \sqrt{\left\lfloor \frac{n^2}{4} \right\rfloor}.$$

For $l = 4$, a solution is given in Theorem 2.44 where the corresponding bound reduces to

$$\lambda_1(\lambda_1 - 1) \leq n - 1,$$

and also in the text below the same theorem.

For $l = 2k+2 \geq 4$, [166] offers

$$\lambda_1^2 - (k-1)\lambda_1 \leq k(n-1).$$

Concerning $c_3(n)$, Lin, Ning and Wu [143] proved that every non-bipartite graph G of order n and size m contains a triangle if one of the following is true:

(i) $\lambda_1 \geq \sqrt{m-1}$ and $G \ncong C_5 \cup (n-5)K_1$;
(ii) $\lambda_1 \geq \lambda_1(S(K_{\lfloor (n-1)/2 \rfloor, \lceil (n-1)/2 \rceil}))$ and $G \ncong S(K_{\lfloor (n-1)/2 \rfloor, \lceil (n-1)/2 \rceil})$,

where $S(K_{\lfloor (n-1)/2 \rfloor, \lceil (n-1)/2 \rceil})$ is obtained from $K_{\lfloor (n-1)/2 \rfloor, \lceil (n-1)/2 \rceil}$ by subdividing an edge.

Returning to signed graphs, we highlight the result of Wang and Lin [250], which asserts that any unbalanced signed graph of order $n \geq 5$ contains a negative quadrangle provided that (i) its spectral radius is at least as large as that of the signed graph obtained by identifying a vertex of K_{n-2} with a vertex of a negative triangle, and (ii) $\dot{G}$ is not isomorphic to this particular signed graph. This result can be viewed as a counterpart to Theorem 2.44 for the parameters $(s,t) = (2,2)$.

2.6 Hoffman Program

The Hoffman program originates from Hoffman's study of limit points of spectral radii of non-negative symmetric integral matrices, as well as from independent investigations by Smith and by Lemmens and Seidel into geometric problems concerning equiangular lines and root systems (cf. [253]).

Let $\mathscr{M}$ denote the set of non-negative symmetric integral matrices. For any $M \in \mathscr{M}$, its spectral radius is denoted by $\rho(M)$. By the Perron–Frobenius theorem, $\rho(M)$ coincides with the largest eigenvalue of M.

In 1972, Hoffman determined the limit points of $\mathscr{S} = \{\rho(M) \colon M \in \mathscr{M}\}$ that are less than $\varepsilon = \sqrt{2+\sqrt{5}}$. Precisely, he proved the following result.

Theorem 2.45 (Hoffman theorem [115]) *Write* $\tau = (\sqrt{5}+1)/2$ *and* $\varepsilon = \sqrt{2+\sqrt{5}}$. *For* $n \in \mathbb{N}$, *define* $\eta_n = \beta_n^{1/2} + \beta_n^{-1/2}$, *where* β_n *is the positive root of* $x^{n+1} - (1+x+x^2+\cdots+x^{n-1})$. *Then* $2 = \eta_1 < \eta_2 < \cdots$ *are precisely the limit points of* $\mathscr{S}$ *less than* $\lim_{n\to\infty} \eta_n = \tau^{1/2} + \tau^{-1/2} = \varepsilon$.

Hoffman reduced the problem to symmetric $(0,1)$-matrices with zero diagonal and showed that for each $M \in \mathscr{M}$ there exists a graph G such that $\rho(A_G) = \rho(M)$. On the other hand, to investigate equiangular lines and the corresponding root systems, Smith [199] and independently Lemmens and Seidel [139] characterized connected graphs with $\rho(A_G) = 2$; they are commonly known as the Smith graphs.

These results mark the historical foundations of what is now referred to as the Hoffman program for graphs. To provide further detail and broaden the context, let us fix M_G as a real or Hermitian matrix associated with a graph G. The M-spectral radius $\rho(M_G)$ is defined as above; notably, it may correspond to the absolute value of the smallest eigenvalue. A real number $\gamma(M)$ is called an *M-limit point* if there exists a sequence of graphs $\{G_k\}_{k\geq 0}$ such that $\rho(M_{G_i}) \neq \rho(M_{G_j})$ for all $i \neq j$, and $\lim_{k\to\infty} \rho(M_{G_k}) = \gamma(M)$.

The *Hoffman program* for graphs consists of:

- determining all M-limit points;
- finding all connected graphs with $\rho(M_G) \leq \gamma(M)$ for fixed limit points.

For ordinary graphs, the Hoffman program began with the adjacency matrix for which $\rho(A_G)$ equals the index, and Theorem 2.45 remains unchanged. Shearer subsequently determined all the remaining limit points for the adjacency matrix.

Theorem 2.46 ([185]) *For* λ *not smaller than the limit value* $\varepsilon = \sqrt{2+\sqrt{5}}$, *there exists a sequence of graphs* $\{G_k\}_{k\geq 0}$ *such that* $\lim_{k\to\infty} \rho(A_{G_k}) = \lambda$.

For the second part of the Hoffman program, the graphs with index at most ε have been established in [46, 71, 115, 199]. For results concerning other matrices associated with ordinary graphs, the reader is referred to the survey in [253].

The focus now shifts to signed graphs, with the goal of extending the aforementioned results. It is evident that ε remains a limit point, and Theorem 2.46 holds in the context of signed graphs. A complete characterization of limit points less than ε was provided by Belardo and Brunetti.

Theorem 2.47 ([27]) *For every non-negative integer k, let γ_k denote the positive root of $x^{k+2}(x^2-x-1)+x^2-1$, and let γ be the positive root of $x^6-2x^5+x^4-x^2+x-1$. Then $\xi_k=\gamma_k^{1/2}+\gamma_k^{-1/2}$ and $\xi=\gamma^{1/2}+\gamma^{-1/2}$ are the limit points of the spectral radii of signed graphs that cannot be obtained from sequences of unsigned graphs. Moreover, the sequence $\{\xi_k\}_{k\geq 0}$ is strictly increasing and* $\lim_{k\to\infty}\xi_k=\varepsilon$.

To complete the Hoffman program for ε, it remains to determine all signed graphs whose spectral radius does not exceed this bound. Signed graphs whose spectral radius does not exceed 2 are determined by McKee and Smyth [154]. We will deal with them in Subsection 3.5.2; as a matter in fact, they are subgraphs of at least one of the signed graphs $\dot{T}_{2k}$, $k\geq 3$, $\dot{Q}_3$ or $\dot{S}_{14}$ illustrated in Figure 3.4 (see page 91). The remaining signed graphs, that is, those with spectral radius in $(2,\varepsilon]$, are established by Wang et al. [245].

2.7 Spectral Distances and Related Functions

A *spectral distance* on the set of finite graphs with equal number of vertices was suggested in a problem of Brualdi (see [237]). He suggested the l_1- or l_2-norm to define such distances, but they can be generalized to

$$\sigma(G_1,G_2)=\Big(\sum_{i=1}^{n}|\lambda_i(G_1)-\lambda_i(G_2)|^p\Big)^{1/p}.$$

Naturally, this definition is not confined solely to the spectrum of the adjacency matrix. Furthermore, it extends straightforwardly to signed graphs, thereby providing a robust framework for comparing signed graphs through their spectra. In many applications within graph theory, graphs exhibiting small spectral distances are often considered interchangeable, which has motivated significant interest in the study of spectral distances (see [1, 126, 127]).

By considering the particular case of the spectral distance between a signed graph $\dot{G}$ and its underlying graph G, we obtain an alternative spectral measure of balance. The value $\sigma(\dot{G},G)$ vanishes if and only if $\dot{G}$ and G have identical spectra, which, according to Theorem 1.25, occurs precisely when they are switching equivalent; in other words, when $\dot{G}$ is balanced.

Furthermore, we propose the quantities $\gamma_A(\dot{G}) = \lambda_1(G) - \lambda_1(\dot{G})$ and $\gamma_L(\dot{G}) = \mu_n(\dot{G})$ as alternative measures of imbalance. By Theorem 1.26, $\gamma_A(\dot{G}) \geq 0$, with equality holding if and only if $\dot{G}$ is balanced. Similarly, Theorem 1.27 ensures that $\gamma_L(\dot{G}) \geq 0$, with equality attained precisely under the same condition.

In what follows, we fix $\beta_A(\dot{G})$ to $\sigma(\dot{G},G)$ with $p=2$, and $\beta_L(\dot{G})$ to the same measure with Laplacian eigenvalues.

Example 2.48 Let C_n^- denote the negative cycle of order n. Taking into account Example 1.11, we deduce that the eigenvalues λ_j (a common notation for both $\lambda_j(C_n^-)$ and $\lambda_j(C_n)$) are ordered as $\lambda_1 \geq \lambda_n \geq \lambda_2 \geq \lambda_{n-1} \geq \lambda_3 \geq \lambda_{n-2} \geq \cdots \geq \lambda_k \geq \lambda_{k+2} \geq \lambda_{k+1}$ for $n = 2k+1$. Similarly, for $n = 2k$ we have $\lambda_1 \geq \lambda_n \geq \lambda_2 \geq \lambda_{n-1} \geq \lambda_3 \geq \lambda_{n-2} \geq \cdots \geq \lambda_{k+2} \geq \lambda_k \geq \lambda_{k+1}$. Therefore,

$$\beta_A(C_n^-) = \sum_{j=0}^{n-1}\left(2\cos\frac{(2j+1)\pi}{n} - 2\cos\frac{2j\pi}{n}\right)^2$$

$$= 8\sin^2\left(\frac{\pi}{2n}\right)\left(n - \sum_{j=0}^{n-1}\cos\frac{(4j+1)\pi}{n}\right) = 8n\sin^2\left(\frac{\pi}{2n}\right),$$

since $\sum_{j=0}^{n-1}\cos((4j+1)\pi/n) = 0$. As the sine function increases in $[0,\pi/2]$, we have $\gamma_A(C_n^-) = 4\sin^2(\pi/2n)$.

Evidently, $\beta_A(\dot{G}) = \beta_L(\dot{G})$ and $\gamma_L(\dot{G}) = \gamma_A(\dot{G})$ hold whenever $\dot{G}$ is regular; this occurs in the previous example.

We know from [108] that $\beta_A(\dot{G}), \beta_L(\dot{G}) \in [0, 2n(n-2)]$. Other bounds include an interplay with the frustration index, like

$$\beta_L(\dot{G}) \leq 8f \ \text{ or } \ \gamma_A(\dot{G}) \leq \frac{4f\Delta}{\Delta + \overline{d}^2},$$

with Δ and $\overline{d}$ in the role of the maximum vertex degree and the average vertex degree.

We proceed with a related topic. The distance between an eigenvalue λ of a signed graph $\dot{G}$ and the spectrum of a signed graph $\dot{H}$ is defined as

$$\min\{|\lambda - \mu| : \mu \text{ is an eigenvalue of } \dot{H}\}.$$

It is not difficult to see that every n-vertex signed graph which has an eigenvalue of multiplicity k contains an induced subgraph with $n-k$ vertices that avoids the mentioned eigenvalue. We estimate the previous distance.

Theorem 2.49 ([208]) *Let λ be an eigenvalue of multiplicity k of a signed graph $\dot{G}$ with n vertices, and let $\dot{H}$ be an induced subgraph with $n-k$ vertices such that λ does not appear in its spectrum. Then*

$$|\lambda||\lambda-\mu| \leq \sqrt{\delta_{\dot{G}\setminus\dot{H}}\Theta} \leq \sqrt{\Delta\Theta},$$

where μ is an eigenvalue of $\dot{H}$ for which $|\lambda-\mu|$ takes its minimum, $\delta_{\dot{G}\setminus\dot{H}}$ is the minimum vertex degree in $V(\dot{G})\setminus V(\dot{H})$,

$$\Theta = \max_{\substack{i\in V(\dot{G})\setminus V(\dot{H})\\ j\in V(\dot{H}),\, i\sim j}} d_{\dot{H}}(i)\,d_{\dot{G}-V(\dot{H})}(j),$$

and Δ is the maximum vertex degree in $\dot{G}$.

Here is a corollary for $k=1$.

Corollary 2.50 ([208]) *Let λ be a simple eigenvalue of a signed graph $\dot{G}$, and let $\dot{G}-i$ be a vertex-deleted subgraph such that λ does not appear in its spectrum. If μ is an eigenvalue of $\dot{G}-i$ for which $|\lambda-\mu|$ takes its minimum, then*

$$|\lambda||\lambda-\mu| \leq d_i \leq \Delta,$$

where d_i is degree of a vertex i and Δ is the maximum vertex degree, each considered in $\dot{G}$.

On the basis of the previous two results, one may introduce the following function:

$$\varsigma_{\dot{H}}(\dot{G}) = \sum_{\lambda\in\dot{G}} \min_{\mu\in\dot{H}} |\lambda-\mu|.$$

Equivalently, $\varsigma_{\dot{H}}(\dot{G}) = \sum_{\lambda\in\dot{G},\lambda\notin\dot{H}} \min_{\mu\in\dot{H}} |\lambda-\mu|$.

It is evident that the function ς is not symmetric, in the sense that $\varsigma_{\dot{H}}(\dot{G}) = \varsigma_{\dot{G}}(\dot{H})$ does not hold in general. For example, the spectrum of the unsigned triangle C_3 is $[2,(-1)^2]$ and the spectrum of the unsigned quadrangle C_4 is $[2,0^2,-2]$, and thus $3=\varsigma_{C_3}(C_4)\neq\varsigma_{C_4}(C_3)=2$. For another example, let K_i, K_j be a pair of complete graphs with $i\geq j$. We have $\varsigma_{K_i}(K_j)=i-j$ and

$$\varsigma_{K_j}(K_i) = \begin{cases} i-j & \text{if } i\geq 2j,\\ j & \text{otherwise.}\end{cases}$$

Since a distance between objects is assumed to be symmetric, ς is not called by this name, but it still measures how much the spectrum of a signed graph deviates from the spectrum of some other signed graph. The term *spectral deviation* was proposed in [208, 231] as a more appropriate designation.

Furthermore, spectral deviation is not restricted to signed graphs of the same order, which means it can be meaningfully applied to comparisons between a signed graph and its induced subgraph or any other related signed graph. Importantly, ς equals zero if and only if the two signed graphs share the same set of eigenvalues allowing for differences in multiplicities. This includes, as a special case, signed graphs sharing the same spectrum. Constructions of such signed graphs are considered in Chapter 7.

The section is concluded with some properties formulated in the following theorem.

Theorem 2.51 *For a signed graph $\dot{G}$ of order n, the following statements hold true:*

(i) $\varsigma_{K_1}(\dot{G}) = \sum_{i=1}^{n} |\lambda_i(\dot{G})|$.
(ii) If $n \geq 2$, then $\varsigma_{\dot{G}-i}(\dot{G}) \leq \lambda_1(\dot{G}) - \lambda_n(\dot{G})$, where i is an arbitrary vertex of $\dot{G}$. Also, $\varsigma_{\dot{G}-i}(\dot{G}) = \sum_{\lambda \in \dot{G}, \lambda \text{ is simple}} \min_{\mu \in \dot{H}} |\lambda - \mu|$.
(iii) If zero is not an eigenvalue of $\dot{G}$, then $\varsigma_{\dot{G}-i}(\dot{G}) \leq \Delta \sum_{j=1}^{n} 1/|\lambda_j(\dot{G})|$.

Proof Part (i) follows since K_1 has a single eigenvalue equal to zero. Both claims of (ii) are consequences of the eigenvalue interlacing. For (iii), by Theorem 2.50 we have $\varsigma_{\dot{G}-i}(\dot{G}) \leq \sum_{j=1}^{n} d_j/|\lambda_j(\dot{G})| \leq \Delta \sum_{j=1}^{n} 1/|\lambda_j(\dot{G})|$. □

2.8 Hoffman Polynomial

A notable result for ordinary graphs, due to Hoffman, establishes the existence of a particular polynomial associated with connected regular graphs [222, Theorem 2.1.6]. Specifically, for a graph G with adjacency matrix A, there exists a polynomial P such that $P(A) = J$, where J is the all-ones matrix, if and only if G is connected and regular. If G is r-regular, then

$$P(x) = n \prod \frac{x - \lambda}{r - \lambda},$$

where n is the number of vertices and the product goes over all distinct eigenvalues excluding r.

The signed counterpart concerns net regularity rather than regularity and is inherently weaker, as it guarantees the equivalence only when the net degree is a simple eigenvalue.

Theorem 2.52 (Hoffman polynomial [210]) *For a signed graph $\dot{G}$, if there exists a polynomial P such that $P(A_{\dot{G}}) = J$ (J being an all-ones matrix of suitable size), then $\dot{G}$ is a connected net-regular signed graph.*

Conversely, if in addition the net degree ϱ is a simple eigenvalue of $\dot{G}$, then the polynomial exists and has the form

$$P(x) = n\prod \frac{x-\lambda}{\varrho-\lambda}, \tag{2.16}$$

where n is the number of vertices and the product goes over all distinct eigenvalues, excluding ϱ.

Proof The proof is similar to that of the original result, along with certain adjustments tailored for signed graphs. Write $A = A_{\dot{G}}$. The existence of P implies the identity $AJ = JA$. Since the (i, j)-entry of AJ is $d_i^+ - d_i^-$ and the same entry of JA is $d_j^+ - d_j^-$, we have that $d_i^+ - d_i^-$ is constant on the vertex set.

Clearly, $\dot{G}$ cannot be disconnected, since for i and j belonging to different components the (i, j)-entry of any power of A would be zero, giving $P(A) \neq J$.

Write $W(x) = \phi(x)/(x-\varrho)$, where ϕ stands for the minimal polynomial of A. From $\phi(A) = O$, we have $(A-\varrho I)W(A) = O$, giving

$$AW(A) = \varrho W(A).$$

Now, the unit vector $\mathbf{j}$ is an eigenvector associated with ϱ. Since this eigenvalue is simple, the dimension of its eigenspace is one and, by the last identity, every column of $W(A)$ is a multiple of $\mathbf{j}$. Moreover, the symmetry of $W(A)$ implies the identity

$$W(A) = cJ \tag{2.17}$$

for some $c \neq 0$. Thus, the polynomial exists and, at this moment, has the form $P(x) = (1/c)W(x)$.

Further, the identity $a_k A^k \mathbf{j} = a_k \varrho^k \mathbf{j}$ (for a real a_k) yields $W(A)\mathbf{j} = W(\varrho)\mathbf{j}$, which together with (2.17) gives $cJ\mathbf{j} = W(\varrho)\mathbf{j}$, that is, $nc = W(\varrho)$, which finally gives (2.16). □

This result applies to net-regular signed graphs, which are not necessarily regular. It also encompasses ordinary graphs, for which the assumption of the simplicity of the spectral radius ρ is superfluous. A notable consequence of (2.16), in the case where the spectrum of $\dot{G}$ consists entirely of integers, is that the number of vertices n divides the product $\prod(\varrho-\lambda)$.

Problems

2.1 Show that the polynomial (2.1) divides the characteristic polynomial of a divisor matrix.

2.2 Prove Theorem 2.12.

2.3 Propose a suitable definition of the complement of a signed graph.

2.4 Compute the eigenvalues of the Cartesian product $\dot{G}_1 \square \dot{G}_2$ and the tensor product $\dot{G}_1 \times \dot{G}_2$ without using Theorem 2.13.

2.5 Prove that the spectrum of every r-dimensional cube with negative quadrangles is symmetric. (In subsequent chapters, this will follow as a special case of a more general result.)

2.6 Construct a complete graph with a symmetric spectrum. (Recall the Seidel matrix of unsigned graphs and identify a graph with two symmetric Seidel eigenvalues.)

2.7 Prove Lemma 2.23.

2.8 Compare the upper bounds (2.7) and (2.8), and decide in which cases each provides a tighter estimate.

2.9 Use Theorem 2.40 to deduce the structure of a signed graph that maximizes the spectral radius among all signed graphs with given order and frustration index.

2.10 Prove that the upper bound for λ_1 in Theorem 2.44, for $(s,t) = (2,2)$, is attained by the friendship graph.

2.11 ([94]) Let $\dot{G}$ be a complete signed graph whose negative edges induce a spanning tree of fixed diameter. Prove that if the index λ_1 of $\dot{G}$ is maximal, then the induced tree is a path with additional vertices attached at a fixed vertex. Is that vertex arbitrary?

2.12 ([142]) Let $\dot{G}$ be a signed graph of order $n \geq 6$ whose negative edges induce a unicyclic graph. Prove that if the least Laplacian eigenvalue of $\dot{G}$ is minimal, then the induced unicyclic graph is a triangle with pendant vertices attached to a fixed vertex.

2.13 Identify the unsigned graphs with index 2, as discussed in [199]. Attempt to extend this classification to signed graphs using the results from the first two chapters of this book. (A complete characterization will be provided later.)

2.14 Compute $\varsigma_{P_{n-1}}(\dot{G})$ (the spectral deviation) when $\dot{G}$ is a positive or a negative cycle of order n.

2.15 Show that if the eigenvalues of a signed graph from Theorem 2.52 are integers, then its order divides the product $\prod(\varrho-\lambda)$.

3

Strong Regularity

In certain contexts, there is no unique or natural way to extend notions from ordinary graphs to signed graphs. For instance, two distinct concepts of regularity coexist: the classical notion and the concept of net regularity introduced at the beginning of Chapter 1. Similarly, Chapter 5 discusses two variants of signed line graphs. Here, we introduce the concept of strong regularity for signed graphs, which generalizes the classical notion of strongly regular graphs in the sense that every strongly regular graph can be regarded as a strongly regular signed graph equipped with an all-positive signature. This concept is relatively new, originating in 2019 [217], yet it has already been explored in several subsequent works.

We say that a signed graph $\dot{G}$ is *strongly regular* (for short, $\dot{G}$ is an *SRSG*) whenever it is regular and satisfies the following conditions:

(SRSG-i) $\dot{G}$ is neither homogeneous complete nor edgeless.

(SRSG-ii) There exists $a \in \mathbb{Z}$ such that $a_{ij}^{(2)} = a$ for all $i \overset{+}{\sim} j$.

(SRSG-iii) There exists $b \in \mathbb{Z}$ such that $a_{ij}^{(2)} = b$ for all $i \overset{-}{\sim} j$.

(SRSG-iv) There exists $c \in \mathbb{Z}$ such that $a_{ij}^{(2)} = c$ for all $i \nsim j$.

Here, $a_{ij}^{(2)}$ is the (i,j)-entry of $A_{\dot{G}}^2$. Clearly, this entry counts the difference between the number of positive and the number of negative i–j walks of length two in $\dot{G}$. Condition (SRSG-i) eliminates particular configurations to maintain consistency with the definition of strongly regular graphs. The definition of an SRSG is purely combinatorial in the sense that switching equivalence does not preserve strong regularity. In conditions (SRSG-ii)–(SRSG-iv), it might occur that the corresponding parameter is not unique; for instance, if $\dot{G}$ does not contain any positive edge, then a is arbitrary, and similarly in the remaining cases. However, if $\dot{G}$ is strongly regular with n vertices and vertex degree r, then the *parameters* of $\dot{G}$ are written as the ordered quintuple (n,r,a,b,c).

The matrix form of the previous definition is

$$A_{\dot{G}}^2 = \frac{a}{2}(A_{\dot{G}} + A_G) - \frac{b}{2}(A_{\dot{G}} - A_G) + cA_{\overline{G}} + rI, \tag{3.1}$$

where $\overline{G}$ denotes the complementary graph. For homogeneous SRSGs, the identity (3.1) reduces to the well-known matrix identity defining strong regularity.

In what follows, our focus is primarily on inhomogeneous non-complete SRSGs. Observe that $\dot{G}$ is an inhomogeneous non-complete SRSG with the above parameters if and only if $-\dot{G}$ is an inhomogeneous non-complete SRSG with parameters (n, r, b, a, c). We mention in passing that inhomogeneous complete SRSGs arise, for example, from the switching class of complete signed graphs that corresponds to a regular two-graph [101, 189].

We note that Zaslavsky defined *very strongly regular signed graphs* to be those whose adjacency matrix satisfies

$$A_{\dot{G}}^2 = aA_{\dot{G}} + cA_{\overline{G}} + rI, \qquad A_{\dot{G}}\mathbf{j} = \varrho\,\mathbf{j}, \tag{3.2}$$

for $a, c, \varrho \in \mathbb{Z}$, $r \in \mathbb{N}$ and $\mathbf{j} \in \mathbb{R}^n$ [260]. Due to the latter defining equality, every very SRSG is net-regular with net degree ϱ. In addition, every very SRSG is an SRSG.

Here is a simple characterization of SRSGs with two eigenvalues.

Theorem 3.1 ([217]) *An inhomogeneous signed graph has two eigenvalues if and only if it is strongly regular with $b = -a$ and, if it is non-complete, $c = 0$.*

Therefore, contrary to the general case, every SRSG with two eigenvalues switches to an SRSG, with the exception of the one that switches to a homogeneous complete signed graph excluded by defining condition (SRSG-i).

3.1 Disconnected and Bipartite SRSGs

The results of this section are taken from [217]. Homogeneous disconnected SRSGs consist of copies of either a complete graph or its negation. The proof of the following characterization is an immediate consequence of the definition.

Theorem 3.2 *An inhomogeneous disconnected signed graph is strongly regular if and only if every component is an SRSG with the same parameters r, a, b and $c = 0$, allowing that a parameter may be undefined in one component but fixed in another.*

We proceed to characterize bipartite SRSGs; the main result is formulated in Theorem 3.5.

Lemma 3.3 *If $\dot{G}$ is a bipartite signed graph, then the following statements hold:*

(i) The square of the adjacency matrix has the form

$$A_{\dot{G}}^2 = \begin{pmatrix} cJ+(r-c)I & O \\ O & cJ+(r-c)I \end{pmatrix} \tag{3.3}$$

if $\dot{G}$ is either strongly regular, complete or edgeless.

(ii) If $\dot{G}$ is strongly regular, but not complete bipartite, then $c=0$.

(iii) If $\dot{G}$ is strongly regular, then it has at most four eigenvalues.

Proof The implication of (i) follows from the definition of strong regularity and fact that $\dot{G}$ is bipartite. For (ii), considering non-adjacent vertices belonging to different colour classes, we arrive at $c=0$.

For (iii), clearly the eigenvalues of $A_{\dot{G}}^2$ are $cn+r-c$ (where n is the number of vertices in one colour class) and $r-c$. Hence, every eigenvalue of $A_{\dot{G}}$ has the form

$$\pm\sqrt{cn+r-c} \quad \text{or} \quad \pm\sqrt{r-c}. \tag{3.4}$$

Thus, the number of eigenvalues is at most four. □

The following definition is needed. A *symmetric balanced incomplete block design* (a *symmetric BIBD*) with parameters (n,r,l) is an arrangement of n points into n blocks in such a way that every point is contained in r blocks, every block has r points and every pair of points occurs together in l blocks. If B is the $(0,1)$-point-block incidence matrix, then

$$BB^\top = lJ+(r-l)I, \qquad JB = rJ. \tag{3.5}$$

The corresponding *incidence graph* is bipartite with the adjacency matrix

$$\begin{pmatrix} O & B \\ B^\top & O \end{pmatrix}. \tag{3.6}$$

Observe that the negation of the incidence graph can also be considered as the incidence (signed) graph of the same design. Moreover, the first identity of (3.5) remains valid if B is replaced by $-B$. Finally, the block design determined by the incidence matrix $J-B$ is also a symmetric BIBD (known as *complementary* to the previous one) [67, Proposition 1.15].

Next, if $\dot{G}$ is an inhomogeneous complete bipartite signed graph, then the graph induced by positive edges $\dot{G}^+$ is the incidence graph of a symmetric

BIBD if and only if the signed graph induced by negative edges $\dot{G}^-$ is the incidence graph of a symmetric BIBD. These BIBDs are complementary.

Let $\dot{G}^*$ stand for either $\dot{G}^+$ or $\dot{G}^-$.

Theorem 3.4 *An inhomogeneous complete bipartite net-regular signed graph $\dot{G}$ is strongly regular if and only if $\dot{G}^*$ is the incidence graph of a symmetric BIBD.*

Proof If $\dot{G}$ is strongly regular, the number of positive paths of length two between any pair of vertices in one colour class is constant. Since $\dot{G}$ is net-regular, the number of paths (between the same vertices) containing two positive edges is also constant, which means that $\dot{G}^+$ is the incidence graph of a BIBD, and so is $\dot{G}^-$.

For the opposite implication, let P (resp. N) be the $(0,1)$ (resp. $(0,-1)$) incidence matrix of the design determined by $\dot{G}^+$ (resp. $\dot{G}^-$). Then

$$A_{\dot{G}} = \begin{pmatrix} O & P+N \\ (P+N)^{\intercal} & O \end{pmatrix}.$$

It follows that $A_{\dot{G}}^2$ has the form

$$\begin{pmatrix} (2(l_+ + l_-) - r)J + 2(r - l_+ - l_-)I & O \\ O & (2(l_+ + l_-) - r)J + 2(r - l_+ - l_-)I \end{pmatrix}.$$

Therefore, $\dot{G}$ is strongly regular. □

Under the notation of Theorem 3.4, if the parameters of a symmetric BIBD associated with $\dot{G}^*$ are given by (n, r_*, l_*), then for the parameters of $\dot{G}$, we have

$$r = n \quad \text{and} \quad c = n - 4(r_* - l_*), \tag{3.7}$$

which, together with (3.4), gives the eigenvalues expressed in terms of the parameters:

$$\pm\sqrt{n^2 + 4(1-n)(r_* - l_*)} \quad \text{and} \quad \pm\sqrt{4(r_* - l_*)}. \tag{3.8}$$

We now prove a characterization of bipartite SRSGs.

Theorem 3.5 *A connected bipartite SRSG has at most four eigenvalues. In addition, the following statements hold:*

(i) Apart from K_2, every bipartite signed graph with two eigenvalues is strongly regular.

(ii) A bipartite signed graph with three eigenvalues is strongly regular if and only if it is a positive quadrangle, complete bipartite graph or its negation.

(iii) A bipartite signed graph $\dot{G}$ with four eigenvalues is strongly regular if and only if it is inhomogeneous, complete bipartite and net-regular, such that the subgraph induced by edges of the same sign is the incidence graph of a symmetric BIBD.

Proof The limit on the number of eigenvalues follows from Lemma 3.3(iii). Part (i) follows from Theorem 3.1.

For part (ii), since the spectrum of a bipartite signed graph is symmetric, due to (3.4), we have either $cn+r-c=0$ or $r=c$.

In the former case we obtain $r/(n-1)=-c$, yielding the following two possibilities: $n=2$, $r=2$, $c=-2$ or $r=n-1$, $c=-1$. The first scenario corresponds to the balanced quadrangle whose adjacent edges have opposite signs. The second scenario admits no solution, as the existence of non-adjacent vertices in different colour classes contradicts the condition $c=-1$.

The second possibility (so, $r=c$) implies that our signed graph must be homogeneous. Furthermore, the non-zero eigenvalues are $\pm\sqrt{rn}$ (again, by (3.4)). Since the spectrum lies in $[-r, r]$ we get $r=n$, giving the desired result.

The converse implication can be verified straightforwardly.

For part (iii), if $\dot{G}$ is a bipartite SRSG with four eigenvalues, then the eigenvalues are given by (3.4), along with $r=n$. Indeed, $r<n$, by Lemma 3.3(ii), gives $c=0$, but then by (3.4) $\dot{G}$ cannot have four eigenvalues. Thus, $\dot{G}$ is complete bipartite.

We prove that $\dot{G}$ is net-regular. An eigenvector associated with the eigenvalue $cn+r-c$ of $A_{\dot{G}}^2$ has the form $(s,s,\dots,s,t,t,\dots,t)^\intercal$, $s,t\in\mathbb{R}$, with the first n entries equal to s; this follows directly from the block structure of the matrix in (3.3). It follows that the eigenvector associated with $\sqrt{cn+r-c}$ of $A_{\dot{G}}$ has the same form. Examining the eigenvalue equation reveals that the net degree is constant on each colour class. Coupled with the regularity condition, this implies that the net degree is constant across the entire vertex set.

Thus, the signed graph under consideration is complete bipartite and net-regular. Moreover, it must be inhomogeneous; otherwise, it would possess fewer than four eigenvalues. Both directions of the statement are established by Theorem 3.4. □

By the previous result, if a signed graph of Theorem 3.4 has three eigenvalues, then it must be the unbalanced quadrangle with $c=-2$. The remaining ones have either two or four eigenvalues, where those with two satisfy $c=0$. Using the second equality of (3.7), we immediately get the following result.

Corollary 3.6 *If $\dot{G}$ is an inhomogeneous complete bipartite signed graph with two eigenvalues, then $\dot{G}^*$ gives rise to the symmetric BIBD with parameters $(4(r_* - l_*), r_*, l_*)$.*

Here is an example inspired by Corollary 3.6.

Example 3.7 A *Menon design* is a symmetric BIBD with parameters

$$(4s^2, s(2s+1), s(s+1)), \quad s \in \mathbb{Z}\backslash\{0\}.$$

Obviously, these parameters fit into those of Corollary 3.6. Such a design exists if and only if there exists a regular Hadamard matrix of order $4s^2$ [222, Theorem 3.8.10]. It is conjectured that Menon designs exist for all values of s.

By taking $\dot{G}^+$ to be the incidence graph of a Menon design, we arrive at an inhomogeneous complete bipartite signed graph $\dot{G}$. By (3.8), the eigenvalues of $\dot{G}$ are $\pm 2s$.

Here are some examples with four eigenvalues.

Example 3.8 There is exactly one symmetric BIBD with parameters $(7,3,1)$, $(11,5,2)$ or $(13,4,1)$ [67, Section II.1]. The first is known as the *Fano plane*. The eigenvalues of complete bipartite SRSGs that are associated (in the sense of Theorem 3.5(iii)) with these designs are $\pm 4, \pm 1$ (Fano plane), $\pm\sqrt{12}, \pm 1$ (the second design) and $\pm 5, \pm\sqrt{12}$ (the third).

We proceed with an infinite family. A *Hadamard design* is a symmetric BIBD with parameters $(4s-1, 2s-1, s-1)$ for $s \geq 1$. Such a design exists if and only if there exists a Hadamard matrix of order $4s$ (see [222, Theorem 3.8.8]). If H is a Hadamard matrix of order $4s$ whose last row and last column consist entirely of ones (such a matrix is equivalent to a Hadamard matrix of the same order), and if B is obtained by deleting the last row and last column of H, then the corresponding complete bipartite SRSG arises from the matrix (3.6). Its eigenvalues are $\pm 2s$ and ± 1.

3.2 Relations with Association Schemes

Let d be a positive integer and X a non-empty set. A *symmetric d-class association scheme* on X consists of a partition of $X \times X$ into $d+1$ non-empty symmetric binary relations $R_0, R_1, \dots, R_d$ satisfying the following conditions:

- $R_0 = \{(x,x) : x \in X\}$.
- If $(x,y) \in R_h$, then the number $z \in X$ such that $(x,z) \in R_i$ and $(y,z) \in R_j$ is a constant p_{ij}^h depending on i, j, h, but not on the choice of x, y.

One may observe that symmetric BIBDs as defined in the previous section fall into this definition.

The numbers p_{ij}^h are called the *intersection numbers* of the scheme. The ith binary relation R_i, also known as the ith associate class, can be represented by a $(0,1)$-matrix A_i of order $|X|$ whose rows and columns are indexed by the elements of X, and $(A_i)_{xy} = 1$ if and only if $(x,y) \in R_i$. Accordingly,

$$A_0 = I, \qquad A_i = A_i^\top, \qquad \sum_{h=0}^{d} A_h = J, \qquad A_iA_j = \sum_{k=0}^{d} p_{ij}^h A_h.$$

These matrices are also known as the *adjacency matrices* of the association scheme.

It follows that the real span $\langle \mathscr{A} \rangle$ of $A_0, A_1, \ldots, A_d$ creates a $(d+1)$-dimensional commutative algebra of symmetric matrices, known as the *Bose–Mesner algebra* of the association scheme. Since these matrices are symmetric and commute pairwise, they can be diagonalized simultaneously and the entire space $\mathbb{R}^{|X|}$ is expressed as a direct sum of their common eigenspaces. As the dimension of $\langle \mathscr{A} \rangle$ is $d+1$, there are exactly $d+1$ (common) eigenspaces of dimension $f_j, 0 \le j \le d$. Note that one of the eigenspaces has dimension 1, as J belongs to $\langle \mathscr{A} \rangle$ and $|X|$ is its simple eigenvalue. Without loss of generality, we may assume that $f_0 = 1$. Denote by $E_0, E_1, \ldots, E_d$ the orthogonal projections of $\mathbb{R}^{|X|}$ onto those $d+1$ eigenspaces. We have $E_0 = |X|^{-1}J$ and $E_iE_j = \delta_{ij}E_i$. The sets $\{A_0, A_1, \ldots, A_d\}$ and $\{E_0, E_1, \ldots, E_d\}$ are bases for $\langle \mathscr{A} \rangle$ (cf. [44, Theorem 2.6.1]), so there exist numbers $p_i(j)$ such that $A_i = \sum_j p_i(j)E_j$, and there exist numbers $q_j(i)$ such that $E_j = |X|^{-1}\sum_i q_j(i)A_i$. The numbers $p_i(j)$ are known as the *eigenvalues* of the association scheme. If P and Q are the $(d+1) \times (d+1)$ matrices defined by $(P)_{i,j} = p_j(i)$ and $(Q)_{i,j} = q_j(i)$, then $PQ = QP = |X|I$. It also holds that $A_iE_j = p_i(j)E_j$, so $p_i(j)$ is an eigenvalue of A_i with multiplicity $f_j = \operatorname{rank}(E_j)$.

The *d-class Johnson scheme* $J(v,d)$ is defined on the d-subsets of a v-set, with $d \le v/2$. Two d-subsets are in relation R_i if and only if they intersect in $d-i$ elements. The *d-class Hamming scheme* $H(d,q)$ has the set of words of length d over an alphabet of q symbols as its vertex set. Two words are in relation R_i if and only if the Hamming distance between them is i (if and only if they differ in exactly i coordinates).

The eigenvalues of the Johnson scheme are

$$p_i(j) = \sum_{r=0}^{i} (-1)^{i-r} \binom{d-r}{i-r} \binom{v-d+r-j}{r} \binom{d-j}{r} \tag{3.9}$$

and their multiplicities are $m_i = \binom{v}{i} - \binom{v}{i-1}$, along with the convention that

$\binom{v}{-1} = 0$ [67]. From the same reference, the eigenvalues of the Hamming scheme are

$$p_i(j) = \sum_{r=0}^{i} (-1)^r (q-1)^{i-r} \binom{d-j}{i-r} \binom{j}{r} \tag{3.10}$$

and their multiplicities are $m_i = (q-1)^i \binom{d}{i}$.

If m and n are positive integers, then the *rectangular scheme* $R(m,n)$ has as vertices the ordered pairs (i,j), where $1 \leq i \leq m$, $1 \leq j \leq n$. Two such vertices are related according to the following criteria: they are first (resp. second) associates if they agree in the first (resp. second) coordinate. They are third associates if they do not agree in any coordinate. Computing the eigenvalues is left as an exercise.

3.2.1 SRSGs from 3-Class Association Schemes

Here we start from a signed graph and arrive at symmetric binary relations that are similar to those of the previous subsection.

Obviously, every signed graph $\dot{G}$ defines the four symmetric binary relations on its vertex set V:

$$\begin{aligned}
R_0 &= \{(x,y) \in V \times V \colon x = y\},\\
R_1 &= \{(x,y) \in V \times V \colon \overset{+}{\sim} y\},\\
R_2 &= \{(x,y) \in V \times V \colon \overset{-}{\sim} y\},\\
R_3 &= \{(x,y) \in V \times V \colon \nsim y, x \neq y\}.
\end{aligned}$$

We have $V \times V = R_0 \cup R_1 \cup R_2 \cup R_3$. Any relation R_1, R_2 or R_3 can be empty; for example, if only R_2 is empty, then $\dot{G}$ is all-positive.

For vertices $x, y \in V$ satisfying $(x,y) \in R_h$, with $h \in \{0,1,2,3\}$, we define the parameters

$$\begin{aligned}
p_{11}^h(x,y) &= |\{z \in V \colon (x,z) \in R_1, (y,z) \in R_1\}|,\\
p_{22}^h(x,y) &= |\{z \in V \colon (x,z) \in R_2, (y,z) \in R_2\}|,\\
p_{12}^h(x,y) &= |\{z \in V \colon (x,z) \in R_1, (y,z) \in R_2\}|,\\
p_{21}^h(x,y) &= |\{z \in V \colon (x,z) \in R_2, (y,z) \in R_1\}|.
\end{aligned}$$

Note that $p_{12}^0(x,y) = p_{21}^0(x,y) = 0$. Now, $\dot{G}$ is strongly regular if at most one of R_1, R_2, R_3 is empty and if, for all $x, y \in V$ and all $h \in \{0,1,2,3\}$ such that $(x,y) \in R_h$, the parameter $p_{11}^h(x,y) + p_{22}^h(x,y) - p_{12}^h(x,y) - p_{21}^h(x,y)$ depends on h, but not on x or y. Clearly, the conditions given in the definition of a symmetric 3-class association scheme are stronger than the previous ones. Con-

Table 3.1 *Eigenvalues of $\dot{G}_{i,j}$ that arise from the three 3-class schemes.*

	$\dot{G}_{1,2}$	$\dot{G}_{1,3}$	$\dot{G}_{2,3}$	Multiplicity
Johnson scheme $J(v,3)$ for $v \geq 6$				
λ_0	$-\frac{1}{2}3(v-3)(v-6)$	$-\frac{1}{6}(v-3)(v^2-9v+2)$	$-\frac{1}{6}(v-3)(v-4)(v-14)$	1
λ_1	$-\frac{1}{2}v^2-17v+54$	$\frac{1}{2}(v-2)(v-3)-2$	$(v-4)(v-7)$	$v-1$
λ_2	$3v-18$	-2	$16-3v$	$\frac{1}{2}v(v-3)$
λ_3	-6	-2	4	$\frac{1}{6}v(v-1)(v-5)$
Hamming scheme $H(3,q)$				
λ_0	$-3(q-1)(q-2)$	$3q-(q-1)^3-3$	$-(q-1)^2(q-4)$	1
λ_1	$-q^2+6q-6$	q^2-2	$2(q-1)(q-2)$	$3(q-1)$
λ_2	$3q-6$	-2	$4-3q$	$3(q-1)^2$
λ_3	-6	-2	4	$(q-1)^3$
Rectangular scheme $R(m,n)$				
λ_0	$n-m$	$(n-1)(2-m)$	$(m-1)(2-n)$	1
λ_1	0	-2	-2	$(m-1)(n-1)$
λ_2	$-m$	$m-2$	$2m-2$	$n-1$
λ_3	n	$2n-2$	$n-2$	$m-1$

sequently, starting from any symmetric 3-class association scheme, one can readily construct an SRSG as follows. Select one of the relations R_1, R_2, R_3 to represent positive edges, choose another to represent the negative edges, and let the remaining relation correspond to non-edges. The resulting signed graph is strongly regular because R_1, R_2, R_3 form a partition of the vertex pairs into disjoint non-empty relations, and the intersection numbers p_{ij}^h depend solely on the indices i, j, h. If R_i and R_j are chosen to represent positive and negative edges respectively, with the corresponding matrices A_i and A_j, then the adjacency matrix of $\dot{G}$ is given by $A_{\dot{G}} = A_i - A_j$, while the parameters a, b, c of $\dot{G}$ are

$$a = p_{11}^i + p_{22}^i - 2p_{12}^i, \quad b = p_{11}^j + p_{22}^j - 2p_{12}^j, \quad c = p_{11}^k + p_{22}^k - 2p_{12}^k, \tag{3.11}$$

where we use that in a symmetric association scheme $p_{ij}^h = p_{ji}^h$ holds [44, Lemma 2.1.1]. Note that an SRSG constructed in this way is net-regular with net degree $p_{ii}^0 - p_{jj}^0$.

We proceed with SRSGs constructed on the basis of either a 3-class Johnson scheme $J(v,3)$, or a 3-class Hamming scheme $H(3,q)$, or a rectangular scheme $R(m,n)$. The data is given in Table 3.1, where $\dot{G}_{i,j}$ denotes the SRSG with the adjacency matrix $A_{\dot{G}} = A_i - A_j$. Only SRSGs with $i < j$ are considered. We remark that the corresponding SRSGs have at most four eigenvalues. Those with three eigenvalues are extracted in [135].

3.2.2 Johnson Signed Graphs and Hamming Signed Graphs

Let $J(v,d)$, $2 \leq d \leq v/2$, be the Johnson scheme, and let $A_1, A_2, \ldots, A_d$ be its adjacency matrices. The *Johnson signed graph* $J(v,d)_{m,n}$ for $1 \leq m, n \leq d$ and $m \neq n$ is the signed graph with the adjacency matrix $A_{J(v,d)_{m,n}} = A_m - A_n$.

Similarly, let $H(d,q)$, $d,q \geq 2$, be the Hamming scheme with unchanged notation for the adjacency matrices. The *Hamming signed graph* $H(d,q)_{m,n}$ for $1 \leq m, n \leq d$ and $m \neq n$ is the signed graph with the adjacency matrix $A_{H(d,q)_{m,n}} = A_m - A_n$.

Note that for a signed graph $\dot{G}$ as defined above, the adjacency matrix of $\dot{G}^+$ (resp. $\dot{G}^-$) is A_m (resp. A_n). Since $m \neq n$, $\dot{G}$ is never edgeless. None of the adjacency matrices is the zero matrix, ensuring that $\dot{G}$ is always inhomogeneous. Moreover, interchanging the parameters m and n results in $-\dot{G}$, the negation of $\dot{G}$. Also, $\dot{G}$ is net-regular with net degree $p^0_{mm} - p^0_{nn}$ and its underlying graph is regular with vertex degree $p^0_{mm} + p^0_{nn}$.

For a fixed d, there are exactly $\binom{d}{2}$ Johnson (Hamming) signed graphs defined as above. Thus, if $d = 2$ only one Johnson signed graph $J(v,2)_{1,2}$ and one Hamming signed graph $H(2,q)_{1,2}$ can be defined. The adjacency matrix of $J(v,2)_{1,2}$ is the Seidel matrix of the complement of the triangular graph $T(v)$, and the adjacency matrix of $H(2,q)_{1,2}$ is the Seidel matrix of the complement of the $q \times q$ grid graph. For $d \geq 3$, we get non-complete Johnson and Hamming signed graphs. Those with the smallest number of vertices are obtained for $(v,d) = (6,3)$ and $(d,q) = (3,2)$, and they have 20 and 8 vertices, respectively. Examples are illustrated in Figure 3.1.

Remark 3.9 We remind the reader that the vertices of an unsigned Johnson graph are identified with the d-subsets of a v-set, and two vertices are adjacent if and only if the size of the intersection of the corresponding subsets is $d-1$. In other words, the adjacency matrix of a Johnson graph is the matrix A_1 of the d-class Johnson scheme.

However, there is a generalization of a Johnson graph in which two vertices are adjacent if and only if the intersection size is equal to i, where i is a fixed integer from 1 to $d-1$ [5]. Many concepts in signed graph theory encompass their unsigned counterparts as special cases. To achieve a similar unification for Johnson signed graphs, one should allow for the possibility that exactly one of the matrices A_m, A_n is the zero matrix, or equivalently, define a Johnson signed graph as either a generalized Johnson graph, its negation, or as having adjacency matrix $A_m - A_n$.

With slight modifications, these remarks also apply to Hamming signed graphs.

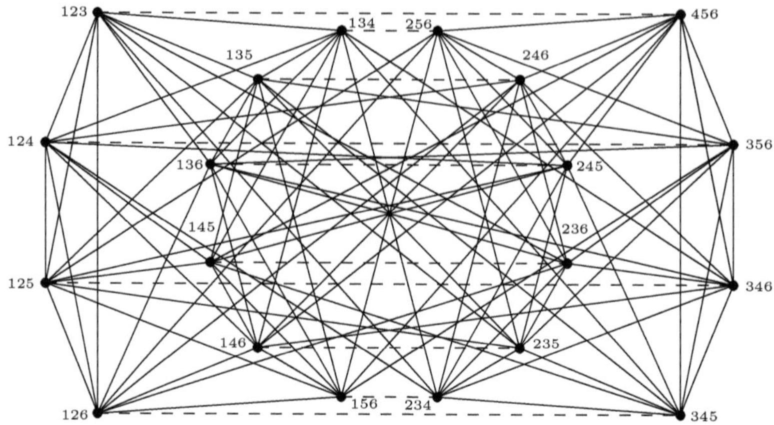

$J(6,3)_{1,3}$

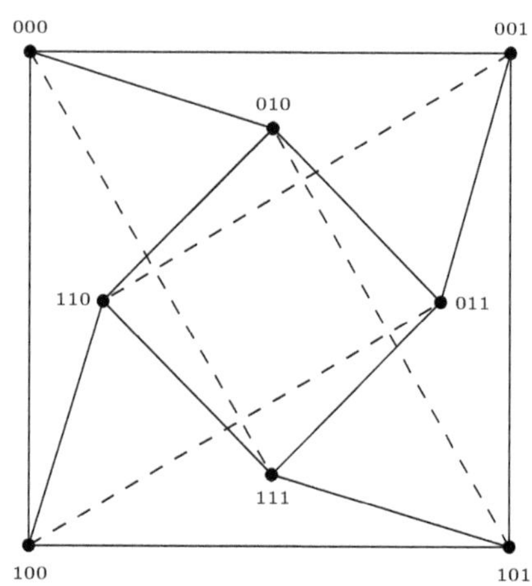

$H(3,2)_{1,3}$

Figure 3.1 The Johnson signed graph $J(6,3)_{1,3}$ and the Hamming signed graph $H(3,2)_{1,3}$. Reproduced from [135], licensed under CC BY 4.0.

We proceed to compute the eigenvalues of these signed graphs.

Theorem 3.10 *The following statements hold:*

(i) The eigenvalues λ_j, $0 \leq j \leq d$, of $J(v,d)_{m,n}$ are

$$\lambda_j = \sum_{r=0}^{m} (-1)^{m-r} \binom{d-r}{m-r} \binom{v-d+r-j}{r} \binom{d-j}{r}$$

$$-\sum_{r=0}^{n}(-1)^{n-r}\binom{d-r}{n-r}\binom{v-d+r-j}{r}\binom{d-j}{r}.$$

(ii) The eigenvalues λ_j, $0 \leq j \leq d$, of $H(d,q)_{m,n}$ are

$$\lambda_j = \sum_{r=0}^{m}(-1)^r(q-1)^{m-r}\binom{d-j}{m-r}\binom{j}{r}$$
$$-\sum_{r=0}^{n}(-1)^r(q-1)^{n-r}\binom{d-j}{n-r}\binom{j}{r}.$$

Proof The result follows from the definitions of the corresponding signed graphs, the facts that the matrices of an association scheme can be simultaneously diagonalized and the formulas (3.9) and (3.10). □

We now consider the question of strong regularity.

Theorem 3.11 *A Johnson signed graph $J(v,d)_{m,n}$ and a Hamming signed graph $H(d,q)_{m,n}$ are strongly regular if and only if $p^h_{mm}+p^h_{nn}-2p^h_{mn}$ is a constant not depending on h, for $1 \leq h \leq d$ and $h \notin \{m,n\}$.*

Proof Denote the signed graph under consideration by $\dot{G}$, and let i,j be its vertices.

We record the following observations. First, $i \overset{+}{\sim} j$ holds precisely when i and j are in relation m. It follows that the difference of the numbers of positive and negative i–j walks of length 2 is $p^m_{mm}+p^m_{nn}-p^m_{mn}-p^m_{nm}=p^m_{mm}+p^m_{nn}-2p^m_{mn}$. By the definition of a symmetric association scheme, this is a constant, say a, not depending on the choice of i or j. Similarly, $i \overset{-}{\sim} j$ holds precisely when i and j are in relation n and the mentioned difference is now equal to $p^n_{mm}+p^n_{nn}-2p^n_{mn}$. This is a constant, say b, not depending on the choice of i or j.

Assume now that $\dot{G}$ is strongly regular. According to the previous observations, it remains to consider the case $i \nsim j$. Clearly, $i \nsim j$ holds precisely if i,j are in a relation $h \notin \{m,n\}$. The difference of the numbers of positive and negative i–j walks of length 2 is $p^h_{mm}+p^h_{nn}-2p^h_{mn}$. Since $\dot{G}$ is strongly regular, the last expression has the same value for every $h \notin \{m,n\}$, and we are done.

Conversely, if $p^h_{mm}+p^h_{nn}-2p^h_{mn}$ is a constant not depending on $h \notin \{m,n\}$, then the mentioned difference is also a constant for $i \nsim j$, which together with the initial observations implies the strong regularity. □

The parameters of a strongly regular Johnson signed graph are

$$\left(\binom{v}{d},\, p^0_{mm}+p^0_{nn},\, p^m_{mm}+p^m_{nn}-2p^m_{mn},\, p^n_{mm}+p^n_{nn}-2p^n_{mn},\, p^h_{mm}+p^h_{nn}-2p^h_{mn}\right),$$

where $1 \leq h \leq d$ and $h \notin \{m,n\}$. Similarly, for a strongly regular Hamming signed graph we have

$$\left(q^d,\ p^0_{mm}+p^0_{nn},\ p^m_{mm}+p^m_{nn}-2p^m_{mn},\ p^n_{mm}+p^n_{nn}-2p^n_{mn},\ p^h_{mm}+p^h_{nn}-2p^h_{mn}\right),$$

where $1 \leq h \leq d$ and $h \notin \{m,n\}$.

The sets of parameters are derived based on the arguments established in the proof of Theorem 3.11.

3.3 A Classification of Inhomogeneous SRSGs

The results of this section are transferred from [136]. It is convenient to rewrite the identity (3.1) in the following way:

$$A^2_{\dot{G}}+\frac{b-a}{2}A_{\dot{G}}=\frac{a+b}{2}A_G+c(J-A_G-I)+rI=\frac{a+b-2c}{2}A_G+cJ+(r-c)I.$$

Condition (SRSG-i) ensures that $A_{\dot{G}}$ is neither the zero matrix nor $J-I$, thereby excluding degenerate cases from (3.3).

Here is a classification of inhomogeneous SRSGs:

$\mathscr{C}_1$: SRSGs with $a=-b$, which are either complete or non-complete with $c \neq 0$. The matrix identity reduces to

$$A^2_{\dot{G}}+bA_{\dot{G}}=cA_{\overline{G}}+rI. \tag{3.12}$$

$\mathscr{C}_2$: SRSGs with $a=-b$, which are non-complete with $c=0$. Then

$$A^2_{\dot{G}}+bA_{\dot{G}}=rI.$$

$\mathscr{C}_3$: SRSGs with $a \neq -b$, which are either complete or non-complete with $c=(a+b)/2$. Then

$$A^2_{\dot{G}}+\frac{b-a}{2}A_{\dot{G}}=\frac{a+b}{2}J+\left(r-\frac{a+b}{2}\right)I. \tag{3.13}$$

$\mathscr{C}_4$: SRSGs with $a \neq -b$, which are non-complete with $c=0$. Then

$$A^2_{\dot{G}}+\frac{b-a}{2}A_{\dot{G}}=\frac{a+b}{2}A_G+rI. \tag{3.14}$$

$\mathscr{C}_5$: SRSGs with $a \neq -b$, which are non-complete with $c \neq (a+b)/2$ and $c \neq 0$. Then

$$A^2_{\dot{G}}+\frac{b-a}{2}A_{\dot{G}}=\frac{a+b-2c}{2}A_G+cJ+(r-c)I. \tag{3.15}$$

Note that if $\dot{G}$ belonging to $\mathscr{C}_1$ is complete, then $A_{\overline{G}}$ is the zero matrix, and we have $A_{\dot{G}}^2 + bA_{\dot{G}} = rI$. Obviously, $\dot{G}$ has two eigenvalues. Moreover, $A_{\dot{G}}$ coincides with the Seidel matrix of a graph induced by negative edges.

Here is a short result concerning non-complete graphs of $\mathscr{C}_1$.

Theorem 3.12 *If $\dot{G}$ belonging to $\mathscr{C}_1$ is non-complete and the complement of its underlying graph is connected, then $\dot{G}$ is net-regular.*

Proof Since $A_{\dot{G}}$ commutes with the left-hand side of (3.12), it also commutes with $A_{\overline{G}}$, which means that $A_{\dot{G}}$ and $A_{\overline{G}}$ are simultaneously diagonalizable. Since $\overline{G}$ is a connected regular graph, $n-r-1$ is its eigenvalue with multiplicity one and $\mathbf{j}$ is the corresponding eigenvector. Thus, $\mathbf{j}$ is an eigenvector of $A_{\dot{G}}$, and so $\dot{G}$ is net-regular. □

A signed graph satisfying the conditions of Theorem 3.12 is very strongly regular in the sense of (3.2). Examples with disconnected $\overline{G}$ appear in Section 3.1, where they correspond to inhomogeneous complete bipartite signed graphs whose positive and negative subgraphs, $\dot{G}^+$ and $\dot{G}^-$, are incidence graphs of complementary symmetric BIBDs. These signed graphs are also net-regular, which demonstrates that the converse of the previous theorem does not hold; specifically, for non-complete signed graphs in $\mathscr{C}_1$, net regularity does not imply the connectedness of the complement of the underlying graph.

By the defining identity, every SRSG of $\mathscr{C}_2$ has two eigenvalues.

3.3.1 SRSGs of $\mathscr{C}_3$

As previously observed, every connected inhomogeneous signed graph with two eigenvalues (appearing in either $\mathscr{C}_1$ or $\mathscr{C}_2$) must necessarily be strongly regular. Moreover, it is well known that a connected regular (unsigned) graph is strongly regular if and only if it has three eigenvalues, with the largest among them being simple [222, Theorem 3.4.7].

In what follows, we concentrate on the class $\mathscr{C}_3$, which corresponds naturally to the class of strongly regular graphs since, by Theorem 3.14, it comprises SRSGs with three distinct eigenvalues, one of which is simple.

Theorem 3.13 *If $\dot{G} \in \mathscr{C}_3$, then $\dot{G}$ is net-regular.*

The proof follows the line of the proof of Theorem 3.12. By Theorem 3.12, the subgraphs $\dot{G}^+$ and $\dot{G}^-$ are regular. If ϱ is the net degree of $\dot{G}$, then $-\varrho$ is the net degree of $-\dot{G}$, which means that many considerations may be restricted to SRSGs with non-negative net degree.

Theorem 3.14 *If $\dot{G} \in \mathscr{C}_3$, then $\dot{G}$ has three eigenvalues. One of them is the net degree with multiplicity* 1.

Proof From (3.13), we deduce that ϱ satisfies the equality

$$\varrho^2 + \frac{b-a}{2}\varrho = \frac{a+b}{2}n + r - \frac{a+b}{2}. \tag{3.16}$$

Since J and $A_{\dot{G}}$ commute, each eigenvector of $A_{\dot{G}}$ which is not in the eigenspace of $\mathbf{j}$ is a 0-eigenvector of J. Thus, any eigenvalue associated with an eigenvector other than $\mathbf{j}$ satisfies the quadratic

$$x^2 + \frac{b-a}{2}x + \frac{a+b}{2} - r = 0. \tag{3.17}$$

Thus, if the multiplicity of ϱ is greater than 1, then it also satisfies the equality (3.17), which, together with (3.16), implies $\frac{1}{2}a + bn = 0$, that is, $a + b = 0$. Since this last equality does not hold for strongly regular signed graphs in $\mathscr{C}_3$, it follows that ϱ must be a simple eigenvalue.

It remains to show that $\dot{G}$ has three eigenvalues. Apart from ϱ, there are at most two other eigenvalues. If $\dot{G}$ has two eigenvalues, the simple one ϱ (associated with $\mathbf{j}$) and another one, say μ, with multiplicity $n-1$, then $A_{\dot{G}} - \mu I = [(\varrho - \mu)/n]\mathbf{j}\mathbf{j}^\intercal$. It can now be seen that $(\varrho - \mu)/n = 1$ or $(\varrho - \mu)/n = -1$, but then $\dot{G}$ is homogeneous and complete, which contradicts its strong regularity. □

We denote the eigenvalues determined by the quadratic equation (3.17) as λ and μ. Note that regularity combined with net regularity in a signed graph having three eigenvalues does not necessarily imply strong regularity. For instance, the Seidel matrix of the three-dimensional cube can be viewed as the adjacency matrix of a complete net-regular signed graph with spectrum $[5^1, 1^4, (-3)^3]$, yet this signed graph is not strongly regular. The following theorem provides sufficient conditions under which a regular signed graph with three eigenvalues is strongly regular.

Theorem 3.15 *Let $\dot{G}$ be a connected inhomogeneous regular and net-regular signed graph with three eigenvalues. If its net degree is a simple eigenvalue, then $\dot{G}$ is strongly regular.*

Proof We resume the introduced notation on eigenvalues and exploit the result of Theorem 2.52. The corresponding polynomial reads

$$P(x) = n\frac{(x-\lambda)(x-\mu)}{(\varrho - \lambda)(\varrho - \mu)}.$$

By using the first three spectral moments, we can express the multiplicities of the eigenvalues in terms of n, r, ϱ, λ and μ. In particular, for the multiplicity of ϱ, we have

$$1 = m_\varrho = \frac{(r+\lambda\mu)n}{(\varrho-\lambda)(\varrho-\mu)}.$$

Thus, $P(x) = (x-\lambda)(x-\mu)/(r+\lambda\mu)$, and then

$$\frac{(A_{\dot{G}}-\lambda I)(A_{\dot{G}}-\mu I)}{r+\lambda\mu} = J.$$

(Note that $r+\lambda\mu \neq 0$, since otherwise we would have $m_\varrho = 0$.)

The last identity gives $A_{\dot{G}}^2 - (\lambda+\mu)A_{\dot{G}} = (r+\lambda\mu)J - \lambda\mu I$. By setting $a = r+\lambda\mu+\lambda+\mu$ and $b = r+\lambda\mu-\lambda-\mu$, we obtain that the adjacency matrix of $\dot{G}$ satisfies the identity (3.13). Since $r+\lambda\mu \neq 0$, we have $a \neq -b$, and so $\dot{G}$ is an SRSG (belonging to $\mathscr{C}_3$). □

A signed graph is *walk-regular* if, for each vertex i, the difference between the number of positive walks and the number of negative walks starting and terminating at i, and traversing k edges, is constant for all non-negative integers k. By setting $k = 2$, it follows that every walk-regular signed graph is regular. It is shown in [13] that every regular signed graph with at most three eigenvalues must be walk-regular. Together with Theorem 3.14, this implies the following result.

Corollary 3.16 *If $\dot{G} \in \mathscr{C}_3$, then $\dot{G}$ is walk-regular.*

Example 3.17 Examples of $\mathscr{C}_3$ SRSGs are easily constructed by considering the Seidel matrix of a strongly regular graph. Indeed, if G is a strongly regular graph with eigenvalues (of the adjacency matrix) r, θ_1 and θ_2 (r being the vertex degree), then the eigenvalues of its Seidel matrix are $n-2r-1$, $-2\theta_1-1$ and $-2\theta_2-1$. Thus, if θ_1 and θ_2 are both distinct from $r-n/2$, the Seidel matrix of G is also the adjacency matrix of the desired SRSG of $\mathscr{C}_3$.

We now provide further details on complete SRSGs belonging to $\mathscr{C}_3$; non-complete SRSGs will be discussed subsequently.

There is a natural relation between the parameters of a strongly regular graph G and the parameters of an SRSG determined by the Seidel matrix of G. Namely, if G is a strongly regular graph with parameters (n,r,a',b') (these are standard parameters associated with a strongly regular graph: order, degree, the number of common neighbours of adjacent vertices and the number of common neighbours of non-adjacent vertices), then its Seidel matrix is the adjacency matrix of a complete inhomogeneous SRSG with $a = n+4b'-4r-2$ and $b = n+4a'-4r+2$, which can be verified by a straightforward exercise.

Conversely, a simple algebraic computation gives that if $\dot{G}$ belonging to $\mathscr{C}_3$ is complete, then $A_{\dot{G}}$ is the Seidel matrix of a strongly regular graph with parameters

$$\left(n,\ \frac{n-1-\varrho}{2},\ \frac{n-2\varrho+b-4}{4},\ \frac{n-2\varrho+a}{4}\right),$$

where ϱ is the net degree of $\dot{G}$.

We proceed with some sufficient conditions for $\dot{G}$ belonging to $\mathscr{C}_3$ to be complete.

Theorem 3.18 *Let $\dot{G}\in\mathscr{C}_3$. If its net degree is* 0*, then $\dot{G}$ is complete.*

The proof follows by considering the defining identity (3.13). Here is another condition.

Theorem 3.19 *Let $\dot{G}\in\mathscr{C}_3$. If the eigenvalues distinct from the net degree are equal in multiplicity, then $\dot{G}$ is complete with $a=b=-1$ and its net degree is* 0.

Proof The multiplicities of λ and μ are

$$m_\lambda=\frac{(r+\mu\varrho)n}{(\lambda-\varrho)(\lambda-\mu)},\qquad m_\mu=\frac{(r+\lambda\varrho)n}{(\mu-\varrho)(\mu-\lambda)},\tag{3.18}$$

which follows by considering spectral moments (and also appears in the problem section). From $m_\lambda=m_\mu$, we deduce $(r+\mu\varrho)(\mu-\varrho)=(r+\lambda\varrho)(\varrho-\lambda)$, which can be rewritten as

$$(\lambda+\mu)\varrho^2+(2r-\lambda^2-\mu^2)\varrho-r(\lambda+\mu)=0.\tag{3.19}$$

Recall that λ and μ are the roots of the quadratic equation (3.17). Computing the roots and inserting them in (3.19), we get

$$\frac{a-b}{2}\varrho^2+\left(a+b-\frac{(a-b)^2}{4}\right)\varrho-r\frac{a-b}{2}=0.\tag{3.20}$$

Assume, by the way of contradiction, that $a\neq b$. Then, using (3.20), we get

$$\varrho^2+\left(\frac{2(a+b)}{a-b}-\frac{a-b}{2}\right)\varrho-r=0.$$

This equality, in conjunction with that of (3.16), leads to

$$\frac{2(a+b)}{a-b}\varrho=-\frac{a+b}{2}(n-1),\quad \text{i.e.}\ \varrho=\frac{b-a}{4}(n-1).$$

Observing ϱ, which can be assumed to be non-negative, and taking into account that a and b have the same parity, we arrive at two possibilities: $b-a=2$ or $b-a=4$.

The first scenario further gives $\varrho = (n-1)/2$. Using (3.16) once again,

$$r = \frac{n-1}{2}\left(\frac{n+1}{2} - 2(a+1)\right).$$

The second factor $(n+1)/2 - 2(a+1)$ is obviously a positive integer not greater than 2, so it can be either 2 or 1. If $(n+1)/2 - 2(a+1) = 2$, then $(n+1)/4 = a+2$, so $n+1$ is divisible by 4. In this case, we have $r = n-1$ and $\varrho = (n-1)/2$.

Denote by r_1 the vertex degree of $\dot{G}^+$ and by r_2 the vertex degree of $\dot{G}^-$. Then $r_1 + r_2 = r = n-1$ and $r_1 - r_2 = \varrho = (n-1)/2$, so $2r_2 = (n-1)/2$. This means that $n-1$ is also divisible by 4, which is impossible. If $(n+1)/2 - 2(a+1) = 1$, we have $\varrho = r$, but this contradicts the assumption that $\dot{G}$ is inhomogeneous.

The second scenario (so, $b - a = 4$) leads to $\varrho = n-1$, that is, $\varrho = r = n-1$, which as before contradicts the assumption that $\dot{G}$ is inhomogeneous.

Therefore, we have $a = b$. By (3.20), we get $(a+b)\varrho = 0$. Since $\dot{G} \in \mathscr{C}_3$, it holds that $a \neq -b$, and so $\varrho = 0$. It follows by Theorem 3.18 that $\dot{G}$ is complete with $a + b = -2$, so we must have $a = b = -1$. □

Observe now that if S is the Seidel matrix of a conference graph (with parameters $(n, (n-1)/2, (n-5)/4, (n-1)/4)$), then S also serves as the adjacency matrix of an inhomogeneous complete SRSG, say $\dot{G}$. Moreover, the parameters a and b are mutually equal and non-zero, implying that $\dot{G} \in \mathscr{C}_3$. Obviously, the net degree of $\dot{G}$ is zero and the remaining eigenvalues are equal in multiplicity. Conversely, if $\dot{G} \in \mathscr{C}_3$ and the eigenvalues distinct from the net degree are equal in multiplicity, then Theorem 3.19 guarantees that $\dot{G}$ is complete, and hence $A_{\dot{G}}$ is the Seidel matrix of a strongly regular graph. By Theorem 3.19, we also have $\varrho = 0$ and $a = b = -1$, which leads to the parameters of the conference graph mentioned above.

We record these observations as the following theorem.

Theorem 3.20 *Let $\dot{G} \in \mathscr{C}_3$. The following statements are equivalent:*

(i) $m_\lambda = m_\mu$.
(ii) $\dot{G}$ is complete and $A_{\dot{G}}$ is the Seidel matrix of a conference graph.

All the $\mathscr{C}_3$ SRSGs encountered so far have been complete. Here, we turn our attention to the possible parameters of non-complete SRSGs in $\mathscr{C}_3$. We utilize the fact that the eigenvalues (ϱ, λ and μ) must be integers; otherwise, the multiplicities satisfy $m_\lambda = m_\mu$, which by Theorem 3.20 implies that $\dot{G}$ is inhomogeneous and complete. In addition, if $\dot{G}$ (or $-\dot{G}$) is switching equivalent to its underlying graph, then we have $\varrho = r$ (or $\varrho = -r$), which means that $\dot{G}$

(or $-\dot{G}$) is homogeneous. Therefore, we may assume that the eigenvalues of $\dot{G}$ lie in $[-r+1, r-1]$.

Now, for a fixed r, we consider all the possible triples (ϱ, λ, μ) for which m_λ and m_μ, given in (3.18), are non-negative integers and $n = m_\lambda + m_\mu + 1 \geq r+2$ holds.

By a straightforward computation based on the previously mentioned restrictions and the equalities derived from spectral moments of small order, we find that the vertex degree of a non-complete SRSG of $\mathscr{C}_3$ cannot be less than 6. Moreover, for $r \geq 6$, there exists a triple of eigenvalues $(\varrho, \lambda, \mu) = (r-2, 1, -(r-2))$ that passes all the numerical tests and leads to the quintuple $(n, r, a, b, c) = ((r-2)(r-3), r, 5-r, r-1, 2)$. The non-existence of SRSGs with these parameters is proved in [136]. For $r \leq 10$, apart from these parameters, we get one additional set: $(n, r, a, b, c) = (18, 8, 1, 3, 2)$. We record this as the following theorem.

Theorem 3.21 *If the vertex degree of $\dot{G} \in \mathscr{C}_3$ does not exceed* 10*, then $\dot{G}$ is complete or (eventually) has the parameters* $(18, 8, 1, 3, 2)$.

Of course, for large r, there exist triples which produce feasible parameters. For example, $(\varrho, \lambda, \mu) = (20, 5, -2)$ gives $(n, r, a, b, c) = (60, 38, 16, -20, 18)$, but the question of existence remains open.

At this point the reader might ask whether $\mathscr{C}_3$ contains any non-complete signed graph, and the answer to this question arises from the previous section. In the corresponding notation, these are the SRSGs $\dot{G}_{1,2}$ obtained from 3-class Johnson schemes $J(9,3)$ and $J(14,3)$, respectively. By considering Table 3.1 and applying the parameter conditions from (3.11), we can easily identify more examples. The 3-class Johnson schemes yield the infinite family of SRSGs $\dot{G}_{1,3}$ obtained from $J(v,3)$ with $v = 8$ or $v \geq 10$, and two SRSGs $\dot{G}_{2,3}$ from $J(6,3)$ and $J(8,3)$. Similarly, the 3-class Hamming schemes produce $\dot{G}_{1,2}$ from $H(3,6)$ and the infinite family $\dot{G}_{1,3}$ from $H(3,q)$ with $q \notin \{2,3\}$. The rectangular schemes give the infinite family $\dot{G}_{1,3}$ from $R(2n,n)$ with $n \neq 2$, and another family $\dot{G}_{2,3}$ from $R(m,2m)$ with $m \neq 2$.

3.4 Net Regularity, Strong Regularity and Walk Regularity

In this section we consider the relationships between net-regular, strongly regular and walk-regular signed graphs. As illustrated in Figure 3.2, there is no inclusion relation between any two of these classes. We investigate conditions under which a signed graph lies in the intersection of two or all three classes.

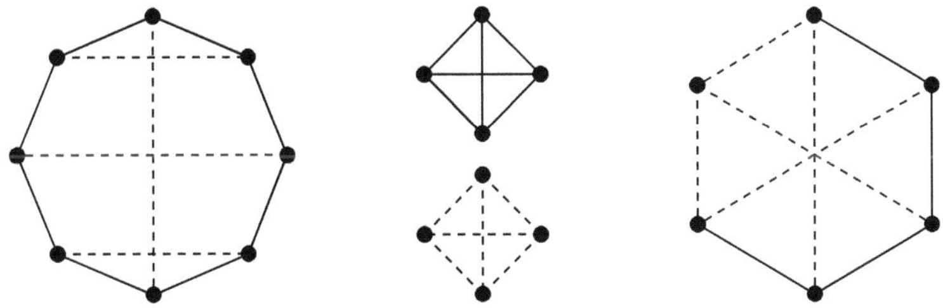

Figure 3.2 Examples of signed graphs: (a) net-regular that is neither strongly regular nor walk-regular; (b) strongly regular that is neither net-regular nor walk-regular; (c) walk-regular that is neither net-regular nor strongly regular. Reproduced from [14].

In this context, we continue examining the five classes introduced in the previous section to identify which strongly regular signed graphs are also net-regular. Additionally, we provide several sufficient conditions under which a walk-regular signed graph is net-regular, strongly regular, or both. The results in this section are drawn from [14].

3.4.1 Net Regularity of SRSGs

If $\dot{G}$ belonging to $\mathscr{C}_1$ is non-complete and the complement of its underlying graph is connected, then $\dot{G}$ must be net-regular as stated in Theorem 3.12. Examples of complete SRSGs of $\mathscr{C}_1$ that are not net-regular can be found in the regular two-graph equivalence class. Constructions of non-complete SRSGs of $\mathscr{C}_1$ that are not net-regular are given in Subsection 3.4.2; see the discussion after Theorem 3.25 concerning walk-regular SRSGs that are not net-regular.

Even if $\dot{G} \in \mathscr{C}_2$ is net-regular, an arbitrary switching need not be net-regular. In addition, if $\dot{G}$ is net-regular then its eigenvalues are integers since the net degree is one of the eigenvalues. The class $\mathscr{C}_3$ is completely resolved in Subsection 3.3.1; see Theorems 3.14 and 3.15.

We proceed with $\mathscr{C}_4$ and $\mathscr{C}_5$.

Theorem 3.22 *Every connected SRSG of $\mathscr{C}_4$ is net-regular.*

Proof We consider the identity (3.14). Since $A_{\dot{G}}$ commutes with the left-hand side, it also commutes with A_G. Then $A_{\dot{G}}$ and A_G are simultaneously diagonalizable. Since G is connected and regular, its degree r appears as its simple eigenvalue associated with $\mathbf{j}$. Hence, $\mathbf{j}$ is an eigenvector of $A_{\dot{G}}$, and so $\dot{G}$ is net-regular. □

A simple way to construct disconnected net-regular SRSGs of $\mathscr{C}_4$ is to take k $(k \geq 2)$ copies of a connected SRSG from the same class. In the notation of Subsection 3.2.2, $kJ(8,3)_{1,2}$ is such a signed graph with parameters $(n,r,a,b,c) = (56k,45,6,-3,0)$ and spectrum $[(-15)^k,(9)^{7k},6^{20k},(-6)^{28k}]$. Such SRSGs can also be constructed by taking multiple copies of a complete SRSG from $\mathscr{C}_3$. Specifically, if $\dot{H} \in \mathscr{C}_3$ is complete, then $k\dot{H}$ is disconnected and shares the parameters a and b with $\dot{H}$. It also has $c = 0$, so it belongs to $\mathscr{C}_4$. It is also net-regular, inheriting the net degree of $\dot{H}$, but now with multiplicity k in the spectrum.

Examples of SRSGs of $\mathscr{C}_4$ that are not net-regular are disjoint unions of m copies of K_n and l copies of $-K_n$, where m,l are positive integers and $n \geq 3$. Evidently, these signed graphs are not net-regular with $a = b = n-2$ and $c = 0$, so they belong to $\mathscr{C}_4$. The spectrum is comprised of four eigenvalues: $[(n-1)^m,(1-n)^l,1^{(n-1)l},(-1)^{m(l-1)}]$.

We proceed with a combinatorial property of $\mathscr{C}_5$.

Theorem 3.23 *An SRSG of $\mathscr{C}_5$ is net-regular whenever either*

$$a+b \equiv 1 \pmod 2 \quad \text{or} \quad \frac{2c}{a+b-2c} \in \mathbb{R} \setminus (-1,0).$$

Proof Assume that $\dot{G}$ is not net-regular. We show that this assumption implies the negation of the constraints stated in the theorem. The starting point is the defining matrix identity (3.15). Additionally, we use the three parameter conditions that hold for every signed graph in $\mathscr{C}_5$ (given in the definition): (i) $a \neq -b$, (ii) $c \neq (a+b)/2$ and (iii) $c \neq 0$.

Due to (ii), the right-hand side of (3.15) can be rewritten as

$$\frac{a+b-2c}{2}\left(A_G + \frac{2c}{a+b-2c}J\right).$$

Observe that the term in parentheses has $\mathbf{j}$ as an eigenvector associated with the eigenvalue $r + (2c/(a+b-2c))n$. As $A_{\dot{G}}$ commutes with the left-hand side of (3.15), $\mathbf{j}$ is an eigenvector for $A_{\dot{G}}$ (and consequently $\dot{G}$ is net-regular) whenever the aforementioned eigenvalue is simple. Therefore, this eigenvalue is non-simple, which implies the existence of an eigenvalue λ of G satisfying $\lambda = r + (2c/(a+b-2c))n$.

From (iii), we deduce that G is connected, and thus $-r \leq \lambda < r$, which implies that $-2 \leq 2c/(a+b-2c) < 0$. If $-2 \leq 2c/(a+b-2c) < -1$, then $-A_G - (2c/(a+b-2c))J$ is a positive matrix with an all-positive eigenvector, which necessarily corresponds to a simple eigenvalue. Also, $2c/(a+b-2c) \neq -1$, because of (i). Hence, $2c/(a+b-2c) \in (-1,0)$.

From (3.15), we deduce that solutions of the quadratic $x^2 + \frac{1}{2}(b-a)x =$

$\frac{1}{2}(a+b-2c)r+cn+r-c$ $(=\frac{1}{2}(a+b-2c)\lambda+r-c)$ must be distinct and both must be the eigenvalues of $\dot{G}$; otherwise, there would be only one eigenvalue of $\dot{G}$ corresponding to the eigenvalue of $A_G+(2c/(a+b-2c))J$. Therefore, the polynomial

$$x^2+\frac{b-a}{2}x-\left(\frac{a+b-2c}{2}r+cn+r-c\right)$$

divides the characteristic polynomial of $\dot{G}$. Since both polynomials are monic and the latter has integral coefficients, the same holds for the former [78, Lemma 1.3.5]. In particular, this implies that $a+b\equiv 0 \pmod 2$. □

3.4.2 Walk Regularity of SRSGs

Every SRSG belonging to $\mathscr{C}_2\cup\mathscr{C}_3$ is walk-regular, since it has at most three eigenvalues. Taking into account that an SRSG of $\mathscr{C}_3$ is net-regular, we deduce that every signed graph belonging to this class is simultaneously net-regular, strongly regular and walk-regular.

We continue with the inspection of net-regular SRSGs from $\mathscr{C}_1\cup\mathscr{C}_4\cup\mathscr{C}_5$. The following lemma is needed.

Lemma 3.24 ([14]) *For a net-regular SRSG $\dot{G}$, the difference between the number of positive and the number of negative triangles containing a fixed vertex is constant.*

For the proof, one may express $A^3_{\dot{G}}$ from the defining identity (3.1) and take into account that the (i,i)-entry of this matrix is twice the number of the considered difference.

Observe that, if a signed graph $\dot{G}$ has exactly s eigenvalues then, for $k\geq s$,

$$A^k_{\dot{G}}\in\langle A^{k-1}_{\dot{G}},A^{k-2}_{\dot{G}},\dots,A^{k-s}_{\dot{G}}\rangle. \tag{3.21}$$

This assertion follows from the identity $A^{k-s}_{\dot{G}}\phi(A_{\dot{G}})=O$, where ϕ represents the minimal polynomial.

Theorem 3.25 *The following statements hold:*

(i) Every net-regular SRSG with at most four eigenvalues is walk-regular.
(ii) Every net-regular SRSG with at most five eigenvalues, for which the net degree is a simple eigenvalue, is walk-regular.

Proof (i) By setting $s=4$ in (3.21), it follows that $A^4_{\dot{G}}$ has constant diagonal entries. Fixing $k\geq s+1$ and assuming that the diagonal entries of $A^{k-1}_{\dot{G}}$ are mutually equal, the claim follows directly from (3.21) by an induction argument.

(ii) In view of part (i), it suffices to consider the case where $\dot{G}$ has exactly five eigenvalues. Substituting $A_{\dot{G}}$ for x in (2.16) and invoking Lemma 3.24, we obtain that the diagonal entries of $A^4_{\dot{G}}$ are mutually equal. The desired conclusion then follows by setting $s = 5$ in (3.21) and applying the same induction argument as above. □

Here are some examples illustrating Theorem 3.25. Examples for part (i) include the Hamming signed graphs $H(4,3)_{2,3}$, $H(4,2)_{2,4}$ and $H(4,4)_{1,2}$. These graphs are strongly regular with three eigenvalues. For the first two, the net degree corresponds to a simple eigenvalue, whereas in the third example it corresponds to a multiple eigenvalue. Further examples arise from Johnson and Hamming signed graphs with four eigenvalues where the net degree is simple, such as $J(v,3)_{1,2}$, $v \notin \{6,7,9,14\}$; $J(v,3)_{2,3}$, $v \notin \{6,8,11\}$; $H(3,q)_{1,2}$, $q \notin \{2,3,6\}$; and $H(3,q)_{2,3}$, $q \neq 3$.

Examples for part (ii) are $J(12,4)_{1,2}$ and $J(8,4)_{3,4}$. Both are strongly regular, have five eigenvalues and their net degree is a simple one.

On the other hand, walk regularity combined with strong regularity and having a small number of eigenvalues does not guarantee net regularity. Obvious examples are SRSGs with two eigenvalues, such as those in the class $\mathscr{C}_2$ (many of which are not net-regular), as well as a SRSGs of $\mathscr{C}_1$ whose adjacency matrices correspond to the Seidel matrices of non-regular graphs with two eigenvalues.

When we deal with at most four eigenvalues, the walk regularity for each class defined in Section 3.3 is summarized as follows.

Theorem 3.26 *Every SRSG of $\mathscr{C}_2 \cup \mathscr{C}_3$ is walk-regular. Let $\dot{G}$ be an SRSG with four eigenvalues.*

(i) If $\dot{G} \in \mathscr{C}_1$ and the complement of its underlying graph is connected, then $\dot{G}$ is walk-regular.
(ii) If $\dot{G}$ is a connected signed graph of $\mathscr{C}_4$, then $\dot{G}$ is walk-regular.
(iii) If the parameters of $\dot{G} \in \mathscr{C}_5$ satisfy either

$$a+b \equiv 1 \pmod 2 \quad \text{or} \quad \frac{2c}{a+b-2c} \in \mathbb{R} \setminus (-1,0),$$

then $\dot{G}$ is walk-regular.

Proof Classes $\mathscr{C}_2$ and $\mathscr{C}_3$ were already resolved in the beginning of this subsection.

If $\dot{G}$ has four eigenvalues, it belongs to some of the remaining three classes, and then the assertion follows from Theorem 3.25(i) used in conjunction with Theorem 3.12 for (i), Theorem 3.22 for (ii) and Theorem 3.23 for (iii). □

We observe the existence of SRSGs in $\mathscr{C}_1$ that have four eigenvalues and are walk-regular, yet the complement of their underlying graph is disconnected. In other words, the converse of Theorem 3.26(i) does not hold. Namely, according to Theorem 3.5, every complete bipartite SRSG with four eigenvalues is net-regular and has parameters $a = b = 0$, $c \neq 0$, so it belongs to $\mathscr{C}_1$. However, the complement of its underlying graph is obviously disconnected.

Disconnected and walk-regular SRSGs which have four eigenvalues and belong to $\mathscr{C}_4$ are the ones that are net-regular. An example is $kJ(8,3)_{1,2}$, $k \geq 2$.

Disconnected but not walk-regular SRSGs which have four eigenvalues and belong to $\mathscr{C}_4$ are not net-regular. Recall from Subsection 3.4.1 a disjoint union of m copies of K_n and l copies of its negation. This SRSG has four eigenvalues, but the diagonal of $A^k_{\dot{G}}$ is not constant for odd k.

3.4.3 Walk Regularity of Net-Regular Signed Graphs

Walk-regular signed graphs can be easily constructed using adjacency matrices A_i, $1 \leq i \leq d$, of a symmetric d-class association scheme. We simply set

$$A_{\dot{G}} = \sum_{s \in S} A_s - \sum_{t \in T} A_t, \quad \text{where } S,T \subseteq \{1,2,\ldots,d\},\ S \cap T = \emptyset.$$

Since the adjacency matrices form a commutative (Bose–Mesner) algebra, $A^k_{\dot{G}}$ $(k \geq 1)$ is a linear combination of $A_0 = I, A_1, A_2, \ldots, A_d$, and therefore has a constant diagonal. In this way we have proved the following result.

Theorem 3.27 *Let $\dot{G}$ be a signed graph with adjacency matrix $A_{\dot{G}} = \sum_{s \in S} A_s - \sum_{t \in T} A_t$, where A_i, $i \in \{0,1,\ldots d\}$, are the adjacency matrices of a d-class symmetric association scheme, $S,T \subseteq \{1,2,\ldots,d\}$ and $S \cap T = \emptyset$. Then $\dot{G}$ is walk-regular.*

Taking into account the definitions of Johnson and Hamming signed graphs, together with the preceding theorem, we arrive at the following corollary.

Corollary 3.28 *Every Johnson signed graph and every Hamming signed graph is walk-regular.*

Examples of walk-regular (and also net-regular) signed graphs that are not strongly regular can be constructed on the basis of this corollary. For instance, the signed graphs $J(v,d)_{1,2}$ and $H(d,q)_{1,2}$ are walk-regular, but they are not strongly regular (see Subsection 3.2.2).

In what follows, we focus on walk regularity of signed graphs with at least four eigenvalues, as those with fewer eigenvalues are necessarily walk-regular. Additionally, we assume net regularity as part of the framework.

Theorem 3.29 *The following statements hold:*

(i) Let $\dot{G}$ be a regular and net-regular signed graph with at most four eigenvalues. If its net degree is a simple eigenvalue then $\dot{G}$ is walk-regular.
(ii) Let $\dot{G}$ be a bipartite regular and net-regular signed graph with at most six eigenvalues. If the net degree occurs as a simple eigenvalue then $\dot{G}$ is walk-regular.

Proof The proof of (i) is similar to the proof of the first item of Theorem 3.25, with the only critical case occurring when the signed graph $\dot{G}$ has four eigenvalues. In this case, we can use the polynomial identity (2.16) to express $A^3_{\dot{G}}$ as a linear combination of $A_{\dot{G}}$, $A^2_{\dot{G}}$, J and I. This implies that diagonal entries of $A^3_{\dot{G}}$ are constant. From here, the assertion follows by an induction argument applied to the span in (3.21).

For (ii), we may assume that $\dot{G}$ has five or six eigenvalues. Since $\dot{G}$ is bipartite, we immediately get that $A^k_{\dot{G}}$ has a constant diagonal whenever k is odd. By setting $x = A_{\dot{G}}$ in (2.16) and splitting the corresponding polynomial into the even-degree part and the odd-degree part, we obtain that $A^4_{\dot{G}}$ is a linear combination of $A^2_{\dot{G}}$, I and

$$J^* = \begin{pmatrix} J & O \\ O & J \end{pmatrix},$$

so we can write $A^4_{\dot{G}} = \alpha A^2_{\dot{G}} + \gamma J^* + \delta I$ for some α, β, γ. Thus, $A^4_{\dot{G}}$ has a constant diagonal.

The previous identity also gives $A^{2k}_{\dot{G}} = \alpha A^{2k-2}_{\dot{G}} + \beta J^* A^{2k-4}_{\dot{G}} + \gamma A^{2k-4}_{\dot{G}}$. We know that the net degree ϱ is an eigenvalue of $A_{\dot{G}}$ associated with $\mathbf{j}$, so $\mathbf{j}$ is an eigenvector of A^{2k-4} for the eigenvalue ϱ^{2k-4}. This means that row sums in A^{2k-4} are all equal to ϱ^{2k-4}, and thus $J^* A^{2k-4}_{\dot{G}} = \varrho^{2k-4} J^*$. Hence, $A^{2k}_{\dot{G}} = \alpha A^{2k-2}_{\dot{G}} + \beta \varrho^{2k-4} J^* + \gamma A^{2k-4}_{\dot{G}}$. By fixing k ($k \geq 3$), assuming that diagonal entries are mutually equal for all powers of A^{2k-2} and considering the right-hand side of the last identity, we deduce that the diagonal of $A^{2k}_{\dot{G}}$ is constant for every k. □

The converses do not hold. If, under the assumptions of (i) (or (ii)) of the previous theorem, a signed graph is walk-regular then its net degree does not need to be a simple eigenvalue.

3.5 Net-Regular SRSGs with Degree at Most Four

Following [11], we classify all signed graphs with the specified properties. Unless otherwise stated, throughout this section we restrict attention to connected signed graphs. The only 1-regular signed graphs are K_2 and $-K_2$, so they are not strongly regular because both are homogeneous and complete. For 2-regular graphs, that is, cycles, the only signatures yielding net-regular signed graphs occur on cycles of even length.

Theorem 3.30 *There are no connected 1-regular SRSGs. The unique connected 2-regular net-regular SRSG is the four-vertex cycle in which negative edges form a perfect matching.*

We prove a result that is used for both cubic and quartic cases.

Theorem 3.31 *Let $\dot{G}$ be a connected non-complete net-regular SRSG belonging to $\mathscr{C}_1 \cup \mathscr{C}_4 \cup \mathscr{C}_5$. Then, for any two vertices i and j of $\dot{G}$, the number of negative walks of length 2 between i and j is even. Moreover, if this number is $2k$, $k \in \mathbb{N}$, then there are exactly k common neighbours of i and j that are joined to i by a positive edge and to j by a negative edge, and exactly k common neighbours that are joined to i by a negative edge and to j by a positive edge.*

Proof If $\dot{G} \in \mathscr{C}_1 \cup \mathscr{C}_4 \cup \mathscr{C}_5$ is connected and non-complete, then the coefficient of A_G in (3.1) does not vanish. In addition, $\dot{G}$ is net-regular, and thus $A_{\dot{G}}$ commutes with J, which implies that $A_{\dot{G}}$ commutes with A_G. This yields that the matrices $A_{G^+} = \frac{1}{2}(A_G + A_{\dot{G}})$ and $A_{G^-} = \frac{1}{2}(A_G - A_{\dot{G}})$ mutually commute.

Now, the number of common neighbours of i and j that are joined to i by a positive edge and to j by a negative edge is the (i,j)-entry of $A_{G^+}A_{G^-}$. Similarly, the remaining number given in the statement formulation is the (i,j)-entry of $A_{G^-}A_{G^+}$. The desired result follows by commutativity of A_{G^+} and A_{G^-}. □

3.5.1 Cubic Case

If $\dot{G}$ is a cubic net-regular signed graph with a positive net degree, then G^+ and G^- are regular of degree 2 and 1, respectively. Hence, G^- induces a perfect matching in $\dot{G}$.

If either $\dot{G} \in \mathscr{C}_2$ or $\dot{G}$ is a complete SRSG from $\mathscr{C}_1$, then the defining identity (3.1) reduces to $A_{\dot{G}}^2 + bA_{\dot{G}} = 3I$. Obviously, $\dot{G}$ has two eigenvalues, where one of them is $\varrho = 1$, and so the other one must be -3 (as the product of eigenvalues is the negation of the common vertex degree). Then the eigenvalues of $-\dot{G}$ are 3 and $\varrho = -1$. Now, Theorem 1.26 yields that $-\dot{G}$ is switching equivalent to K_4, and then $\dot{G} \in \mathscr{C}_1$. It is the net-regular SRSG $\dot{H}_1$, with parameters $(n,r,a,b) = (4,3,-2,2)$, obtained by reversing the sign of a perfect matching in K_4; see Figure 3.3.

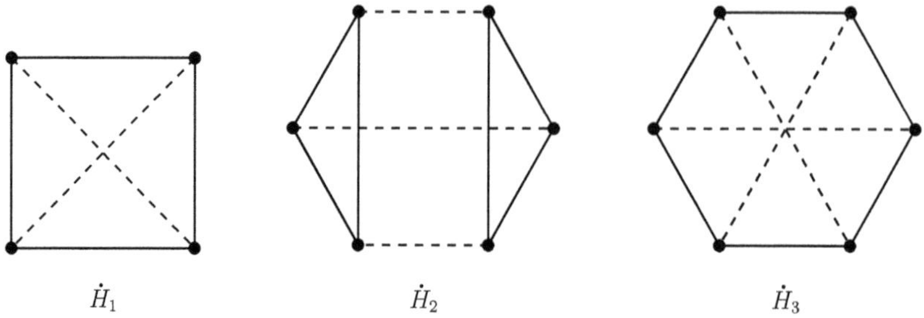

Figure 3.3 Connected cubic net-regular SRSGs with net degree 1. Reproduced from [11].

Suppose that $\dot{G} \in \mathscr{C}_3$. We know from Theorem 3.21 that every SRSG of $\mathscr{C}_3$ whose degree is at most 6 must be complete. However, it can be easily verified that there is only one signature on K_4 that satisfies the above conditions, and it produces the already obtained SRSG $\dot{H}_1$.

It remains to consider the case of non-complete SRSGs of $\mathscr{C}_1 \cup \mathscr{C}_4 \cup \mathscr{C}_5$.

Lemma 3.32 *There are no non-complete cubic net-regular ($\varrho = 1$) SRSGs that belong to $\mathscr{C}_1 \cup \mathscr{C}_4 \cup \mathscr{C}_5$ and contain a negative triangle.*

Proof Let the vertices i, j, k, with $i \overset{+}{\sim} j$, $j \overset{-}{\sim} k$, $k \overset{+}{\sim} i$, form a negative triangle in $\dot{G}$. According to Theorem 3.31, i and j must have a common neighbour, say l, such that $i \overset{-}{\sim} l$ and $j \overset{+}{\sim} l$. By the same theorem, since $k \overset{+}{\sim} i$, there is a positive edge between k and l. Now, all four vertices have degree 3 and the obtained graph is the complete SRSG $\dot{H}_1$. □

Suppose now that a non-complete cubic net-regular SRSG $\dot{G}$ from $\mathscr{C}_1 \cup \mathscr{C}_4 \cup \mathscr{C}_5$ does not contain a negative triangle. Then the parameter b equals 0. Similarly, a is either 1 or 0. Observe that $\mathbf{j}$ is an eigenvector for all four matrices that appear in the defining identity; it corresponds to the eigenvalue 1 of $A_{\dot{G}}^2$ and $A_{\dot{G}}$, 3 of A_G and $n-4$ of $A_{\overline{G}}$. Thus, we have $1 = 2a + b + c(n-4) + 3$, that is, $c(n-4) = -4$ for $a = 1$, $b = 0$, and $c(n-4) = -2$ for $a = 0$, $b = 0$. Obviously, $c < 0$ holds, leading to the following possibilities:

$$(8,3,1,0,-1), \quad (6,3,1,0,-2), \quad (6,3,0,0,-1), \quad (5,3,0,0,-2).$$

A straightforward analysis eliminates the first two parameter sets, and produces the signed graphs $\dot{H}_2$ and $\dot{H}_3$ of Figure 3.3. Here is the statement.

Theorem 3.33 *There are exactly six connected cubic net-regular SRSGs: $\dot{H}_1$, $\dot{H}_2$, $\dot{H}_3$ (illustrated in Figure 3.3) and their negations.*

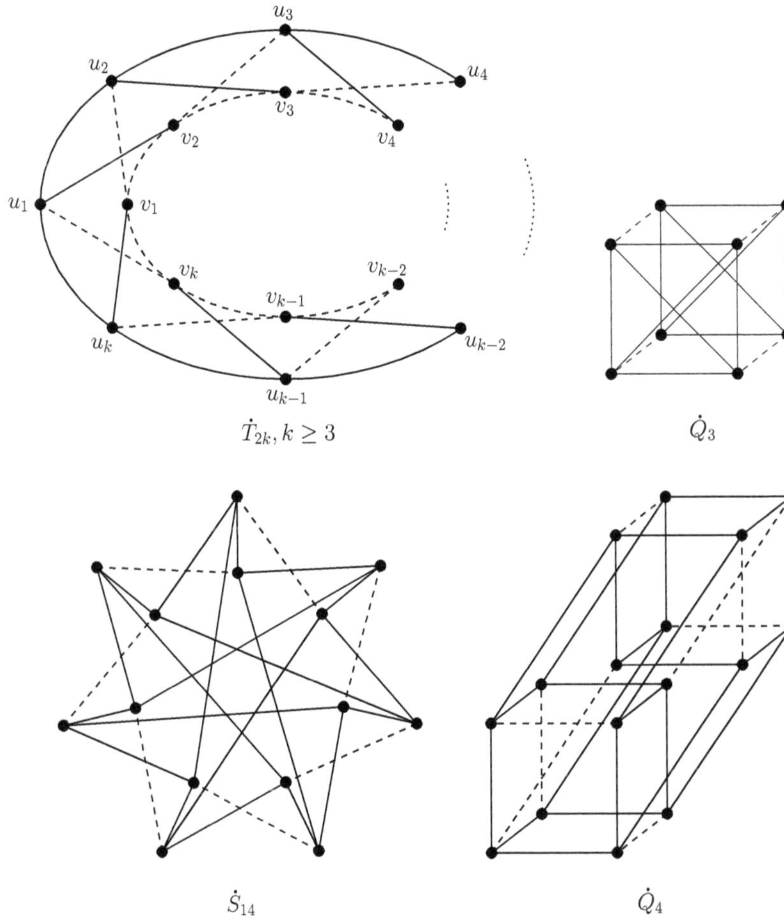

Figure 3.4 Signed graphs $\dot{T}_{2k}$, $k \geq 3$, $\dot{Q}_3$, $\dot{S}_{14}$ and $\dot{Q}_4$. Reproduced from [11].

3.5.2 Quartic Case and Recapitulation

The quartic case is considered in a similar way, but requires intensive computation. We omit the details and just quote the main results. The first one deals with net degree 2 or -2.

Theorem 3.34 *There are exactly six connected quartic net-regular SRSGs with net degree 2 or -2 ($\dot{Q}_3$, $\dot{Q}_4$, $\dot{S}_{14}$, illustrated in Figure 3.4, and their negations) and one infinite family comprised of signed graphs switching equivalent to $\dot{T}_{4j}$, $j \geq 2$, of Figure 3.4 and whose negative edges are $v_{2i-1} \stackrel{-}{\sim} v_{2i}$ and $u_{2i-1} \stackrel{-}{\sim} u_{2i}$ for $1 \leq i \leq j$.*

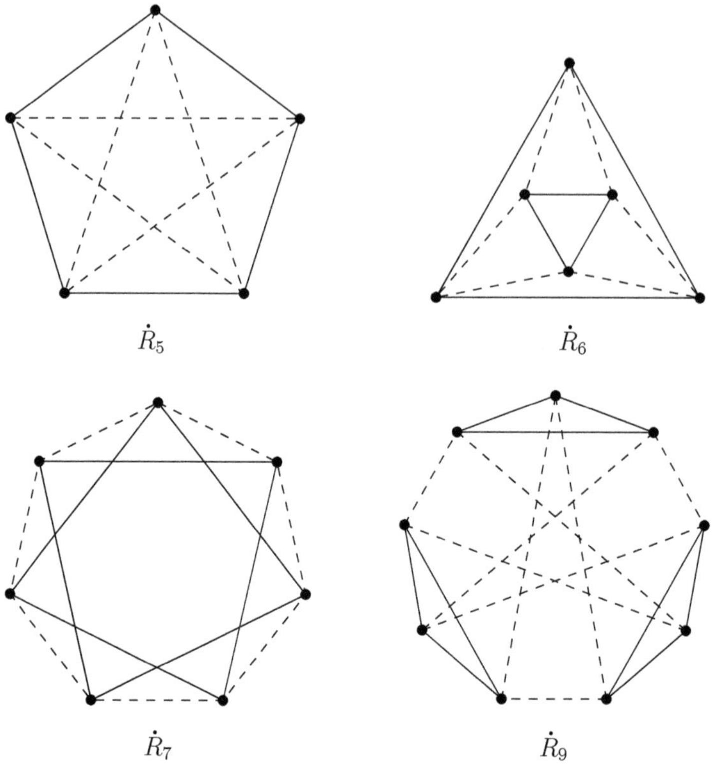

Figure 3.5 Connected quartic net-regular SRSGs with net degree 0 and $a \geq b$. Reproduced from [11].

For $\varrho = 0$, we have the following theorem.

Theorem 3.35 *There are exactly six connected quartic net-regular SRSGs with net degree* 0*:* $\dot{R}_5$, $\dot{R}_6$, $-\dot{R}_6$, $\dot{R}_7$, $-\dot{R}_7$ *and* $\dot{R}_9$ *of Figure* 3.5.

The final result recapitulates the entire section.

Theorem 3.36 *The following statements hold:*

(i) There are no connected 1*-regular net-regular SRSGs (Theorem* 3.30*).*

(ii) The four-vertex cycle with two negative non-adjacent edges is the unique connected 2*-regular net-regular SRSG (Theorem* 3.30*).*

(iii) There are exactly six connected cubic net-regular SRSGs (Theorem 3.33*).*

(iv) There is one infinite family and an additional six connected quartic net-regular SRSGs with net degree $\neq 0$ *(Theorem* 3.34*).*

(v) *There are exactly six connected quartic net-regular SRSGs with net degree* 0 *(Theorem* 3.35*).*

In total, there is an infinite family and an additional 19 connected net-regular SRSGs with vertex degree at most 4.

Problems

3.1 Use the catalogue of BIBDs from [67] to construct all bipartite SRSGs with at most 20 vertices and four eigenvalues.

3.2 Use the result of the previous problem to determine all SRSGs with at most 20 vertices.

3.3 Show that every strongly regular (unsigned) graph remains strongly regular when regarded as a signed graph with the all-positive signature.

3.4 Compute the eigenvalues of the rectangular association scheme $R(m,n)$.

3.5 ([217]) Show that if $\dot{G}$ is an SRSG satisfying $w_2(i,j)=0$ for all i,j, and $c \neq 0$, then

$$n \leq \frac{r(r-1)}{c} + r + 1,$$

where n and r denote the order and vertex degree, respectively.

3.6 Determine necessary and sufficient conditions for an SRSG arising from the Johnson, Hamming or rectangular association schemes to have three eigenvalues. Can any such SRSGs have only two eigenvalues?

3.7 Show that every signed graph in $\mathscr{C}_3$ is net-regular; that is, prove Theorem 3.13.

3.8 Prove Theorem 3.18.

3.9 Prove or disprove the existence of an SRSG with parameters $(18,8,1,3,2)$.

3.10 Construct an example of a connected SRSG that is neither net-regular nor walk-regular.

3.11 Derive structural conditions under which a signed graph is simultaneously net-regular, strongly regular and walk-regular, beyond those that follow directly from the definitions.

3.12 Prove or disprove the existence of SRSGs in $\mathscr{C}_5$ satisfying

$$\frac{2c}{a+b-2c} \in \mathbb{R} \setminus (-1,0).$$

3.13 Investigate whether a signed Johnson graph or a signed Hamming graph is strongly regular or walk-regular.

3.14 ([135]) Prove that the Johnson signed graphs $J(12,4)_{1,2}$ and $J(8,4)_{3,4}$ are strongly regular with five eigenvalues each.

3.15 Find an upper bound for the number of eigenvalues of an SRSG in terms of its parameters.

4

Star Complements

We focus on particular decompositions of signed graphs into two induced subgraphs, one of which serves as a star complement for a specified eigenvalue. This notion, originally developed in the theory of unsigned graphs, is adapted here to the framework of signed graphs. It will feature prominently in the forthcoming chapters, occasionally serving as a central tool in the spectral analysis.

In Chapter 2, we introduced matrices that define orthogonal projections of the Euclidean space $\mathbb{R}^n$ onto the eigenspaces corresponding to the eigenvalues of a signed graph $\dot{G}$, with respect to the standard basis. Let P denote such a projection matrix associated with a fixed eigenvalue μ of $\dot{G}$. Then there exists a subset

$$S \subseteq V(\dot{G}) = \{1, 2, \dots, n\}$$

such that the vectors $P\mathbf{e}_i$, for $i \in S$, form a basis of the eigenspace $\mathscr{E}(\mu)$. The set S is called a *star set* for μ, while the induced subgraph on the complementary vertex set $\overline{S} = V(\dot{G}) \setminus S$ is known as a *star complement* for μ.

By construction, μ is not an eigenvalue of the star complement, and the multiplicity of μ in the spectrum of $\dot{G}$ is precisely $|S|$. Denoting this multiplicity by k and letting $t = n - k$, we refer to t as the *co-dimension* of μ. Importantly, a star complement exists for every eigenvalue of any signed graph, which is a straightforward observation. Also, a signed graph $\dot{G}$ contains a star complement $\dot{H}$ for μ if and only if $-\dot{G}$ contains a star complement $-\dot{H}$ for $-\mu$.

Since the adjacency matrix $A_{\dot{G}}$ and the matrix P commute, we have the following identity:

$$\lambda P\mathbf{e}_i = PA_{\dot{G}}\mathbf{e}_i = \sum_{j \sim i} \sigma(ij) P\mathbf{e}_j. \tag{4.1}$$

The following results are, with slight modifications, transferred from the field of graphs.

Theorem 4.1 ([202, 77]) *Let S be a star set for μ in $\dot{G}$.*

(i) If $\mu \neq 0$, then every vertex of S is adjacent to at least one vertex of $\overline{S} = V(\dot{G}) \setminus S$.
(ii) If $\mu \notin \{1, 0, -1\}$, then each two vertices of S have distinct neighbourhoods in $\overline{S}$.

Proof (i): From (4.1) we have that, for $\mu \neq 0$ and $u \in S$, the vectors $P\mathbf{e}_u, P\mathbf{e}_j$ $(j \sim u)$ are linearly dependent. Since the vectors $P\mathbf{e}_i$ $(i \in S)$ are linearly independent (see the definition of S), it follows that u is adjacent to a vertex in $\overline{S}$.

(ii): If $u, v \in S$ are the vertices sharing the same neighbourhood in $\overline{S}$, then by (4.1) we have

$$\lambda(P\mathbf{e}_u - P\mathbf{e}_v) - \sum_{i \in S,\, i \sim u} \sigma(ui) P\mathbf{e}_i + \sum_{i \in S,\, i \sim v} \sigma(vi) P\mathbf{e}_i = \mathbf{0}.$$

Linear independence of the vectors on the left-hand side yields

$$\begin{cases} \mu = 0 & \text{and} \quad u \nsim v, \\ \mu = -1 & \text{and} \quad u \overset{+}{\sim} v, \\ \mu = 1 & \text{and} \quad u \overset{-}{\sim} v, \end{cases}$$

completing the assertion. □

The next result establishes the basic properties of star complements.

Theorem 4.2 (Reconstruction theorem [77]) *Given a signed graph $\dot{G}$ with the adjacency matrix*

$$\begin{pmatrix} A_S & B^\intercal \\ B & A_{\dot{H}} \end{pmatrix}, \tag{4.2}$$

where A_S is the $k \times k$ adjacency matrix of the subgraph induced by a vertex set S, while $\dot{H}$ is the subgraph induced by $V(\dot{G}) \setminus S$, then S is a star set for μ if and only if μ is not an eigenvalue of $\dot{H}$. In this case

$$\mu I - A_S = B^\intercal(\mu I - A_{\dot{H}})^{-1} B$$

and the eigenspace of μ consists of the eigenvectors

$$\begin{pmatrix} \mathbf{x} \\ (\mu I - A_{\dot{H}})^{-1} B\mathbf{x} \end{pmatrix},$$

where $\mathbf{x} \in \mathbb{R}^k$.

If $\dot{H}$ is a star complement for an eigenvalue μ and $\mathbf{x}, \mathbf{y} \in \mathbb{R}^t$, then we define the following bilinear form:

$$\langle \mathbf{x}, \mathbf{y} \rangle = \mathbf{x}^\intercal (\mu I - A_{\dot{H}})^{-1} \mathbf{y}.$$

Here is a direct consequence.

Corollary 4.3 (cf. [77]) *If μ is not an eigenvalue of a signed graph $\dot{H}$, then there is a signed graph $\dot{G}$ with $\dot{H}$ as a star complement for μ if and only if*

$$\langle \mathbf{b}_i, \mathbf{b}_i \rangle = \mu \quad and \quad \langle \mathbf{b}_i, \mathbf{b}_j \rangle \in \{1, 0, -1\},$$

for all distinct $i, j \in S = V(\dot{G}) \setminus V(\dot{H})$, where $\mathbf{b}_i$ and $\mathbf{b}_j$ determine neighbourhoods of i and j in $\dot{H}$, respectively.

Clearly, the vectors $\mathbf{b}_i$ $(1 \leq i \leq k)$ form the submatrix B of the reconstruction theorem. Also, if $\langle \mathbf{b}_i, \mathbf{b}_j \rangle = 0$ (resp. $\langle \mathbf{b}_i, \mathbf{b}_j \rangle = -1$, $\langle \mathbf{b}_i, \mathbf{b}_j \rangle = 1$), then $i \nsim j$ (resp. $i \overset{+}{\sim} j$, $i \overset{-}{\sim} j$).

Theorem 4.1(ii) asserts that, under the given constraints, distinct columns of B satisfy $\mathbf{b}_i \neq \mathbf{b}_j$. A sharper version rules out the case $\mathbf{b}_i = -\mathbf{b}_j$, and establishing this refinement is an elementary exercise.

The matrix $(\mu I - A_{\dot{H}})^{-1}$ can be computed by the following method.

Theorem 4.4 (cf. [77, Proposition 5.1.11]) *If the minimal polynomial of the adjacency matrix A of a signed graph $\dot{G}$ is $\phi(x) = \sum_{i=0}^{d} c_i x^i + x^{d+1}$ and μ is not an eigenvalue of $\dot{G}$, then*

$$\phi(\mu)(\mu I - A)^{-1} = \sum_{i=0}^{d} a_i A^i,$$

with $a_d = 1$ and $a_{d-j} = \sum_{i=0}^{j-1} c_{d-i} \mu^{j-1-i} + \mu^j$ for $1 \leq j \leq d$.

We proceed with some additional characterizations, transferred from the unsigned domain without significant changes.

Corollary 4.5 (cf. [75, 77]) *If μ is a non-main eigenvalue of $\dot{G}$, then, in the notation of Corollary 4.3, $\langle \mathbf{b}_i, \mathbf{j} \rangle = -1$ holds for all $i \in S$.*

The proof follows since $\mathbf{j}$ is orthogonal to the eigenspace of μ which, together with the eigenspace specification of the reconstruction theorem, yields $\langle \mathbf{b}_i, \mathbf{j} \rangle = -1$.

Theorem 4.6 (cf. [75, 77]) *If μ is an eigenvalue of a connected signed graph $\dot{G}$, then $\dot{G}$ contains a connected star complement for μ.*

If $\lambda_1, \lambda_2, \ldots, \lambda_s$ are distinct eigenvalues of a signed graph $\dot{G}$, then a *star partition* in $\dot{G}$ is the partition $V(\dot{G}) = X_1 \sqcup X_2 \sqcup \cdots \sqcup X_s$ such that X_i is the star set for λ_i. An analogue of [76, Theorem 7.4.5] reads as follows.

Theorem 4.7 *Let μ be an eigenvalue of multiplicity k in a signed graph $\dot{G}$ and $S \subset V(\dot{G})$ a set of vertices such that $|S| = k$ and μ is not an eigenvalue of*

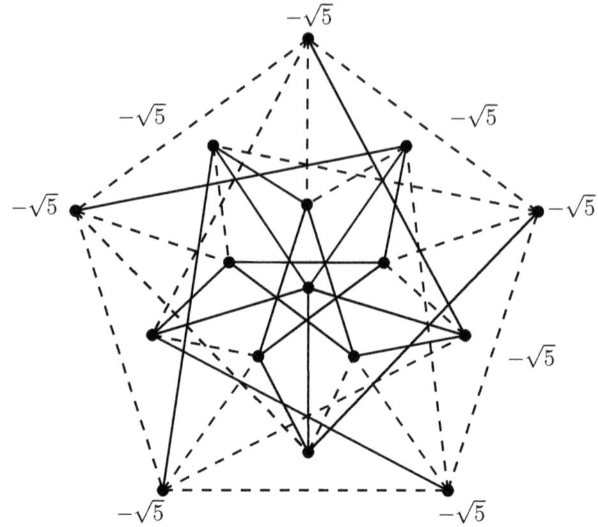

Figure 4.1 A star partition of a signed graph with spectrum $[\sqrt{5}^8, (-\sqrt{5})^8]$.

$G-S$. Then there is a star partition $X_1 \sqcup X_2 \sqcup \cdots \sqcup X_s$ such that $S = X_i$ for some $i \in \{1,2,\ldots,s\}$.

Here is an example.

Example 4.8 A positive triangle, viewed as a subgraph of the signed graph in Figure 1.3, serves as a star complement for the eigenvalue -2. According to Theorem 4.7, the subgraph induced by the remaining vertices forms a star complement for the eigenvalue 2.

Similarly, Figure 4.1 depicts another signed graph with two eigenvalues, $\pm\sqrt{5}$, together with a corresponding star partition.

The previous results provide a spectral tool known as the *star complement technique*, which enables the construction of signed graphs from their smaller components – namely, star complements. Specifically, any signed graph possessing an eigenvalue μ can be reconstructed from its star complement corresponding to μ by analysing all possible neighbourhoods within the star complement. These neighbourhoods are represented by the columns of the matrix B in the reconstruction theorem, that is, by the characteristic vectors $\mathbf{b}_i$. According to Theorem 4.1, if $\mu \notin \{1,0,-1\}$, then only distinct characteristic vectors are relevant; moreover, no two of them are negatives of each other. Furthermore, Corollary 4.3 gives a restriction on the matrix B stating that for every pair of columns the inner product must satisfy $\langle \mathbf{b}_i, \mathbf{b}_j \rangle \in \{1,0,-1\}$.

Every signed graph that admits a star complement $\dot{H}$ for an eigenvalue μ is called a *μ-extension* of $\dot{H}$. A *maximal extension* is a μ-extension to which no additional vertex can be added without violating the μ-extension property. It can be verified that the signed graphs in Example 4.8 are maximal extensions with respect to all of their eigenvalues. Maximal extensions are particularly important, as every μ-extension is an induced subgraph of some maximal one. Finally, if two star complements $\dot{H}$ and $\dot{H}'$ have unique maximal μ-extensions $\dot{G}$ and $\dot{G}'$, respectively, then $\dot{H} \cong \dot{H}'$ implies $\dot{G} \cong \dot{G}'$ [180].

The following example illustrates the technique.

Example 4.9 Consider the complete graph K_6 in the role of a star complement for -3. In the notation of Corollary 4.3, by direct computation we obtain that $\langle \mathbf{b}_i, \mathbf{b}_i \rangle = -3$ if and only if $\mathbf{b}_i$ is a $(1,-1)$-vector with exactly three 1s (and three -1s), having no zero entry. This yields exactly ten possibilities when those differing only by sign are excluded. For such a pair of vectors we compute the bilinear form:

$$\langle \mathbf{b}_i, \mathbf{b}_j \rangle = \begin{cases} -1 & \text{if } \mathbf{b}_i \text{ and } \mathbf{b}_j \text{ match in exactly four positions,} \\ 1 & \text{if } \mathbf{b}_i \text{ and } \mathbf{b}_j \text{ match in exactly two positions,} \\ 0 & \text{never.} \end{cases}$$

This means that the corresponding vertices are by either a positive or a negative edge (depending on whether the previous bilinear form is -1 or 1), and leads to the unique maximal extension (up to the switching operation). Its adjacency matrix is

$$\left(\begin{array}{rrrrrrrrrr|c}
0 & 1 & 1 & 1 & -1 & -1 & 1 & 1 & -1 & -1 & \\
1 & 0 & 1 & -1 & 1 & -1 & 1 & -1 & 1 & -1 & \\
1 & 1 & 0 & -1 & -1 & 1 & -1 & 1 & 1 & -1 & \\
1 & -1 & -1 & 0 & 1 & 1 & 1 & 1 & -1 & -1 & \\
-1 & 1 & -1 & 1 & 0 & 1 & 1 & -1 & 1 & -1 & B^{\mathsf{T}} \\
-1 & -1 & 1 & 1 & 1 & 0 & -1 & 1 & 1 & -1 & \\
1 & 1 & -1 & 1 & 1 & -1 & 0 & 1 & 1 & 1 & \\
1 & -1 & 1 & 1 & -1 & 1 & 1 & 0 & 1 & 1 & \\
-1 & 1 & 1 & -1 & 1 & 1 & 1 & 1 & 0 & 1 & \\
-1 & -1 & -1 & -1 & -1 & -1 & 1 & 1 & 1 & 0 & \\
\hline
1 & 1 & -1 & 1 & 1 & -1 & 1 & -1 & -1 & -1 & \\
1 & -1 & 1 & 1 & -1 & 1 & -1 & 1 & -1 & -1 & \\
-1 & 1 & 1 & -1 & 1 & 1 & -1 & -1 & 1 & -1 & \\
1 & 1 & 1 & -1 & -1 & -1 & 1 & 1 & 1 & 1 & A_{K_6} \\
-1 & -1 & -1 & 1 & 1 & 1 & 1 & 1 & 1 & 1 & \\
-1 & -1 & -1 & -1 & -1 & -1 & -1 & -1 & -1 & 1 &
\end{array}\right),$$

and the spectrum is $[5^6, (-3)^{10}]$.

4.1 Eigenvalue Multiplicity

A natural question in the context of star complements is: what is the maximum possible size k of a star set S for a given star complement $\dot{H}$ and a fixed eigenvalue μ? Theorem 4.1 offers insight into this matter. In particular, if $\mu = 0$, the size of S is unbounded, as any collection of p mutually non-adjacent vertices with identical neighbourhoods in $\dot{H}$ yields the eigenvalue 0 with multiplicity $p-1$. A similar observation holds for $\mu \in \{1,-1\}$. In other words, if t denotes the co-dimension of $\mu \in \{1,0,-1\}$, then the order of the corresponding signed graph $\dot{G}$ is not bounded by any function of t. However, for $\mu \notin \{1,0,-1\}$, Theorem 4.1 implies that all vertices in S must have pairwise distinct neighbourhoods in $\dot{H}$, leading to an intuitive exponential upper bound on the order of $\dot{G}$ in terms of t.

However, [172] offers a sharp cubic bound that reads

$$n \leq \binom{t+2}{3}. \tag{4.3}$$

In the notation of the reconstruction theorem, we denote the columns of $(B\,|\,A_{\dot{H}} - \mu I)$ by $\mathbf{s}_u$, $1 \leq u \leq n$. The key step in the proof of the upper bound (4.3) is the fact that the functions $F_u\colon \mathbb{R}^t \to \mathbb{R}$, $F_u(\mathbf{x}) = \langle \mathbf{s}_u, \mathbf{x} \rangle$, defined for every vertex $u \in V(\dot{G})$, are linearly independent. Indeed, since these functions belong to the space $\mathscr{H}_3$ of homogeneous cubic functions on $\mathbb{R}^t$ and the dimension of this space is $\binom{t+2}{3}$, we get the desired inequality.

Example 4.10 The negative quadrangle has the eigenvalues $\sqrt{2}$ and $-\sqrt{2}$, both with multiplicity 2. Thus, the upper bound of (4.3) is attained (for both eigenvalues and $(n,t) = (4,2)$). Another example is given later in Remark 5.7.

If μ is a non-main eigenvalue, then

$$n \leq \binom{t+2}{3} - 1. \tag{4.4}$$

The proof uses the result of Corollary 4.5 to show that $F(\mathbf{x}) = \langle \mathbf{j}, \mathbf{x} \rangle$ and F_u, $1 \leq u \leq n$, make a system of linearly linearly independent functions.

If $\dot{G}$ is homogeneous and μ is still non-main, by [37], the upper bound (4.3) reduces to $n \leq \binom{t+1}{2} - 1$, and the case of equality is completely characterized in the same reference (see also [75, Theorem 5.3.3]), where certain strongly regular graphs are identified as extreme examples that achieve the bound. Following [172], we consider the structure of inhomogeneous signed graphs that attain the equality in (4.4). If equality is attained, then the functions $F, F_1, \ldots, F_n$

form a basis for $\mathscr{H}_3$. In particular, we have

$$\langle \mathbf{x},\mathbf{x}\rangle\langle \mathbf{j},\mathbf{x}\rangle = \sum_{u=1}^{n} \beta_u F_u(\mathbf{x}) + \gamma F(\mathbf{x})$$

for some $(\beta_1,\beta_2,\ldots,\beta_n)^\intercal = \mathbf{b} \in \mathbb{R}^n$ and some $\gamma \in \mathbb{R}$.

By considering $\mathbf{x}+\mathbf{y}$ and $\mathbf{x}-\mathbf{y}$ instead of $\mathbf{x}$, we obtain

$$\langle \mathbf{x},\mathbf{x}\rangle\langle \mathbf{j},\mathbf{y}\rangle + 2\langle \mathbf{x},\mathbf{y}\rangle\langle \mathbf{j},\mathbf{x}\rangle = 3\sum_{u=1}^{n} \beta_u \langle \mathbf{s}_u,\mathbf{x}\rangle^2\langle \mathbf{s}_u,\mathbf{y}\rangle + 3\gamma\langle \mathbf{j},\mathbf{x}\rangle^2\langle \mathbf{j},\mathbf{y}\rangle. \quad (4.5)$$

By setting $\mathbf{x}=\mathbf{j}$, $\mathbf{y}=\mathbf{s}_i$ and also $\mathbf{x}=\mathbf{y}=\mathbf{s}_i$, we obtain

$$(s^2\gamma - s)\mathbf{j}_n = (\mu I - A_{\dot{G}})\mathbf{b}$$

with $s = \langle \mathbf{j},\mathbf{j}\rangle$ and

$$(\gamma - \mu)\mathbf{j}_n = (\mu^3 I - A_{\dot{G}})\mathbf{b},$$

respectively. It follows that $\mathbf{b} = \beta\mathbf{j}_n$, where

$$\beta = \frac{s^2\gamma - s - \gamma + \mu}{\mu(1-\mu^2)},$$

that is, $\mathbf{b}$ is a constant vector.

Now, if we consider $\mathbf{x}+\mathbf{z}$ and $\mathbf{x}-\mathbf{z}$ instead of $\mathbf{x}$ in (4.5), we arrive at

$$\begin{aligned}\langle \mathbf{x},\mathbf{y}\rangle\langle \mathbf{j},\mathbf{z}\rangle + \langle \mathbf{x},\mathbf{z}\rangle\langle \mathbf{j},\mathbf{y}\rangle + \langle \mathbf{y},\mathbf{z}\rangle\langle \mathbf{j},\mathbf{x}\rangle &= 3\beta\sum_{u=1}^{n} \langle \mathbf{s}_u,\mathbf{x}\rangle\langle \mathbf{s}_u,\mathbf{y}\rangle\langle \mathbf{s}_u,\mathbf{z}\rangle \\ &\quad + \gamma\langle \mathbf{j},\mathbf{x}\rangle\langle \mathbf{j},\mathbf{y}\rangle\langle \mathbf{j},\mathbf{z}\rangle.\end{aligned}$$

By setting $\mathbf{x}=\mathbf{s}_i$, $\mathbf{y}=\mathbf{z}=\mathbf{j}$, we obtain $-s = \beta\mu - \beta d_i^{\pm} - \gamma s^2$, while for $\mathbf{x}=\mathbf{y}=\mathbf{s}_i$, $\mathbf{z}=\mathbf{j}$, we have $\mu s + 2 = -3\beta\mu^2 - 3\beta d_i + 3\gamma s$. This equality implies that $\dot{G}$ is regular. Finally, for $\mathbf{x}=\mathbf{s}_i$, $\mathbf{y}=\mathbf{s}_j$, $\mathbf{z}=\mathbf{j}$, we find

$$\begin{aligned}\langle \mathbf{s}_i,\mathbf{s}_j\rangle\langle \mathbf{j},\mathbf{j}\rangle + \langle \mathbf{s}_i,\mathbf{j}\rangle\langle \mathbf{j},\mathbf{s}_j\rangle + \langle \mathbf{s}_j,\mathbf{j}\rangle\langle \mathbf{j},\mathbf{s}_i\rangle &= 3\beta\sum_{u=1}^{n} \langle \mathbf{s}_u,\mathbf{s}_i\rangle\langle \mathbf{s}_u,\mathbf{s}_j\rangle\langle \mathbf{s}_u,\mathbf{j}\rangle \\ &\quad + 3\gamma\langle \mathbf{j},\mathbf{s}_i\rangle\langle \mathbf{j},\mathbf{s}_j\rangle\langle \mathbf{j},\mathbf{j}\rangle,\end{aligned}$$

which together with the previous two equalities implies the following:

$$\text{for } i \nsim j, \quad 3\beta \sum_{u\in N(i)\cap N(j)} \sigma(ui)\sigma(uj) = 3\gamma s - 2; \quad (4.6)$$

$$\text{for } i \sim j, \quad 3\beta \sum_{u\in N(i)\cap N(j)} \sigma(ui)\sigma(uj) = 6\sigma(ij)\beta\mu - 2 + \sigma(ij)s + 3\gamma s. \quad (4.7)$$

Since $\dot{G}$ is inhomogeneous, the last equality leads to $\beta \neq 0$. Gathering the previous results, we arrive at the following theorem.

Theorem 4.11 *If equality in* (4.4) *is attained for a signed graph* $\dot{G}$, *then* $\dot{G}$ *is net-regular and strongly regular with parameters*

$$a = \frac{6\beta\mu - 2 + s + 3\gamma s}{3\beta}, \quad b = \frac{-6\beta\mu - 2 - s + 3\gamma s}{3\beta}, \quad c = \frac{3\gamma s - 2}{3\beta}.$$

Proof The upper bound is not attained for $t = 2$, while for $t \geq 3$, $\dot{G}$ must be inhomogeneous (as $\binom{t+1}{2} < \binom{t+2}{3}$). Then, by the foregoing discussion, we have $\beta \neq 0$. The net regularity follows since $\dot{G}$ is regular and satisfies (4.6) and (4.7). □

In the previous theorem we have $c = \frac{1}{2}(a+b)$. From Theorems 3.1 and 3.14, it follows that under this condition, either $\dot{G}$ has two eigenvalues, or it has three eigenvalues with one of them being the net degree which appears with multiplicity one. Conversely, if a net-regular signed graph of order n has spectrum $[\varrho, \lambda^f, \mu^g]$ $(f \leq g)$, then a star complement for the eigenvalue μ has order $f+1$. Applying inequality (4.4), we thus obtain the bound $n \leq \binom{f+3}{3} - 1$, which serves as an analogue of the classical Seidel absolute bound for strongly regular graphs (cf. [190]).

4.2 Decompositions of Signed Graphs with Two Eigenvalues

We consider connected signed graphs $\dot{G}$ that have two eigenvalues and admit a vertex partition into one of the following types: (i) an edgeless star complement $\dot{H}$ and the signed subgraph $\dot{S}$ induced by the corresponding star set S, or (ii) two induced subgraphs, each having two eigenvalues. In case (i), we will show that the subgraph induced by the star set also has at most two eigenvalues, matching the eigenvalue count of $\dot{G}$. In case (ii), we identify conditions under which one of these induced subgraphs must be a star complement within $\dot{G}$. A detailed investigation of signed graphs possessing two eigenvalues is undertaken in Chapter 6.

We begin by examining case (i) in a more general setting that does not require assuming anything about the number of eigenvalues of $\dot{G}$. Adopting the notation introduced in Theorem 4.2, we derive the following result.

Theorem 4.12 ([225]) *If* tK_1 *is a star complement for* μ *in a signed graph* $\dot{G}$ *with* n *vertices and adjacency matrix* (4.2) *(with* $A_{\dot{H}} = O$*), then:*

(i) $B^{\intercal}B = \mu(\mu I - A_{\dot{S}})$.

(ii) The $(n-t) \times (n-t)$ *top-left block* $X^{(p)}$ *of* $A^p_{\dot{G}}$ $(p \geq 3)$ *is determined by*

the recurrence

$$\begin{cases} X^{(p)} = X^{(p-1)}A_{\dot{S}} + \mu X^{(p-2)}(\mu I - A_{\dot{S}}), \\ X^{(1)} = A_{\dot{S}}, \\ X^{(2)} = A_{\dot{S}}^2 - \mu A_{\dot{S}} + \mu^2 I. \end{cases} \tag{4.8}$$

(iii) *The number of eigenvalues of $\dot{S}$ does not exceed the number of eigenvalues of $\dot{G}$.*

Proof (i): If $B = (\mathbf{b}_1 \mid \mathbf{b}_2 \mid \cdots \mid \mathbf{b}_{n-k})$, then by Corollary 4.3 we have

$$\langle \mathbf{b}_i, \mathbf{b}_j \rangle = \mathbf{b}_i^\intercal \frac{1}{\mu} I \mathbf{b}_j = \begin{cases} \mu & \text{if } i = j, \\ 0 & \text{if } i \nsim j, \\ -1 & \text{if } i \overset{+}{\sim} j, \\ 1 & \text{if } i \overset{-}{\sim} j. \end{cases}$$

In other words, $B^\intercal B = \mu(\mu I - A_{\dot{S}})$.

(ii): Considering the corresponding top-left block of $A_{\dot{G}}$, we immediately get $X^{(1)} = A_{\dot{S}}$ and $X^{(2)} = A_{\dot{S}}^2 + B^\intercal B = A_{\dot{S}}^2 - \mu A_{\dot{S}} + \mu^2 I$. In addition, for $p \geq 3$, we have $X^{(p)} = X^{(p-1)}A_{\dot{S}} + X^{(p-2)}B^\intercal B = X^{(p-1)}A_{\dot{S}} + \mu X^{(p-2)}(\mu I - A_{\dot{S}})$, completing the recurrence (4.8).

(iii): Assume that $\dot{G}$ has exactly q eigenvalues. Since the top-left block of $A_{\dot{G}}^p$ is a polynomial in $A_{\dot{S}}$ of degree at most p (by (4.8)), by considering the minimal polynomial of $A_{\dot{G}}$, we deduce that there exists a polynomial ϕ of degree at most q such that $\phi(A_{\dot{S}}) = O$, which concludes the proof. □

Assume now that the signed graph $\dot{G}$ has two eigenvalues, λ and μ. It follows that exactly one of them is positive, and the other is negative. By computing the minimal polynomial, we deduce that $\dot{G}$ is regular with vertex degree $-\lambda\mu$.

Lemma 4.13 *If tK_1 is a star complement for μ in a signed graph $\dot{G}$ with n vertices and two eigenvalues, then $2t \leq n$.*

Proof If the adjacency matrix of $\dot{G}$ is given by (4.2), then by inspecting the minimal polynomial, we get $BB^\intercal = -\lambda\mu I$, that is, $\operatorname{rank}(BB^\intercal) = t$ (as λ and μ are non-zero). On the other hand, $\operatorname{rank}(BB^\intercal) \leq \operatorname{rank}(B) \leq n - t$, and the desired result follows. □

By Theorem 4.12(iii), if $\dot{G}$ has two eigenvalues, then $\dot{S}$ has at most two eigenvalues. We consider them in more detail.

Theorem 4.14 ([225]) *Let a signed graph $\dot{G}$ with n vertices and two eigenvalues, λ and μ, be decomposed into the star complement tK_1 for μ and the signed graph $\dot{S}$ induced by the star set. Then either*

(i) $\dot{S}$ is isomorphic to tK_1 and $\lambda = -\mu$, or
(ii) $\dot{S}$ has two eigenvalues: $\lambda+\mu$ with multiplicity t and μ with multiplicity $n-2t$.

Proof For $1 \leq i \leq t$, let $\mathbf{e}_i$ denote the ith vector of the canonical basis of $\mathbb{R}^t$. Observe that $B^\top\mathbf{x} \neq \mathbf{0}$ holds for every non-zero vector $\mathbf{x} \in \mathbb{R}^t$, as otherwise we would have $-\lambda\mu I\mathbf{x} = BB^\top\mathbf{x} = \mathbf{0}$, which is impossible.

The minimal polynomial of $A_{\dot{G}}$ gives $A_{\dot{S}}B^\top = (\lambda+\mu)B^\top$; hence,

$$A_{\dot{S}}B^\top\mathbf{e}_i = (\lambda+\mu)B^\top\mathbf{e}_i,$$

that is, $\lambda+\mu$ is an eigenvalue of $A_{\dot{S}}$. As the kernel of $B^\top$ trivially intersects the space $\mathbb{R}^t$, the multiplicity of $\lambda+\mu$ is at least t.

Denote by $\overline{\mu}$ the average of the $n-2t$ remaining eigenvalues (counted with multiplicities). Considering that $\operatorname{trace}(A_{\dot{G}}) = 0$ and $\operatorname{trace}(A_{\dot{S}}) = 0$, we have the equalities $t\lambda + (n-t)\mu = 0$ and $t(\lambda+\mu) + (n-2t)\overline{\mu} = 0$, respectively. Subtracting these yields $(n-2t)(\mu-\overline{\mu}) = 0$. Therefore, either $2t = n$ or $\mu = \overline{\mu}$. In the former case, we immediately obtain conclusion (i). In the latter case, since $\dot{S}$ is an induced subgraph of $\dot{G}$, the eigenvalue interlacing argument implies that μ is the unique remaining eigenvalue of $\dot{S}$ with multiplicity $n-2t$, yielding conclusion (ii). □

Here is a simple example.

Example 4.15 If B is a Hadamard matrix or an adjacency matrix of a signed graph with two symmetric eigenvalues, then the signed graph $\dot{G}$ with adjacency matrix

$$\begin{pmatrix} O & B^\top \\ B & O \end{pmatrix}$$

has two symmetric eigenvalues and admits a decomposition into two isomorphic edgeless signed graphs. Each features as a star complement for both eigenvalues of $\dot{G}$.

We now proceed with a decomposition earlier denoted by (ii). In the first result, we omit any assumption on the number of eigenvalues of two subgraphs. If $\dot{G}$ has two eigenvalues, λ and μ, then their multiplicities are

$$m_{\dot{G}}(\lambda) = \frac{\mu n}{\mu-\lambda} \quad \text{and} \quad m_{\dot{G}}(\mu) = \frac{\lambda n}{\lambda-\mu}. \tag{4.9}$$

Theorem 4.16 ([202]) *Let a signed graph $\dot{G}$ with two eigenvalues, λ and μ, be decomposed into signed graphs $\dot{H}_1$ and $\dot{H}_2$. If ν is an eigenvalue of $\dot{H}_1$ and $\nu \notin \{\lambda, \mu\}$, then $\lambda + \mu - \nu$ is an eigenvalue of $\dot{H}_2$. In addition, $m_{\dot{H}_1}(\nu) = m_{\dot{H}_2}(\lambda + \mu - \nu)$.*

Proof Expressing $A_{\dot{G}}$ in the form

$$A_{\dot{G}} = \begin{pmatrix} A_{\dot{H}_1} & B^\intercal \\ B & A_{\dot{H}_2} \end{pmatrix}, \tag{4.10}$$

and using $A_{\dot{G}}^2 = (\lambda + \mu)A_{\dot{G}} - \lambda\mu I$, we get

$$A_{\dot{H}_2}B = -BA_{\dot{H}_1} + (\lambda + \mu)B.$$

Now, if ν is an eigenvalue of $\dot{H}_1$ and $\mathbf{x}$ is an associated eigenvector, then

$$\begin{aligned} A_{\dot{H}_2}B\mathbf{x} &= -BA_{\dot{H}_1}\mathbf{x} + (\lambda + \mu)B\mathbf{x} = -\nu B\mathbf{x} + (\lambda + \mu)B\mathbf{x} \\ &= (\lambda + \mu - \nu)B\mathbf{x}, \end{aligned} \tag{4.11}$$

which means that $\lambda + \mu - \nu$ is an eigenvalue of $A_{\dot{H}_2}$, unless $B\mathbf{x} = \mathbf{0}$. In the last case, the identity

$$A_{\dot{G}}\begin{pmatrix} \mathbf{x} \\ \mathbf{0} \end{pmatrix} = \begin{pmatrix} A_{\dot{H}_1}\mathbf{x} \\ \mathbf{0} \end{pmatrix} = \nu\begin{pmatrix} \mathbf{x} \\ \mathbf{0} \end{pmatrix},$$

gives $\nu \in \{\lambda, \mu\}$.

Finally, if the eigenspace $\mathscr{E}_{\dot{H}_1}(\nu)$ of $\nu \notin \{\lambda, \mu\}$ is spanned by $\mathbf{x}_1, \mathbf{x}_2, \ldots, \mathbf{x}_k$, then, since the kernel of B trivially intersects the space $\mathscr{E}_{\dot{H}_1}(\nu)$, the vectors $B\mathbf{x}_1, B\mathbf{x}_2, \ldots, B\mathbf{x}_k$ are linearly independent, and so

$$\dim(\mathscr{E}_{\dot{H}_1}(\nu)) \le \dim(\mathscr{E}_{\dot{H}_2}(\lambda + \mu - \nu)).$$

By interchanging $\dot{H}_1$ and $\dot{H}_2$, we arrive at the opposite inequality, which concludes the proof. □

An immediate corollary of the preceding theorem is that, under its hypotheses, the number of eigenvalues of $\dot{H}_1$ absent from the spectrum of $\dot{G}$ coincides with the corresponding number for $\dot{H}_2$.

From this point onward, we assume that the subgraphs $\dot{H}_1$ and $\dot{H}_2$ each possess two eigenvalues. The next result establishes relations between these eigenvalues and those of $\dot{G}$.

Theorem 4.17 ([202]) *Assume that a connected signed graph $\dot{G}$ with n vertices and eigenvalues λ and μ is decomposed into signed graphs $\dot{H}_1$ (with n_1 vertices and eigenvalues λ_1 and μ_1) and $\dot{H}_2$ (with n_2 vertices and eigenvalues λ_2 and μ_2).*

(i) *The case* $\lambda_1, \mu_1 \in \{\lambda, \mu\}$ *cannot occur.*

(ii) *If* $\lambda_1, \mu_1 \notin \{\lambda, \mu\}$, *then the spectrum of* $\dot{G}$ *is symmetric,* $\dot{H}_2 \cong -\dot{H}_1$ *and, for* $i, j \in V(\dot{H}_1)$, *the columns of* B *(from* (4.10)*) satisfy*

$$\mathbf{b}_i^\intercal \mathbf{b}_j = \begin{cases} 0 & \text{if } i \nsim j, \\ -\lambda_1 - \mu_1 & \text{if } i \overset{+}{\sim} j, \\ \lambda_1 + \mu_1 & \text{if } i \overset{-}{\sim} j. \end{cases}$$

(iii) *If* $\lambda_1 = \lambda$ *and* $\mu_2 = \mu$, *then* $n_1\lambda_1 = -n_2\mu$, $\mu_1 + \lambda_2 = \lambda + \mu$, $m_{\dot{H}_1}(\mu_1) = m_{\dot{H}_2}(\lambda_2)$, $m_{\dot{G}}(\lambda) = n_1$ *and* $m_{\dot{G}}(\mu) = n_2$.

(iv) *If* $\lambda_1 = \lambda$ *and* $\lambda_2 = \lambda$, *then* $n_1(\lambda - \mu_2) = n_2(\lambda - \mu_1)$, $\mu_1 + \mu_2 = \lambda + \mu$, *and* $m_{\dot{H}_1}(\mu_1)$, $m_{\dot{H}_2}(\mu_2)$ *and* $m_{\dot{G}}(\mu)$ *are mutually equal.*

Proof (i): Otherwise, $\dot{H}_1$ and $\dot{G}$ would have the same vertex degree which, since $\dot{G}$ is connected, means that $\dot{H}_2$ is edgeless, so has exactly one eigenvalue.

(ii): It follows that $\dot{G}$ has $2n_1$ vertices. If the spectrum of $\dot{G}$ is non-symmetric, then there would exist an eigenvalue whose multiplicity exceeds n_1. However, this contradicts the fact that $\dot{G}$ contains an induced signed subgraph on n_1 vertices that does not possess this eigenvalue in its spectrum.

Since the spectrum of $\dot{G}$ is symmetric, from (4.10), we obtain

$$A_{\dot{H}_2}B = -BA_{\dot{H}_1}. \tag{4.12}$$

According to the assumption that $\dot{H}_2$ does not share any eigenvalue with $\dot{G}$, there are n_1 independent eigenvectors of the form $B\mathbf{x}$ (see (4.11)) associated with its eigenvalues. As these eigenvectors lie in the image of B, it follows that B is invertible. Then, by employing (4.12), we get $A_{\dot{H}_2} = -BA_{\dot{H}_1}B^{-1}$, giving $\dot{H}_2 \cong -\dot{H}_1$.

Observe that $\dot{H}_2$ is a star complement in $\dot{G}$, for both λ and $\mu = -\lambda$. By Corollary 4.3, for $i, j \in V(\dot{H}_1)$, the columns $\mathbf{b}_i$ and $\mathbf{b}_j$ of B satisfy $\langle \mathbf{b}_i, \mathbf{b}_j \rangle \in \{1, 0, -1\}$, where the bilinear form is formed by $A_{\dot{H}_2}$ and each of $\lambda, -\lambda$. This gives $\mathbf{b}_i^\intercal(A_{\dot{H}_2} + (\pm\lambda - \lambda_2 - \mu_2)I)\mathbf{b}_j = p\phi_{\dot{H}_2}(\pm\lambda)$, where $\phi_{\dot{H}_2}$ stands for the minimal polynomial of $\dot{H}_2$, whereas p is 0, -1 or 1 depending on whether $i \nsim j$, $i \overset{+}{\sim} j$ or $i \overset{-}{\sim} j$. This yields $\lambda \mathbf{b}_i^\intercal \mathbf{b}_j = p\lambda(\lambda_2 + \mu_2)$. Since $\lambda \neq 0$ and the eigenvalues of $\dot{H}_1$ are obtained by reversing the sign of those of $\dot{H}_2$, part (ii) is completed.

(iii), (iv): By Theorem 4.16, the eigenvalues of $\dot{H}_1$ and $\dot{H}_2$ which do not appear in the spectrum of $\dot{G}$ have the same multiplicity, and their sum is equal to $\lambda + \mu$.

Counting the number of edges between $\dot{H}_1$ and $\dot{H}_2$, we get

$$n_1(-\lambda\mu + \lambda_1\mu_1) = n_2(-\lambda\mu + \lambda_2\mu_2),$$

which gives $n_1\lambda_1 = -n_2\mu$ for (iii), and $n_1(\lambda-\mu_2) = n_2(\lambda-\mu_1)$ for (iv).

To conclude (iii), it remains to compute the multiplicities of λ and μ in $\dot{G}$. First, by the eigenvalue interlacing, the multiplicity of each of them is not less than the multiplicity of the same eigenvalue in $\dot{H}_1$ and $\dot{H}_2$, respectively. Hence, we may write

$$m_{\dot{G}}(\lambda) = \frac{\mu_1 n_1}{\mu_1 - \lambda} + a, \qquad m_{\dot{G}}(\mu) = \frac{\lambda_2 n_2}{\lambda_2 - \mu} + b$$

for $a, b \geq 0$; if necessary, compare (4.9). Using the identities $\mathrm{trace}(A_{\dot{G}}) = 0$ and $\mathrm{trace}(A_{\dot{G}}^2) = -\lambda\mu(n_1+n_2)$, we arrive at $a = b = \lambda n_1/(\lambda-\mu_1)$, giving $m_{\dot{G}}(\lambda) = n_1$, so $m_{\dot{G}}(\mu) = n_2$.

For (iv), by setting

$$m_{\dot{G}}(\lambda) = \frac{\mu_1 n_1}{\mu_1 - \lambda} + \frac{\mu_2 n_2}{\mu_2 - \lambda} + a$$

and using $\mathrm{trace}(A_{\dot{G}}) = 0$, we get $a = \lambda(n_1+n_2)/(\lambda-\mu)$. Using the equalities of (iv) we already proved, we get

$$m_{\dot{G}}(\lambda) = n_1 + n_2 - \frac{\lambda n_1}{\lambda - \mu_1},$$

giving $m_{\dot{G}}(\mu) = \lambda n_1/(\lambda-\mu_1)$, meaning that $m_{\dot{G}}(\mu)$ is equal to the multiplicity of μ_1 in $\dot{H}_1$. This completes (iv) and the entire proof. □

We conclude this section with the following result, which serves as an illustrative example of Theorem 4.17(ii) and carries important implications. Additional examples will be presented in the subsequent chapters.

Theorem 4.18 ([202]) *For every $r \geq 2$, there exists a unique r-dimensional signed cube with negative quadrangles $\dot{Q}_r$. Its eigenvalues are $\pm\sqrt{r}$.*

Proof Obviously, $\dot{Q}_2$ is isomorphic to a negative quadrangle, all negative quadrangles are switching isomorphic and their common eigenvalues are $\sqrt{2}$ and $-\sqrt{2}$.

Assume the claim holds for $\dot{Q}_{r-1}$, with colour classes denoted by U_1 and U_2. It then follows by a straightforward verification that $\dot{Q}_r$ can be constructed by taking two copies of $\dot{Q}_{r-1}$ and inserting positive edges between corresponding vertices in U_1 and negative edges between corresponding vertices in U_2. The uniqueness of $\dot{Q}_r$, up to switching isomorphism, is a direct consequence of this construction.

It remains to consider the eigenvalues. The adjacency matrix of $\dot{Q}_r$ is the first of the following matrices, where the 2×2 top-left block and its diagonal counterpart both stand for the adjacency matrix of $\dot{Q}_{r-1}$.

$$\begin{pmatrix} O & N^\intercal & I & O \\ N & O & O & -I \\ I & O & O & N^\intercal \\ O & -I & N & O \end{pmatrix} \qquad \begin{pmatrix} O & N^\intercal & I & O \\ N & O & O & I \\ I & O & O & -N^\intercal \\ O & I & -N & O \end{pmatrix}$$

The second matrix is obtained by applying the switching with respect to the vertices of one colour class of the second copy of $\dot{Q}_{r-1}$. Now, this matrix has the form

$$A_{\dot{G}} = \begin{pmatrix} A_{\dot{H}} & B \\ B & -A_{\dot{H}} \end{pmatrix},$$

which gives $|xI - A_{\dot{G}}| = |(xI - A_{\dot{H}})(xI + A_{\dot{H}}) - B^2| = |x^2 I - (A_{\dot{H}}^2 + B^2)|$. The final determinant equals $|(x^2 - (a+b))I|$, where a and b denote the row sums of $|A_{\dot{H}}|$ and $|B|$, which leads to the desired assertion. □

Here is a corollary taken from the same reference.

Corollary 4.19 *If H is a subgraph of Q_r induced by any set of $2^{r-1}+1$ vertices, then the maximum vertex degree Δ of H satisfies $\Delta \geq \sqrt{r}$.*

Proof If $\dot{Q}_r$ is the r-dimensional signed cube with negative quadrangles and $\dot{H}$ is its subgraph sharing the vertex set with H, then by Theorem 4.18 and the interlacing argument, the index of $\dot{H}$ is $\geq \sqrt{r}$, which means that the index of its underlying graph H is also $\geq \sqrt{r}$ (by Theorem 1.26). Hence, $\Delta \geq \sqrt{r}$. □

The inequality of Corollary 4.19 was a fundamental ingredient for the resolution of the sensitivity conjecture posed by Nisan and Szegedy in 1992 [168]. The resolution can be found in [122].

Problems

4.1 Use Theorem 4.4 to compute $(\mu I - A)^{-1}$, where A is the adjacency matrix of a signed graph with two eigenvalues. Perform the same computation for the adjacency matrix of a strongly regular (unsigned) graph.

4.2 Assume that, in the notation of Theorem 4.2, $\mu \notin \{1, 0, -1\}$. Prove that the columns of B are distinct, and no column is the negative of another.

4.3 Show that every connected signed graph contains a connected star complement for any of its eigenvalues; that is, prove Theorem 4.6.

4.4 In the setting of unsigned graphs, determine all regular extensions of an edgeless graph for a given eigenvalue μ.

4.5 Determine all maximal extensions of the path P_t for the eigenvalue -2. Perform the same analysis for a signed cycle in cases where -2 is not an eigenvalue.

4.6 Use the results of Section 2.6 to identify all graphs whose index is less than 2. Use these graphs as target star complements for the eigenvalue 2 in signed graphs.

4.7 (cf. [160]) Determine a minimal family of signed graphs such that each signed graph in Figure 3.4 is a maximal extension, with respect to both the eigenvalues 2 and -2, of at least one member of the family.

4.8 The underlying graph of the signed graph in Figure 4.1 is the strongly regular graph known as the Clebsch graph. Determine a star partition of this graph.

4.9 Extract star complements for $\sqrt{5}$ and $-\sqrt{5}$ in the signed graph shown in Figure 4.1. Show that this signed graph is the unique maximal extension for each eigenvalue and the corresponding star complement.

4.10 Show that $n \leq \binom{t+2}{3}$ holds for every signed graph of order n, where t is the co-dimension of a given eigenvalue.

5

Signed Line Graphs

The concept of the line graph plays a central role in spectral graph theory. In the context of signed graphs, an analogue of this construction was first introduced by Zaslavsky in [260], where the line graph of a signed graph was defined through its incidence matrix, as outlined in Subsection 1.3.2. This formulation provides a natural extension of the classical line graph to the signed setting and preserves certain combinatorial features. However, as will be demonstrated in subsequent sections, this definition does not align optimally with the goals of spectral analysis. Accordingly, a slight modification is introduced to improve compatibility with spectral methods and to facilitate more effective investigation.

5.1 Definitions of a Signed Line Graph

The *line graph* $L(G)$ of an unsigned graph $G = (V, E)$ is the graph whose vertex set is E, and two vertices are adjacent if and only if the corresponding edges in G are adjacent. In this context, G is referred to as the *root* of $L(G)$. The root is unique, except in the case where $L(G) \cong K_3$. To define the line graph of a signed graph $\dot{G}$, denoted by $\mathscr{L}(\dot{G})$, we require that its underlying graph is $L(G)$. The incidence matrix B_η of $\dot{G}$, as defined in Subsection 1.3.2, can be used to assign a signature to $\mathscr{L}(\dot{G})$ via a column-by-column product construction.

Zaslavsky considered a definition of the line graph of a signed graph in which the adjacency matrix is given by $2I - B_\eta^\intercal B_\eta$. The approach is well-motivated and aligns naturally with the concept of switching equivalence in signed graphs. Different bi-orientations yield switching-equivalent signed line graphs. Likewise, switching-equivalent signed graphs produce switching-equivalent signed line graphs. Therefore, this definition provides a consistent and mean-

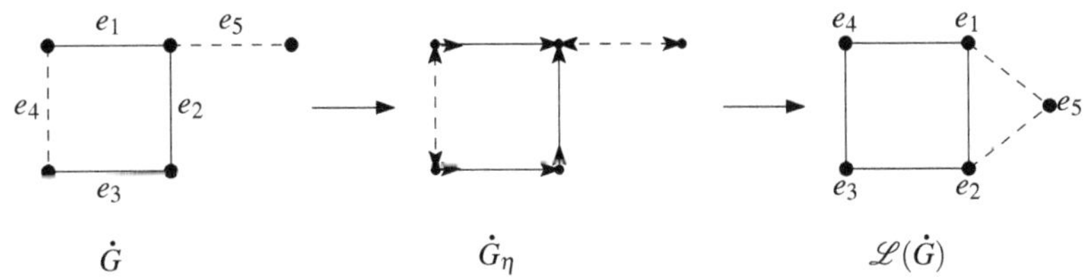

Figure 5.1 A construction of the signed line graph. Reproduced from [35], licensed under CC BY 4.0.

ingful way to define a line graph of a signed graph. We refer to this construction as the *combinatorial signed line graph*; that is, we set

$$A_{\mathscr{L}_C(\dot{G})} = 2I - B_\eta^\intercal B_\eta \tag{5.1}$$

for some incidence matrix B_η of $\dot{G}$. Formally, a signed line graph is a class of switching-equivalent signed graphs obtained by taking different orientations.

Evidently, as $B_\eta^\intercal B_\eta$ is positive semidefinite, the adjacency matrix considered in (5.1) has eigenvalues in the real interval $(-\infty, 2]$, which is the opposite situation to the eigenvalues of the line graph of an unsigned graph (see [77]). Hence, in [34], the authors considered the opposite signature of the combinatorial line graph of Zaslavsky, arriving at the so-called *spectral signed line graph* $\mathscr{L}_S(\dot{G})$:

$$A_{\mathscr{L}_S(\dot{G})} = B_\eta^\intercal B_\eta - 2I. \tag{5.2}$$

Therefore,

$$\mathscr{L}_S(\dot{G}) \cong -\mathscr{L}_C(\dot{G}).$$

Clearly, the adjacency matrix of $\mathscr{L}_S(\dot{G})$ has eigenvalues in the real interval $[-2, +\infty)$, which is consistent with the spectral theory of ordinary line graphs.

For the sake of clarity and consistency – and given that this book is primarily concerned with the spectral properties of signed graphs – we will henceforth adopt the spectral definition of the signed line graph. Accordingly, we will omit the index S and simply write $\mathscr{L}(\dot{G}) := \mathscr{L}_S(\dot{G})$. It is worth noting that all subsequent results can be equivalently formulated in the framework of the combinatorial signed line graph.

Figure 5.1 provides an illustrative example of a signed graph $\dot{G}$, one of its orientations $\dot{G}_\eta$ and the resulting (spectral) signed line graph $\mathscr{L}(\dot{G})$.

Remark 5.1 Which is the most appropriate definition of a line graph for a signed graph? In fact, both definitions introduced earlier – Zaslavsky's combinatorial line graph (5.1) and the spectral line graph (5.2) – have desirable properties and serve different purposes.

Zaslavsky advocates for the combinatorial version (5.1), primarily because it maintains consistency with the foundational relationship between edge signs and vertex–edge orientations in signed graphs. In contrast, the present author favours the spectral version (5.2), as it aligns more naturally with the existing framework of spectral graph theory and is more frequently encountered in the spectral graph literature. For a comprehensive discussion of the relative merits of both approaches, the interested reader is referred to [35].

Additionally, one may wonder why a seemingly more intuitive definition is not considered, namely, that the line graph of a signed graph $\dot{G}$ is a signed graph whose vertices correspond to the edges of $\dot{G}$, with a pair of vertices joined by a positive (resp. negative) edge if and only if the corresponding edges in $\dot{G}$ are adjacent and have the same (resp. different) sign. While this definition appears natural, it lacks the structural and theoretical coherence of the aforementioned versions. First, it does not integrate well with the concept of vertex–edge orientations, which underpin both (5.1) and (5.2), and which extend naturally from unsigned to signed graphs. Second, as shown in [230], a line graph defined in this way always switches to its underlying graph. Consequently, any signed structure imposed on G yields the same signed line graph up to switching, thus trivializing the role of the signature.

These considerations justify the focus on the combinatorial and spectral line graphs in the present study.

It is important to note that neither the combinatorial nor the spectral definition of a signed line graph constitutes a direct generalization of the ordinary line graph. Indeed, if one embeds the class of unsigned graphs into the framework of signed graphs by interpreting them as signed graphs with an all-negative signature (rather than the all-positive signature adopted throughout this work), then the combinatorial signed line graph $\mathscr{L}_C(\dot{G})$ does extend the concept of the ordinary line graph in a natural way. For instance, under this interpretation, the combinatorial signed line graph of the unsigned triangle K_3 satisfies $\mathscr{L}_C(K_3) \cong K_3$, as expected. However, if K_3 is treated as a signed graph with an all-positive signature (as is the convention in this book), then the resulting signed line graph becomes $\mathscr{L}_C(K_3) \cong -K_3$. This discrepancy highlights the role that signature choice plays in determining whether a signed line graph definition behaves as a true generalization of the unsigned case.

The identities (5.1) and (5.2) remain valid even in the presence of multiple edges in the signed graph $\dot{G}$. A pair of edges joining the same pair of vertices is referred to as a *digon*, that is, a cycle of length two. The sign of the digon is determined by the signs of its constituent edges: it is *positive* if and only if both edges have the same sign, otherwise it is a *negative digon*.

In the construction of the signed line graph $\mathscr{L}(\dot{G})$, the nature of digons is treated in a manner consistent with their spectral behaviour. In particular, if a digon is negative, then the corresponding vertices in $\mathscr{L}(\dot{G})$ are non-adjacent. This is in agreement with the fact that a negative digon contributes nothing to the adjacency matrix $A_{\dot{G}}$, and hence does not affect the spectral or combinatorial structure in a way that warrants the addition of an edge in $\mathscr{L}(\dot{G})$.

A *simply signed graph* is a signed graph in which negative digons are permitted, but positive ones are not. In particular, a special type of simply signed graph is obtained by replacing every edge of an (unsigned) graph G with a negative digon; equivalently, this amounts to inserting a negative parallel edge for each edge of G. The resulting structure is called a *signed doubled graph*, denoted by $\ddot{G}$.

We postpone a detailed treatment of signed doubled graphs until Section 5.4. However, we remark here that the adjacency matrix of a signed doubled graph of order n is the $n \times n$ zero matrix. This follows from the fact that each pair of parallel edges with opposite signs contributes cancelling entries in the adjacency matrix. The degree of a vertex in $\ddot{G}$ is twice its degree in G, and the Laplacian matrix of $\ddot{G}$ is a diagonal matrix encoding the vertex degrees.

In [55], the authors studied a signed analogue of Beineke's characterization of line graphs via forbidden subgraphs. Clearly, if $\dot{G}$ is the line graph of a signed graph $\dot{H}$, then its underlying graph G must satisfy $G \cong L(H)$; that is, G is a line graph in the classical (unsigned) sense. Consequently, the nine forbidden subgraphs described in [55, 77] for unsigned line graphs are also forbidden in the context of line graphs of signed graphs. This naturally leads to the following question: assuming that $G \cong L(H)$ for some unsigned graph H, under what conditions does a given signature σ of G yield a signed graph $\dot{G} = (G, \sigma)$ that is a line graph in the signed context? The following theorem provides a complete answer to this question, stated here specifically for the case of the spectral signed line graph.

Theorem 5.2 ([55]) *Let $G \cong L(H)$ for some graph H, and $\dot{G} = (G, \sigma)$. The following statements are equivalent:*

(i) $\dot{G}$ is the line graph of a signed graph.
(ii) $\dot{G}$ has no induced subgraph switching isomorphic to any signed graph of Figure 5.2.
(iii) The least eigenvalue of $\dot{G}$ is at least -2.

Alternatively, in [243] Vijayakumar gave a characterization of signed line graphs by means of forbidden subgraphs and proved that a minimal forbidden signed graph for $\lambda_n \geq -2$ has at most 10 vertices.

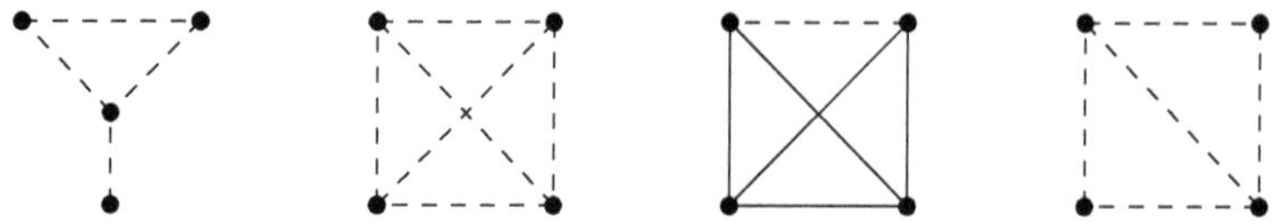

Figure 5.2 Forbidden signed graphs for Theorem 5.2. Reproduced from [55].

In the following, attention is directed to a well-known formula that bridges the Laplacian theory of signed graphs with the adjacency-based framework of their line graph counterparts. Recall from Chapter 1 that, for a signed graph $\dot{G}$ with an incidence matrix B_η,

$$B_\eta B_\eta^\intercal = L_{\dot{G}} \quad \text{and} \quad B_\eta^\intercal B_\eta = A_{\mathscr{L}_S(\dot{G})} + 2I. \tag{5.3}$$

By observing that $B_\eta B_\eta^\intercal$ and $B_\eta^\intercal B_\eta$ share the set of non-zero eigenvalues, we arrive at the following theorem.

Theorem 5.3 *For a signed graph $\dot{G}$ with n vertices and m edges, let $\Phi_{\dot{G}}$ and $\Psi_{\dot{G}}$ be the characteristic polynomials of the adjacency matrix and the Laplacian matrix, respectively. Then $\Phi_{\mathscr{L}(\dot{G})}(x) = (x+2)^{m-n}\Psi_{\dot{G}}(x+2)$.*

The proof is almost identical to the classical one for the unsigned case (cf. [75, Theorem 2.4.1]). There is also the following result.

Theorem 5.4 *If $\dot{G}$ is a regular signed graph of order n and degree r, then*

$$\Phi_{\mathscr{L}(\dot{G})}(x) = -(x+2)^{m-n}\Phi_{\dot{G}}(r-x-2),$$

where $m = rn/2$ is the size of $\dot{G}$.

Proof Recall that the spectrum and the Laplacian spectrum of a regular signed graph differ only by a shift of the indeterminate. In other words, the assertion is a slight manipulation on the identities (5.3). □

5.2 Balance of Signed Line Graphs

The question of balance of signed line graphs is resolved in the next theorem. Observe that all triangles that arise from a star of $\dot{G}$ are positive in $\mathscr{L}(\dot{G})$. Even cycles of $\dot{G}$ keep their signature in $\mathscr{L}(\dot{G})$, while odd cycles of $\dot{G}$ reverse their signature in $\mathscr{L}(\dot{G})$. We analyse the balance and the degree of imbalance of line graphs. Recall that $\dot{G}$ is antibalanced if $-\dot{G}$ is balanced.

Theorem 5.5 ([35]) *Let $\dot{G}$ be a signed graph. The following statements hold:*

(i) $\mathscr{L}(\dot{G})$ *is antibalanced if and only if* $\dot{G}$ *is a disjoint union of paths and positive cycles.*

(ii) $\mathscr{L}(\dot{G})$ *is balanced if and only if* $\dot{G}$ *is antibalanced.*

(iii) $\ell(\mathscr{L}(\dot{G})) \geq f(\mathscr{L}(\dot{G})) = \ell(-\dot{G})$, *where* ℓ *and* f *stand for the frustration index and the frustration number.*

Proof (i): If $\dot{G}$ contains a vertex of degree at least three, then the edges incident to that vertex induce positive triangles in $\mathscr{L}(\dot{G})$, which implies that $\mathscr{L}(\dot{G})$ cannot be antibalanced. Therefore, the maximum degree in $\dot{G}$ must be at most two, and consequently $\dot{G}$ consists of paths and signed cycles. Since the signed line graphs of paths are trees (which are always antibalanced), the remaining case to consider is that of signed cycles. The signed line graph of an even-length negative cycle is itself an even negative cycle, which is not antibalanced. Similarly, the line graph of an odd-length negative cycle results in an odd positive cycle, which is also not antibalanced. On the other hand, an analogous analysis of positive cycles reveals that these are permitted, as their line graphs do not violate antibalance.

(ii): This is a consequence of a straightforward analysis.

(iii): An edge set S in $-\dot{G}$ is a set S of vertices in $\mathscr{L}(\dot{G})$, and $\mathscr{L}(\dot{G}) - S \cong \mathscr{L}(\dot{G} - S)$, where this notation is clear from the context. By virtue of part (ii), $\mathscr{L}(\dot{G}-S)$ is balanced if and only if the negation $-(\dot{G}-S)$ is balanced. Thus, the smallest size of an edge set S such that $-(\dot{G}-S)$ is balanced, which is $\ell(-\dot{G})$, equals the smallest size of a vertex set S such that $\mathscr{L}(\dot{G}-S)$ is balanced, which is $f(\mathscr{L}(\dot{G}))$.

The inequality follows from the general observation that $\ell \geq f$ holds for every signed graph [25]. □

With slight modifications, the preceding theorem can be formulated for the combinatorial signed line graph. Regarding balance, we also present the following result.

Theorem 5.6 ([35]) *Let* G *be an unsigned graph. Then* $\mathscr{L}(-G)$ *is balanced, and therefore switching equivalent to the ordinary line graph of* G.

Proof Choose the orientation for $-G$ in which $\eta(v,e) = 1$ for every incident pair (v,e). Under this orientation, each pair of incident edges in $-G$ contributes a positive edge in the line graph, which establishes the claimed property. Furthermore, any alternative orientation of $-G$ results in a switching-equivalent signed line graph, and thus does not affect the balance or antibalance status of the resulting signed graph. □

5.3 Exceptional Signed Graphs

As established in the preceding sections, the least eigenvalue of a signed line graph is bounded below by -2. A connected signed graph is called *exceptional* if its least eigenvalue is not less than -2, but the signed graph itself is not a line graph of any simply signed graph (defined in Section 5.1). To say more about exceptional signed graphs, we need to introduce some root systems.

5.3.1 Root Systems

If $\mathbf{e}_1, \mathbf{e}_2, \ldots, \mathbf{e}_k$ are the vectors of the canonical basis of $\mathbb{R}^k$, $k \geq 4$, then the root system D_k consists of the vectors (or the *roots*) $\pm\mathbf{e}_i \pm \mathbf{e}_j$, for $1 \leq i < j \leq k$. The root system E_8 consists of the 112 roots of D_8 and the additional 128 roots of the form $\frac{1}{2}\sum_{i=1}^{8} \pm\mathbf{e}_i$, where the number of positive summands is even. The root system E_7 consists of the 126 roots of E_8 orthogonal to a fixed root $\mathbf{x} \in E_8$, and E_6 consists of the 72 roots of E_8 orthogonal to a fixed pair of roots $\mathbf{x}, \mathbf{y} \in E_8$, such that $\mathbf{x} \cdot \mathbf{y} = -1$.

A root system determines a *line system*, that is, a collection of one-dimensional subspaces generated by the corresponding roots. The definitions of the root systems E_7 and E_6 are essentially independent of the particular choice of roots, as different choices yield isometric line systems.

For a root system R, a system of positive roots R^+ is its subset such that: (i) for $\pm\mathbf{x} \in R$, exactly one of $\mathbf{x}, -\mathbf{x}$ belongs to R^+, and (ii) for $\mathbf{x}, \mathbf{y} \in R^+$, if $\mathbf{x} + \mathbf{y} \in R$ then $\mathbf{x} + \mathbf{y} \in R^+$. Observe that R and R^+ produce the same line system.

A *representation* of a signed graph $\dot{G}$ with least eigenvalue ≥ -2 is a mapping $f\colon V(\dot{G}) \to \mathbb{R}^k$, $k \geq 1$, such that

$$f(i) \cdot f(j) = \begin{cases} 2 & \text{if } i = j, \\ 1 & \text{if } i \overset{+}{\sim} j, \\ -1 & \text{if } i \overset{-}{\sim} j, \\ 0 & \text{if } i \nsim j. \end{cases}$$

If $S \subset \mathbb{R}^k$ is the image of f, then we say that $\dot{G}$ is *represented* by S. Clearly, $A_{\dot{G}} + 2I$ is the Gram matrix of S, so it is positive semidefinite and the least eigenvalue of $\dot{G}$ is, indeed, greater than or equal to -2.

It is known that every signed line graph admits a representation in some root system of type D_k [54, 99]. Alternatively, a connected signed graph is called exceptional if it admits a representation in the root system E_8, but does not admit a representation in any D_k. This definition is equivalent to the former.

Table 5.1 *Spectra of signed graphs $\dot{M}_6$, $\dot{M}_7$ and $\dot{M}_8$.*

Signed graph	Spectrum	Spectrum of the underlying graph
$\dot{M}_6$	$[10^6,(-2)^{30}]$	$[20,2^{20},(-4)^{15}]$
$\dot{M}_7$	$[16^7,(-2)^{56}]$	$[32,4^{27},(-4)^{35}]$
$\dot{M}_8$	$[28^8,(-2)^{112}]$	$[56,8^{35},(-4)^{84}]$

5.3.2 Properties of Exceptional Signed Graphs

Every exceptional signed graph has at most 120 vertices, corresponding to the number of positive roots in the root system E_8. The following three exceptional signed graphs play a significant role in the remainder of this chapter: $\dot{M}_6$, represented by E_6^+; $\dot{M}_7$, represented by E_7^+; and $\dot{M}_8$, represented by E_8^+. Each of these signed graphs is unique up to switching equivalence. Their spectra, along with those of the corresponding underlying graphs, are summarized in Table 5.1.

Remark 5.7 The upper bound of (4.3) is attained for $\dot{M}_8$ (with $(\mu,n,t) = (-2,120,8)$). It is known from [180] that $\dot{M}_8$ is the unique signed graph that attains this bound for $\mu = -2$. Naturally, the same bound is also attained by its negation, corresponding to $\mu = 2$.

A natural question arises concerning exceptional signed graphs whose least eigenvalue exceeds -2. This question has been completely resolved through a combination of theoretical analysis and computational methods, as reported by Greaves et al.

Lemma 5.8 ([99]) *An exceptional signed graph whose least eigenvalue is greater than* -2 *is*

(i) *one of* 32 *signed graphs of order* 6,
(ii) *one of* 233 *signed graphs of order* 7 *or*
(iii) *one of* 1242 *signed graphs of order* 8

listed in [99]. Every signed graph of (ii) (resp. (iii)) is a one-vertex extension of some of (i) (resp. (ii)).

As demonstrated in [214] (see also [243]), every signed graph with at most six vertices and least eigenvalue -2 has a representation in D_6, so it is a signed line graph. Consequently, every exceptional signed graph $\dot{G}$ with least eigenvalue -2 has at least seven vertices and contains an exceptional induced subgraph with six vertices and least eigenvalue greater than -2. This implies that the lower bound for the co-dimension of the eigenspace of -2 in an exceptional signed graph $\dot{G}$ is 6. On the other hand, we have $A_{\dot{G}}+2I=S^{\mathsf{T}}S$, where S is obtained by arranging some roots of E_8 as its columns. Since the dimension of the linear span of E_8 is 8, the rank of $A_{\dot{G}}+2I$ is at most 8, which gives the upper bound. We record this as the following lemma.

Lemma 5.9 *An exceptional signed graph has six, seven or eight eigenvalues (with possible repetitions) greater than* -2.

We remind the reader that a maximal extension of a prescribed star complement is defined before Example 4.9. Maximal extensions for -2 of exceptional signed graphs are determined in the next theorem.

Theorem 5.10 ([214]) *For* $6 \leq k \leq 8$, $\dot{M}_k$ *is the unique maximal extension of every exceptional signed graph with k vertices and least eigenvalue greater than* -2.

Proof Every exceptional signed graph with k vertices, where $6 \leq k \leq 8$, and least eigenvalue greater than -2 admits a representation in the root system E_k and, consequently, in the set of positive roots E_k^+. Such a signed graph is naturally an induced subgraph of a signed graph represented by the entire set of positive roots. For $k=8$, the desired result is immediate.

For the case $k=7$, suppose, towards a contradiction, that there exists an extension $\dot{G}$ such that $A_{\dot{G}}+2I=S^{\mathsf{T}}S$, where the columns of S consist of roots from E_7^+ together with at least one root $\mathbf{x} \in E_8^+ \setminus E_7^+$. Since $\mathbf{x}$ is orthogonal to every $\mathbf{y} \in E_7^+$, it follows that the rank of $A_{\dot{G}}+2I$ is 8, implying that $\dot{G}$ has eight eigenvalues (counting multiplicities) greater than -2. This contradicts the assumption that $\dot{G}$ is an extension of the $k=7$ case.

The argument for $k=6$ is analogous. □

Therefore, every signed graph of Lemma 5.8(i) (resp. (ii), (iii)) is a star complement in $\dot{M}_6$ (resp. $\dot{M}_7$, $\dot{M}_8$). In the unsigned domain, there are exactly 473 maximal extensions of exceptional graphs with eight vertices, along with a large number of maximal extensions of exceptional graphs with six or seven vertices [77, Chapter 6]. The extension reported in Theorem 5.10 has a very simple formulation.

5.4 Signed Graph Foundations

For a signed graph $\dot{G}$, let μ be an eigenvalue of its signed line graph. A set of edges S of $\dot{G}$ is a *line star set* for μ if it is a star set for this eigenvalue in $\mathscr{L}(\dot{G})$.

The spanning subgraph of $\dot{G}$ obtained by removing the edges of S is a *line star complement* for μ. In particular, if $\mu = -2$ then the line star complement is known as a *foundation* for $\dot{G}$.

The following classification of connected signed graphs whose least eigenvalue is greater than -2 is a reformulation of [99, Theorem 6].

Theorem 5.11 *Let $\dot{G}$ be a connected signed graph with n vertices and least eigenvalue greater than -2. Then $\dot{G}$ belongs to one of the following classes:*

(i) signed line graphs of trees;
(ii) signed line graphs of unbalanced unicyclic simply signed graphs (so, negative digons are allowed);
(iii) exceptional signed graphs whose least eigenvalue is ≥ -2.

In light of the previous result, foundations of either trees, or unbalanced simply signed graphs, or exceptional signed graphs merit particular attention. The foundations of exceptional signed graphs have already been characterized in terms of maximal extensions for the eigenvalue -2 in Theorem 5.10. In the following, we focus on the non-exceptional foundations, beginning with the following result.

Theorem 5.12 ([180]) *The signed line graph of every tree and every unbalanced unicyclic simply signed graph with at least three vertices is a star complement for -2.*

Here is a sketch of the proof. If a tree T contains a pair of non-adjacent vertices, then by inserting a positive edge e between such a pair we obtain a balanced unicyclic graph $T+e$ such that the least eigenvalue of its signed line graph is -2, by Theorem 5.11. Therefore, $\mathscr{L}(T)$ is a star complement for -2 in $\mathscr{L}(T+e)$. Unbalanced unicyclic simply signed graphs are considered in a similar way.

The previous result is interesting because in the domain of graphs not every line graph of a tree or a unicyclic graph is a star complement for -2. We continue with the following characterization.

A simple algebraic computation gives that spectra of $\mathscr{L}(K_t)$ and $\mathscr{L}(\ddot{K}_t)$ are $\left[(t-2)^{t-1}, (-2)^{\binom{t-1}{2}}\right]$ and $\left[2(t-2)^t, (-2)^{t(t-2)}\right]$, respectively.

Theorem 5.13 ([180]) *Let $\dot{M}$ be a maximal non-exceptional signed graph with a star complement $\mathscr{L}(\dot{G})$ for -2. The following hold:*

(i) If $\dot{G}$ is a tree of order $t \geq 3$, then $\dot{M} \cong \mathscr{L}(K_t)$.
(ii) If $\dot{G}$ is an unbalanced unicyclic simply signed graph of order $t \geq 3$, then $\dot{M} \cong \mathscr{L}(\ddot{K}_t)$.

Proof (i): Since $\dot{G}$ is a tree, we may denote it simply by T. Given that $\dot{M}$ is not exceptional, we may assume that $\dot{M} \cong \mathscr{L}(\dot{F})$ for some simply signed graph $\dot{F}$. Evidently, T spans $\dot{F}$; otherwise, the co-dimension of the eigenspace corresponding to the eigenvalue -2 for $\dot{M}$ would exceed $t-1$.

Moreover, from Theorem 5.11(ii), it follows that $\mathscr{L}(T+e)$, where e is a negative edge that may form a negative digon, has least eigenvalue greater than -2. Consequently, $\dot{F}$ is an unsigned graph spanned by T, implying that $\dot{F}$ is a subgraph of the complete graph K_t.

Since exactly $t-1$ eigenvalues of $\mathscr{L}(K_t)$ differ from -2 and the order of $\mathscr{L}(T)$ is $t-1$, it follows that $\mathscr{L}(T)$ is a star complement for the eigenvalue -2 in $\mathscr{L}(K_t)$. The maximality of $\dot{M}$ then guarantees that $\dot{F} \cong K_t$.

(ii): As before, we assume that $\dot{M} \cong \mathscr{L}(\dot{F})$, where $\dot{F}$ is a simply signed graph. Since $\dot{G}$ spans $\dot{F}$, otherwise the co-dimension of the eigenspace corresponding to the eigenvalue -2 for $\dot{M}$ would not equal t, it follows that $\dot{F}$ has order t. Consequently, $\mathscr{L}(\dot{F})$ is a subgraph of $\mathscr{L}(\ddot{K}_t)$. The remainder of the proof proceeds by a slight modification of the argument given in part (i). □

Theorem 5.12 tells us that every tree and every unbalanced unicyclic simply signed graph (in both cases, with at least three vertices) can appear in the role of $\dot{G}$ in Theorem 5.13.

Lemma 5.9 establishes that if a connected signed graph with least eigenvalue -2 is exceptional, then the co-dimension of the eigenspace associated with -2 is 6, 7 or 8. Combined with Theorems 5.12 and 5.13, this immediately yields the following result.

Corollary 5.14 ([180]) *Let $\dot{G}$ be a tree (resp. an unbalanced unicyclic simply signed graph) of order t. If $t \notin \{7,8,9\}$ (resp. $t \notin \{6,7,8\}$) then $\mathscr{L}(K_t)$ (resp. $\mathscr{L}(\ddot{K}_t)$) is the unique maximal signed line graph with $\dot{G}$ as a star complement for -2.*

Here is an illustration concerning paths and signed cycles.

Example 5.15 The path P_t is an induced subgraph of $\dot{M}_t$ when $t \in \{7,8\}$, and so M_t is the unique maximal extension of $\mathscr{L}(P_{t+1})$ in these cases. On the other hand, P_6 is not an induced subgraph of $\dot{M}_6$, and in this case one can invoke Theorem 4.2 to show that $\mathscr{L}(K_7)$ is the unique maximal extension of $\mathscr{L}(P_7)$.

Similarly, if $\dot{C}_t$ is a negative t-cycle, then we find that $\mathscr{L}(\dot{C}_t)$ has $\mathscr{L}(\ddot{K}_t)$ as the unique maximal extension when $t \neq 8$ and $\dot{M}_8$ as such an extension when $t = 8$.

The next result characterizes the foundations for non-exceptional signed graphs.

Theorem 5.16 ([180]) *Let $\dot{G}$ be a connected balanced (resp. unbalanced) signed graph such that the least eigenvalue of $\mathscr{L}(\dot{G})$ is -2. Then the subgraph $\dot{H}$ of $\dot{G}$ is a foundation of $\dot{G}$ if and only if $\dot{H}$ is a spanning tree of $\dot{G}$ (resp. every component of $\dot{H}$ is an unbalanced unicyclic simply signed graph).*

We highlight the following result, which enumerates the number of star partitions of $\mathscr{L}(K_t)$. In fact, this relates to one of the foundational results in graph theory: the celebrated formula of Cayley for the number of labelled trees [60].

Theorem 5.17 ([180]) *For $t \geq 2$, there are exactly t^{t-2} star partitions of the signed line graph $\mathscr{L}(K_t)$.*

Proof The statement clearly holds for $t = 2$. From Theorem 5.16, if H is a subgraph of K_t such that $\mathscr{L}(H)$ forms a star complement for -2 in $\mathscr{L}(K_t)$, then H must be a spanning tree of K_t. By Theorem 4.7, the vertex set partition $V(\mathscr{L}(H)) \sqcup V(\mathscr{L}(K_t - H))$ constitutes a star partition of $\mathscr{L}(K_t)$. Hence, the number of star partitions corresponds exactly to the number of spanning trees in K_t, which is given by the formula of Cayley as t^{t-2}. □

5.4.1 A Classification of Signed Graphs with Least Eigenvalue ≥ -2

As established in Section 5.3, a connected signed graph whose spectrum is bounded below by -2 is either the signed line graph of a simply signed graph or an exceptional signed graph. In this subsection, we build upon these results to classify such graphs through the notion of their maximal extensions.

Theorem 5.18 ([180]) *Let $\dot{G}$ be a connected signed graph with least eigenvalue $\lambda \geq -2$. If $\dot{G}$ is non-exceptional, then it is the signed line graph of a simply signed graph $\dot{H}$ such that one of the following holds:*

- *(L1) $\dot{H}$ is a tree (with $\lambda > -2$).*
- *(L2) $\dot{H}$ is an unbalanced unicyclic simply signed graph (with $\lambda > -2$).*
- *(L3) $\dot{H}$ is balanced, not a tree and spans the complete graph (with $\lambda = -2$).*
- *(L4) $\dot{H}$ is unbalanced, not a unicyclic simply signed graph and spans the signed doubled complete graph up to switching (with $\lambda = -2$).*

If $\dot{G}$ is exceptional then:

- *(E1) $\dot{G}$ is one of the signed graphs listed in Lemma 5.8 (with $\lambda > -2$).*
- *(E2) $\dot{G}$ has at least $t+1$ vertices and is an induced subgraph of $\dot{M}_t$ (with $t \in \{6,7,8\}$, $\lambda = -2$).*

In addition, every signed graph with $t \geq 3$ *vertices that belongs to (L1) (resp. (L2)) is a foundation for* K_t *(resp.* $\ddot{K}_t$*). Every signed graph that belongs to (E2) is a* (-2)*-extension of some signed graph of (E1).*

Proof Parts (L1) and (L2) follow from Theorem 5.11; (L3) and (L4) follow from Theorem 5.13. Part (E1) follows from Lemma 5.8 and (E2) follows from Theorem 5.10, which also establishes that every signed graph of (E2) acts as a (-2)-extension of some signed graph of (E1). Finally, the fact that every signed graph in (L1) (resp. (L2)) with $t \geq 3$ vertices is a foundation of K_t (resp. $\ddot{K}_t$) follows from Theorem 5.16. □

For an analogous classification in the context of unsigned graphs, see [77, Theorems 2.3.20, 5.4.1, 5.4.3 and Chapter 6].

5.5 Eigenbases for -2 in Signed Line Graphs

The results in this section are taken from [207]; a slightly different presentation appears in [32]. Theorems 5.20 and 5.22 provide the eigenbases corresponding to the eigenvalue -2 in signed line graphs. In this section, we use the notations ij and $\{i,j\}$ interchangeably to denote the edge between vertices i and j.

We first consider the signed line graphs of Theorem 5.18(L3). The following lemma is needed.

Lemma 5.19 *Let* $\dot{C}$ *be a positive signed cycle and* η *a vertex–edge orientation on it. If the vertices of* $\dot{C}$ *are labelled by* $0,1,\dots,k$ *in the natural order, then a function* $f\colon E(\dot{C}) \to \{x,-x\}, x \in \mathbb{R}$*, defined by*

$$\begin{aligned} f(\{0,1\}) &= x, \\ f(\{i,i+1\}) &= -\eta(i,\{i-1,i\})\eta(i,\{i,i+1\})f(\{i-1,i\}), \end{aligned}$$

for $i := i \pmod{k+1}$*, is consistent in the sense that*

$$-\eta(0,\{k,0\})\eta(0,\{0,1\})f(\{k,0\}) = f(\{0,1\}).$$

Proof Assume first that, for every edge $\{i,i+1\}$,

$$\begin{aligned} \eta(i,\{i,i+1\}) &= -1 \quad \text{if } \sigma(\{i,i+1\}) = 1, \\ \eta(i,\{i,i+1\}) &= 1 \quad \text{if } \sigma(\{i,i+1\}) = -1. \end{aligned}$$

Observe that, when moving from the edge $\{i-1,i\}$ to the edge $\{i,i+1\}$, the function f changes sign if and only if $\sigma(\{i,i+1\}) = -1$. Since the signed cycle $\dot{C}$ is positive, it contains an even number of negative edges, which yields

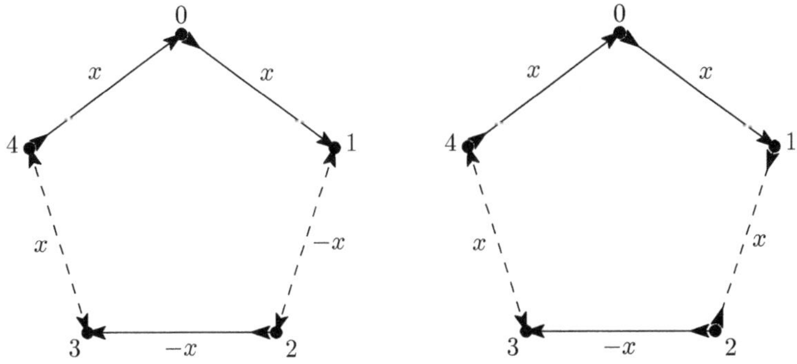

Figure 5.3 An illustration of Lemma 5.19. On the right-hand side, the orientation of the edge $\{1,2\}$ is changed. Reproduced from [207], licensed under CC BY 4.0.

the desired conclusion. Furthermore, changing the orientation on a single edge results in negating f on that edge while leaving it unchanged on all other edges, thereby completing the proof. □

Lemma 5.19 is visualized in Figure 5.3.

We recall that, for a signed graph $\dot{G}$, a vertex–edge orientation η uniquely defines the signature σ on the edge set of the signed line graph $\mathscr{L}(\dot{G})$ as if u is a vertex incident to edges e and f of $\dot{G}$; then $\sigma(ef) = \eta(u,e)\eta(u,f)$.

Theorem 5.20 *Let $\dot{G}$ be a connected balanced signed graph distinct from a tree, $\dot{T}$ a spanning tree and $F = E(\dot{G}) \setminus E(\dot{T})$. For $e_0 \in F$, let $e_0, e_1, \ldots, e_k$ be the edges (listed in the natural order) of the unique cycle obtained by adding e_0 to $\dot{T}$. If σ is the signature on the edge set of $\mathscr{L}(\dot{G})$ defined by a fixed vertex–edge orientation on $\dot{G}$ and $m = |E(\dot{G})|$, then $\mathbf{x} \in \mathbb{R}^m$ that takes*

$$x_{e_i} = \begin{cases} 1 & \text{if } i = 0, \\ -\sigma(e_{i-1}e_i)x_{e_{i-1}} & \text{if } 1 \leq i \leq k \end{cases}$$

and zero on the remaining edges is an eigenvector for -2 in $\mathscr{L}(\dot{G})$. Moreover, the $m - |E(\dot{T})|$ eiegnvectors formed in this way span the eigenspace for -2.

Proof Let e simultaneously denote an edge of $\dot{G}$ and the corresponding vertex of $\mathscr{L}(\dot{G})$, and let x_e be the associated entry of the vector $\mathbf{x}$. Denote by $\dot{C}$ the cycle described in the statement and fix an orientation η. Note that $\dot{C}$ is positive, since $\dot{G}$ is balanced.

If e is non-adjacent to any edge of $\dot{C}$, then $x_e = 0$ and also $(A_{\mathscr{L}(\dot{G})}\mathbf{x})_e = 0$.

Suppose that $e \in \dot{C}$, say $e = e_i$. Let u (resp. v) denote the vertex between e_{i-1}

and e_i (resp. e_i and e_{i+1}). Then

$$\begin{aligned}(A_{\mathcal{L}(\dot{G})}\mathbf{x})_e &= \sigma(e_{i-1}e_i)x_{e_{i-1}} + \sigma(e_ie_{i+1})x_{e_{i+1}} \\ &= \eta(u,e_{i-1})\eta(u,e_i)x_{e_{i-1}} + \eta(v,e_i)\eta(v,e_{i+1})x_{e_{i+1}} \\ &= -(\eta(u,e_{i-1})\eta(u,e_i))^2x_{e_i} - (\eta(v,e_i)\eta(v,e_{i+1}))^2x_{e_i} = -2x_e,\end{aligned}$$

as desired. Lemma 5.19 ensures that the previous equality remains valid for $i \in \{0,k\}$.

Suppose that $e \notin \dot{C}$ and e is incident to a single vertex, say u, of $\dot{C}$. Then $x_e = 0$. On the other hand, if u is located between e_i and e_{i+1}, then

$$\begin{aligned}(A_{\mathcal{L}(\dot{G})}\mathbf{x})_e &= \sigma(ee_i)x_{e_i} + \sigma(ee_{i+1})x_{e_{i+1}} \\ &= \eta(u,e)\eta(u,e_i)x_{e_i} + \eta(u,e)\eta(u,e_{i+1})x_{e_{i+1}} \\ &= \eta(u,e)\eta(u,e_i)x_{e_i} + \eta(u,e)\eta(u,e_{i+1})(-\eta(u,e_i)\eta(u,e_{i+1})x_{e_i}) \\ &= 0.\end{aligned}$$

The remaining case, in which e is not in $\dot{C}$ but lies between two vertices belonging to the same cycle, is considered in a similar way. Therefore, $\mathbf{x}$ is an eigenvector associated with -2.

The eigenvectors constructed in this way are linearly independent because, for each $f \in F$, the f-entry is non-zero in exactly one of them. □

An illustration of Theorem 5.20 is given in Figure 5.3, where x is fixed to 1. We proceed with the signed line graphs of Theorem 5.18(L4).

Lemma 5.21 *Let $\dot{C}$ be a negative signed cycle, η a vertex–edge orientation on it, and let the vertices of $\dot{C}$ be labelled by $0,1,\dots,k$ in the natural order. For $f\colon E(\dot{C}) \to \{x,-x\}$, $x \in \mathbb{R}$, defined by*

$$\begin{aligned}f(\{0,1\}) &= x, \\ f(\{i,i+1\}) &= -\eta(i,\{i-1,i\})\eta(i,\{i,i+1\})f(\{i-1,i\})\end{aligned}$$

for $1 \le i \le k$ and $i := i \pmod{k+1}$, we have

$$\eta(0,\{k,0\})\eta(0,\{0,1\})f(\{k,0\}) = f(\{0,1\}).$$

Proof Assume that the orientation is assigned as in Lemma 5.19. Since $\dot{C}$ is negative, the function f changes its sign an odd number of times when traversing the entire cycle, which yields the desired conclusion. The remainder of the proof, concerning alternative orientations, follows analogously to the corresponding part of the proof of Lemma 5.19. □

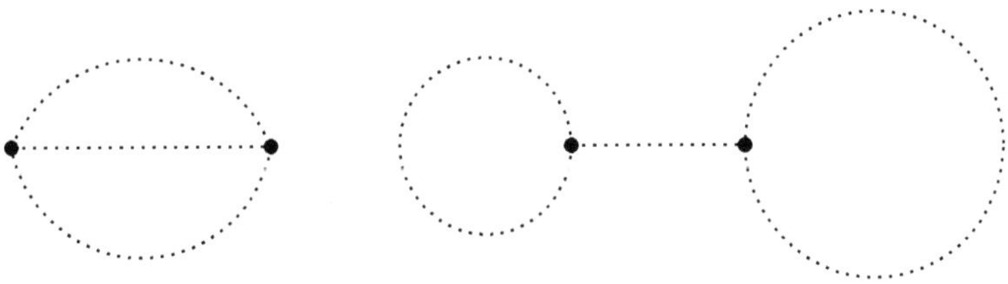

Figure 5.4 A sketch of a simply signed theta-graph and a simply signed dumbbell. In the latter case, the specified vertices may be identified. Reproduced from [207], licensed under CC BY 4.0.

For an illustration, see Figure 5.3, changing the sign of $\{4,0\}$ and setting $f(\{4,0\}) = -x$.

At this point, we introduce two types of simply signed bicyclic graphs. The first type contains a simply signed theta-graph as an induced subgraph, where a *simply signed theta-graph* consists of two vertices connected by three edge-disjoint paths. (Since negative digons are allowed, at most two of these paths have length one.) The second type contains a simply signed dumbbell as an induced subgraph, where a *simply signed dumbbell* consists of two cycles and a path connecting them. These structures are illustrated in Figure 5.4. It is a straightforward observation that a simply signed theta-graph contains either one or three positive cycles.

Theorem 5.22 *Let $\dot{G}$ be a connected unbalanced simply signed graph distinct from a unicyclic simply signed graph, $\dot{U}$ a spanning subgraph whose components are unbalanced unicyclic simply signed graphs and $F = E(\dot{G}) \setminus E(\dot{U})$. Adding e $(e \in F)$ to $\dot{U}$ creates either a single positive cycle or a simply signed dumbbell without positive cycles.*

Let σ be the signature on the edge set of $\mathscr{L}(\dot{G})$ defined by a fixed vertex–edge orientation on $\dot{G}$, and $m = |E(\dot{G})|$.

(i) *If e creates a positive cycle, then $\mathbf{x} \in \mathbb{R}^m$ formed as in Theorem 5.20 is an eigenvector for -2 in $\mathscr{L}(\dot{G})$.*

(ii) *Suppose that e creates a simply signed dumbbell without positive cycles. Let $f_0, f_1, \ldots, f_l$ be edges in the path between two cycles, $e'_0, e'_1, \ldots, e'_{k_1}$ and $e''_0, e''_1, \ldots, e''_{k_2}$ edges of two cycles, all in the natural order, such that f_0 (resp. f_l) is adjacent to e'_0, e'_{k_1} (resp. e''_0, e''_{k_2}). Then $\mathbf{x} \in \mathbb{R}^m$ that takes*

$$x_{f_i} = \begin{cases} 2 & \text{if } i = 0, \\ -\sigma(f_{i-1}f_i)x_{f_{i-1}} & \text{if } 1 \leq i \leq l; \end{cases}$$

$$x_{e'_i} = \begin{cases} -\sigma(f_0e'_i) & \text{if } i \in \{0,k_1\}, \\ -\sigma(e'_{i-1}e'_i)x_{e'_{i-1}} & \text{if } 1 \leq i \leq k_1 - 1 \end{cases}$$

and zero on the remaining edges is an eigenvector for -2 *in* $\mathscr{L}(\dot{G})$.

The $m - |E(\dot{U})|$ *eigenvectors formed in this way span the eigenspace for* -2.

Proof If the edge e lies between two vertices of a fixed cycle, say $\dot{C}$, of $\dot{U}$, then e induces a simply signed theta-graph. Since $\dot{C}$ is negative, this theta-graph contains exactly one positive cycle.

If at most one vertex incident to e belongs to a cycle of $\dot{U}$, then either e connects vertices within the same component of $\dot{U}$ or acts as a bridge between two distinct components. In the former case, e induces either a simply signed theta-graph or a simply signed dumbbell. We conclude that this situation results in exactly one positive cycle. In the latter case, e either induces a single positive cycle or a simply signed dumbbell devoid of positive cycles. Finally, if e forms a link between two unicyclic components, then it creates a simply signed dumbbell. Since both components are unbalanced, the resulting dumbbell contains no positive cycles.

Now, (i) follows from Theorem 5.20, and we proceed with (ii). Let e denote an edge of $\dot{G}$ and the corresponding vertex of $\mathscr{L}(\dot{G})$, and let x_e be the corresponding entry of $\mathbf{x}$. The bicyclic component created by e is denoted by $\dot{B}$ and its cycles by $\dot{C}_1$ and $\dot{C}_2$. Both cycles are negative.

If e is non-adjacent to any edge of $\dot{B}$, then $x_e = (A_{\mathscr{L}(\dot{G})}\mathbf{x})_e = 0$. Further, for $l \geq 1$, we have

$$\begin{aligned}(A_{\mathscr{L}(\dot{G})}\mathbf{x})_{f_0} &= \sigma(f_0e'_0)x_{e'_0} + \sigma(f_0e'_{k_1})x_{e'_{k_1}} + \sigma(f_0f_1)x_{f_1} \\ &= -\sigma(f_0e'_0)^2 - \sigma(f_0e'_{k_1})^2 - 2\sigma(f_0f_1)^2 = -4 = -2x_{f_0}.\end{aligned}$$

Edges f_i $(0 < i < l)$, f_l and f_0 (when $l = 0$) are considered in a similar way. Next, we have

$$\begin{aligned}(A_{\mathscr{L}(\dot{G})}\mathbf{x})_{e_0} &= \sigma(f_0e'_0)x_{f_0} + \sigma(e'_0e'_1)x_{e'_1} + \sigma(e'_0e'_{k_1})x_{e'_{k_1}} \\ &= 2\sigma(f_0e'_0) - \sigma(e'_0e'_1)^2x_{e'_0} + x_{e'_0} \\ &= -2x_{e'_0} - \sigma(e'_0e'_1)^2x_{e'_0} + x_{e'_0} = -2x_{e'_0},\end{aligned}$$

where $\sigma(e'_0e'_{k_1})x_{e'_{k_1}} = x_{e'_0}$ follows from Lemma 5.21. The remaining edges of cycles and the situation with a negative digon are considered in a similar way.

Finally, if u is a vertex of $\dot{B}$, then $\sum \eta(u,f)x_f = 0$ holds by the construction of $\mathbf{x}$, where the summation goes over all edges incident to u. This means that if e is not in $\dot{B}$ but is incident to at least one vertex of $\dot{B}$, then we have $x_e = (A_{\mathscr{L}(\dot{G})}\mathbf{x})_e = 0$. □

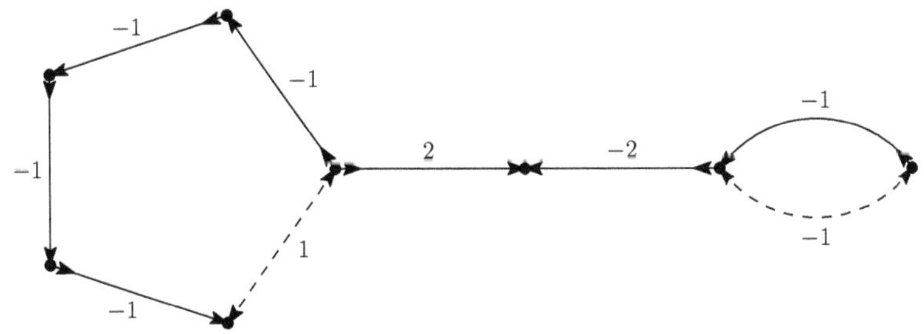

Figure 5.5 An illustration of Theorem 5.22(ii). Reproduced from [207], licensed under CC BY 4.0.

An illustration of Theorem 5.22 is given in Figure 5.5.

Problems

5.1 Prove that different vertex–edge orientations on $\dot{G}$ produce switching-equivalent signed line graphs.

5.2 Determine the signature of $\mathscr{L}_C(\dot{C}_n)$ and $\mathscr{L}_S(\dot{C}_n)$, where $\dot{C}_n$ may be either positive or negative, and n may have any parity.

5.3 Formulate statements analogous to Theorems 5.3 and 5.4 in which the spectral signed line graph is replaced by the combinatorial signed line graph.

5.4 ([230]) Show that the third definition of a signed line graph, as discussed in Remark 5.1, necessarily results in a balanced signed graph, irrespective of the chosen orientation.

5.5 Show that the signed line graph of a simply signed graph has no multiple edges.

5.6 Determine whether the signed line graph of a simply signed graph is a tree.

5.7 Determine whether the signed line graph of a simply signed graph is regular.

5.8 Find an explicit system of positive roots for each of the exceptional root systems E_k, with $6 \leq k \leq 8$.

5.9 Show that every signed line graph is representable in the root system D_k.

5.10 Which signed graph is represented by all positive roots of D_k for $k \geq 4$?

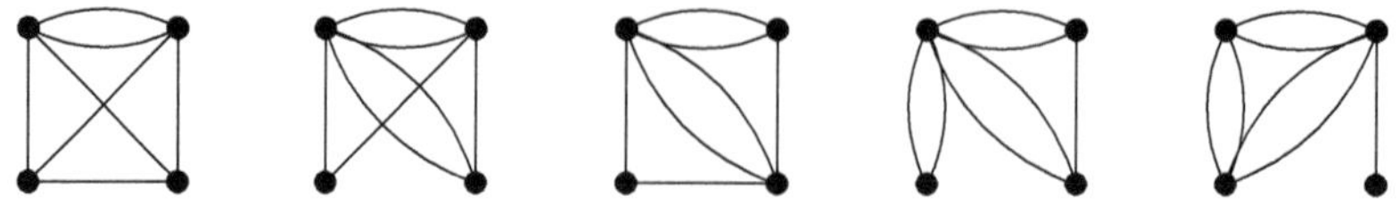

Figure 5.6 Underlying multigraphs of signed graphs that produce switching-isomorphic signed line graphs.

5.11 Show that $\mathscr{L}(K_8)$ has a representation in E_7. (Prove that it is represented by the roots $\mathbf{e}_i - \mathbf{e}_j$ for $1 \leq j < i \leq 8$ from the E_8 root system. Next, find a root from the same system that is orthogonal to all of these roots.)

5.12 Prove that the multigraphs in Figure 5.6 admit signatures producing simply signed graphs $\dot{G}_1, \dot{G}_2, \ldots, \dot{G}_5$ such that $\mathscr{L}(\dot{G}_1), \mathscr{L}(\dot{G}_2), \ldots, \mathscr{L}(\dot{G}_5)$ are mutually switching isomorphic.

5.13 ([221]) Extend the solution of the previous problem by identifying all signed graphs possessing a unique simply signed root. Then, determine the corresponding solutions when simply signed roots are excluded.

5.14 Show that the implication $\dot{G} \not\cong \dot{H} \Rightarrow \mathscr{L}(\dot{G}) \cong \mathscr{L}(\dot{H})$, where $\dot{G}$ and $\dot{H}$ are signed graphs (but not simply signed), holds if and only if one of them is a four-vertex star and the other is a signed triangle.

5.15 Show that the implication $\dot{G} \not\cong \dot{H} \Rightarrow \mathscr{L}(\dot{G}) \cong \mathscr{L}(\dot{H})$, where $\dot{G}$ and $\dot{H}$ are simply signed graphs of order n, does not hold for $n \geq 5$.

5.16 As an exercise, apply the star complement technique to some exceptional signed graphs of order $k \in \{6,7,8\}$ whose least eigenvalue is greater than -2 and construct their extensions for -2.

5.17 ([77, Chapter 6]) Show that an exceptional unsigned graph has at most 36 vertices.

5.18 Prove that the graph of the previous problem is unique, find its representation in E_8 and compute the spectrum of the signed graph represented by the remaining roots.

5.19 ([77, Chapter 4]) Give a characterization of regular exceptional unsigned graphs.

5.20 Prove Theorem 5.11.

5.21 ([235]) Let the adjacency matrix of a connected signed graph $\dot{G}$ be as in the statement of Theorem 4.2. Assume that the spectrum of $\dot{G}$ is $[\theta^t, -\theta^t]$ and that $\dot{G}$ contains a star complement (for both θ and $-\theta$) with spectrum $[\lambda, -\lambda]$. Prove that the eigenspaces corresponding to the eigenvalues $\pm\theta$ are spanned by

$$\begin{pmatrix} \mathbf{e}_i \\ \frac{1}{\theta^2-\lambda^2}(A_{\dot{H}} \pm \theta I)\mathbf{b}_i^\intercal \end{pmatrix}, \ 1 \leq i \leq t, \text{ or } \begin{pmatrix} B^\intercal \mathbf{x}_i \\ \frac{1}{\theta^2-\lambda^2}(A_{\dot{H}} \pm \theta I)BB^\intercal \mathbf{x}_i \end{pmatrix}, \ 1 \leq i \leq t.$$

5.22 Use Theorem 4.18 and the previous problem to compute the eigenspaces of an r-dimensional signed cube with negative quadrangles.

6

Signed Graphs with a Small Number of Eigenvalues

In the context of ordinary graphs, the first non-trivial spectral case arises when the graph has exactly three distinct eigenvalues. It is well known that a connected regular graph has three eigenvalues if and only if it is strongly regular. Strongly regular graphs have been extensively studied over the past several decades. Many such graphs have been explicitly constructed, and numerous existence conditions have been established for cases that remain unresolved. Notable characterizations of non-regular graphs with three eigenvalues can be found in [65, 66, 79], along with further references therein. Additionally, a substantial body of work has addressed the characterization of regular graphs with four eigenvalues; see, for example, [80, 81, 137]. Despite this progress, both the three- and four-eigenvalue cases remain open in full generality.

Motivated by this, the present chapter shifts focus to signed graphs with two eigenvalues, which represents the first non-trivial spectral case in this broader setting, as well as to certain signed graphs with three eigenvalues. Some of these examples have already been encountered in earlier chapters. It is already known that regular signed graphs with at most three eigenvalues are walk-regular, and those with two eigenvalues are necessarily strongly regular. The following sections aim to provide further characterizations and classifications of specific subclasses within this framework.

6.1 Two Eigenvalues

We recall the already used fact that the minimal polynomial of a signed graph with two eigenvalues, λ and μ, is

$$\phi(x) = x^2 - (\lambda + \mu)x + \lambda\mu. \tag{6.1}$$

If $\lambda = -\mu$, then both eigenvalues are square roots of an integer and the adjacency matrix is orthogonal. Otherwise, the eigenvalues are integers.

We proceed with a general result concerning signed graphs with exactly one non-negative eigenvalue.

Lemma 6.1 ([171]) *If $\lambda_1 \geq \lambda_2 \geq \cdots \geq \lambda_n$ are the eigenvalues (with possible repetitions) of a signed graph $\dot{G}$, then $\lambda_2 \leq 0$ if and only if $\dot{G}$ is switching equivalent to a complete multipartite graph with possible isolated vertices. In particular, $\lambda_2 < 0$ holds if and only if $\dot{G}$ is switching equivalent to the complete graph.*

Proof If $\dot{G}$ is unbalanced, then it must contain at least one negative cycle as an induced subgraph. Since the second largest eigenvalue of any negative cycle is strictly positive (see Example 1.11), it follows that $\lambda_2 > 0$. Therefore, if $\lambda_2 \leq 0$, then $\dot{G}$ is switching equivalent to G. In this case, the claim follows from the well-known result stating that a graph has non-positive second largest eigenvalue if and only if it is a complete multipartite graph (possibly with some isolated vertices); see [73, Theorem 6.7].

Furthermore, if $\lambda_2 < 0$, then $\dot{G}$ must be connected and, in fact, complete. Indeed, if $\dot{G}$ were not complete, it would contain $2K_1$ (a pair of isolated vertices) as an induced subgraph, which has $\lambda_2 \geq \lambda_2(2K_1) = 0$, contradicting the assumption. The converse implication is immediate. □

We proceed with some consequences. The first one follows from the last part of the previous proof.

Corollary 6.2 ([171]) *A connected signed graph $\dot{G}$ has two eigenvalues, one of which is simple, if and only if $\dot{G}$ switches to a homogeneous complete signed graph.*

In the following two corollaries we consider vertex-deleted subgraphs of signed graphs with two eigenvalues.

Corollary 6.3 ([171]) *Let $\dot{G}$ be a connected signed graph of order $n \geq 3$, and let $\dot{H}$ be a vertex-deleted subgraph of $\dot{G}$. Then $\dot{G}$ and $\dot{H}$ both have two eigenvalues if and only if $\dot{G}$ switches to a homogeneous complete signed graph.*

Proof If $\dot{G}$ is switching equivalent to K_n or $-K_n$, then $\dot{G}$ has two eigenvalues and every vertex-deleted subgraph is switching equivalent to K_{n-1} or $-K_{n-1}$. Hence, $\dot{H}$ has two eigenvalues.

For the converse, suppose that $\dot{G}$ is not as in the formulation of the statement. By Corollary 6.2, the eigenvalues of $\dot{G}$, say μ and λ, are non-simple, and therefore the same eigenvalues appear in the spectrum of $\dot{H}$. Moreover, since both have two eigenvalues, $\dot{G}$ and $\dot{H}$ are regular and have a common vertex degree (equal to $-\lambda\mu$ by (6.1)), which is impossible. □

6.1.1 Vertex Degree At Most Five

Signed graphs with vertex degree at most 2 and exactly two eigenvalues can be readily classified (for degree-2 cases, compare Example 1.11). Moreover, complete signed graphs may be excluded from consideration, as they correspond to regular two-graphs, and hence represent a well-studied class.

The cases of vertex degrees 3 and 4 have been addressed in previous works; see [121, 152, 233]. In the proof that follows, we will refer to two key results from [92], where the authors classified all triangle-free signed graphs of degree 3 or 4 that possess exactly two eigenvalues.

Theorem 6.4 *The signed cube with negative quadrangles is the unique connected non-complete cubic signed graph with two eigenvalues. This signed graph is illustrated in Figure* 3.4.

A connected non-complete quartic signed graph has two eigenvalues if and only if it is one of the remaining signed graphs of Figure 3.4.

Proof In [92], the signed graph $\dot{Q}_3$ is identified as the unique triangle-free (moreover, bipartite) cubic signed graph with two eigenvalues. Suppose, under the same assumptions, that there exists a signed graph $\dot{G}$ with at least one triangle. Then its bipartite double $\mathrm{bd}(\dot{G}) \cong \dot{G} \times K_2$ is a connected bipartite cubic signed graph that inherits the same eigenvalues as $\dot{G}$; namely, the eigenvalues $\sqrt{3}$ and $-\sqrt{3}$. Since this product graph is bipartite and has two eigenvalues, it must be isomorphic to $\dot{Q}_3$, the aforementioned unique solution. However, this leads to a contradiction as $\dot{Q}_3$ is not a bipartite double, and the possibility is thereby excluded.

We now proceed to the case of quartic signed graphs. Let $\dot{G}$ be such a graph with at least one triangle, satisfying all other conditions of the theorem; in particular, it has eigenvalues 2 and -2. Then the bipartite double $\mathrm{bd}(\dot{G})$ must be one of the connected bipartite quartic signed graphs with the same spectrum classified in [92]. These include the remaining signed graphs in Figure 3.4, where in the family $\dot{T}_{2k}$, $k \geq 4$.

The cases $\mathrm{bd}(\dot{G}) \cong \dot{S}_{14}$ and $\mathrm{bd}(\dot{G}) \cong \dot{Q}_4$ can be readily ruled out, as they admit no preimages under the bipartite double construction that would satisfy the initial assumptions. Turning to the infinite family $\dot{T}_{2k}$, we observe that the presence of a triangle in $\dot{G}$ implies that $\mathrm{bd}(\dot{G})$ contains a hexagon as an induced subgraph (this follows from the structure of the tensor product). Upon examining the family, we find that this condition is met only in the case $\dot{T}_{12}$. Notably, this signed graph switches to $\mathrm{bd}(\dot{T}_6)$, thereby completing the proof. □

We note that the graph $\dot{T}_{2k}$ is isomorphic to the signed line graph of a signed doubled cycle, that is, $\dot{T}_{2k} \cong \mathscr{L}(\ddot{C}_k)$. It is also worth emphasizing that the quartic signed graphs depicted in Figure 3.4 appear as maximal connected signed graphs whose spectra lie entirely within the closed interval $[-2, 2]$. That is, they are not proper subgraphs of any larger signed graphs with the same spectral property; see [154] for further details.

To address the case of vertex degree 5, we first introduce a relevant family of inequivalent *weighing matrices* of weight 5. An $n \times n$ matrix $W = W(n,r)$ is called a *weighing matrix* of order n and weight r if it satisfies the condition $W^{\intercal}W = rI_n$. Two such matrices M and N are said to be *equivalent* if there exist $(1,0,-1)$-monomial matrices P and Q such that $N = PMQ$. The classification result that follows is due to Harada and Munemasa.

Theorem 6.5 ([112]) *Every weighing matrix of weight 5 is equivalent to a matrix which is a direct sum of an arbitrary number of matrices $W(6,5)$, $W(8,5)$, $W_{12,5}$, $W_{14,5}$, $D(16,5)$, $E(4t+2,5)$ and $F(4t+4,5)$, where $t \geq 2$, and the matrices in question are:*

$$\begin{pmatrix}
\cdot & + & + & + & + & + \\
+ & \cdot & + & - & - & + \\
+ & + & \cdot & + & - & - \\
+ & - & + & \cdot & + & - \\
+ & - & - & + & \cdot & + \\
+ & + & - & - & + & \cdot
\end{pmatrix}$$

$W(6,5)$

$$\begin{pmatrix}
+ & + & + & + & + & \cdot & \cdot & \cdot \\
+ & + & - & - & \cdot & + & \cdot & \cdot \\
+ & - & - & + & \cdot & \cdot & + & \cdot \\
+ & - & + & - & \cdot & \cdot & \cdot & + \\
+ & \cdot & \cdot & \cdot & - & - & - & - \\
\cdot & + & \cdot & \cdot & - & - & + & + \\
\cdot & \cdot & + & \cdot & - & + & + & - \\
\cdot & \cdot & \cdot & + & - & + & - & +
\end{pmatrix}$$

$W(8,5)$

$$\begin{pmatrix}
+ & + & + & + & + & \cdot & \cdot & \cdot & \cdot & \cdot & \cdot & \cdot \\
+ & - & \cdot & \cdot & \cdot & + & + & + & \cdot & \cdot & \cdot & \cdot \\
+ & \cdot & - & \cdot & \cdot & - & \cdot & \cdot & + & + & \cdot & \cdot \\
+ & \cdot & \cdot & - & \cdot & \cdot & - & \cdot & - & \cdot & + & \cdot \\
+ & \cdot & \cdot & \cdot & - & \cdot & \cdot & - & \cdot & - & - & \cdot \\
\cdot & + & - & \cdot & \cdot & \cdot & \cdot & + & \cdot & - & \cdot & + \\
\cdot & + & \cdot & - & \cdot & + & \cdot & \cdot & \cdot & + & - & \cdot \\
\cdot & + & \cdot & \cdot & - & \cdot & + & \cdot & \cdot & \cdot & + & - \\
\cdot & \cdot & + & - & \cdot & - & + & \cdot & \cdot & \cdot & \cdot & + \\
\cdot & \cdot & + & \cdot & - & \cdot & - & + & + & \cdot & \cdot & \cdot \\
\cdot & \cdot & \cdot & + & - & \cdot & \cdot & \cdot & - & + & \cdot & + \\
\cdot & \cdot & \cdot & \cdot & \cdot & + & \cdot & - & + & \cdot & + & +
\end{pmatrix}$$

$W_{12,5}$

$$\begin{pmatrix}
+ & + & + & + & + & \cdot & \cdot & \cdot & \cdot & \cdot & \cdot & \cdot & \cdot & \cdot \\
+ & - & \cdot & \cdot & \cdot & + & + & + & \cdot & \cdot & \cdot & \cdot & \cdot & \cdot \\
+ & \cdot & - & \cdot & \cdot & - & \cdot & \cdot & + & + & \cdot & \cdot & \cdot & \cdot \\
+ & \cdot & \cdot & - & \cdot & \cdot & - & \cdot & - & \cdot & + & \cdot & \cdot & \cdot \\
+ & \cdot & \cdot & \cdot & - & \cdot & \cdot & - & \cdot & - & - & \cdot & \cdot & \cdot \\
\cdot & + & - & \cdot & \cdot & + & \cdot & \cdot & \cdot & \cdot & \cdot & + & + & \cdot \\
\cdot & + & \cdot & - & \cdot & \cdot & \cdot & + & \cdot & \cdot & - & \cdot & - & \cdot \\
\cdot & + & \cdot & \cdot & - & \cdot & + & \cdot & \cdot & \cdot & + & - & \cdot & \cdot \\
\cdot & \cdot & + & - & \cdot & \cdot & \cdot & \cdot & + & \cdot & \cdot & \cdot & + & - \\
\cdot & \cdot & + & \cdot & - & \cdot & \cdot & \cdot & \cdot & + & \cdot & + & \cdot & + \\
\cdot & \cdot & \cdot & + & - & \cdot & - & + & \cdot & \cdot & \cdot & \cdot & \cdot & - \\
\cdot & \cdot & \cdot & \cdot & \cdot & + & - & \cdot & + & \cdot & \cdot & - & \cdot & + \\
\cdot & \cdot & \cdot & \cdot & \cdot & + & \cdot & - & \cdot & + & \cdot & \cdot & - & - \\
\cdot & \cdot & \cdot & \cdot & \cdot & \cdot & \cdot & \cdot & + & - & + & + & - & \cdot
\end{pmatrix}$$

$W_{14,5}$

$$\begin{pmatrix}
+ & + & + & + & + & \cdot & \cdot & \cdot & \cdot & \cdot & \cdot & \cdot & \cdot & \cdot & \cdot & \cdot \\
+ & - & \cdot & \cdot & \cdot & + & + & + & \cdot & \cdot & \cdot & \cdot & \cdot & \cdot & \cdot & \cdot \\
+ & \cdot & - & \cdot & \cdot & - & \cdot & \cdot & + & + & \cdot & \cdot & \cdot & \cdot & \cdot & \cdot \\
+ & \cdot & \cdot & - & \cdot & \cdot & - & \cdot & - & \cdot & + & \cdot & \cdot & \cdot & \cdot & \cdot \\
+ & \cdot & \cdot & \cdot & - & \cdot & \cdot & - & \cdot & - & - & \cdot & \cdot & \cdot & \cdot & \cdot \\
\cdot & + & - & \cdot & \cdot & + & \cdot & \cdot & \cdot & \cdot & \cdot & + & + & \cdot & \cdot & \cdot \\
\cdot & + & \cdot & - & \cdot & \cdot & + & \cdot & \cdot & \cdot & \cdot & - & \cdot & + & \cdot & \cdot \\
\cdot & + & \cdot & \cdot & - & \cdot & \cdot & + & \cdot & \cdot & \cdot & \cdot & - & - & \cdot & \cdot \\
\cdot & \cdot & + & - & \cdot & \cdot & \cdot & \cdot & + & \cdot & \cdot & + & \cdot & \cdot & + & \cdot \\
\cdot & \cdot & + & \cdot & - & \cdot & \cdot & \cdot & \cdot & + & \cdot & \cdot & + & \cdot & - & \cdot \\
\cdot & \cdot & \cdot & + & - & \cdot & \cdot & \cdot & \cdot & \cdot & + & \cdot & \cdot & + & + & \cdot \\
\cdot & \cdot & \cdot & \cdot & \cdot & + & - & \cdot & + & \cdot & \cdot & - & \cdot & \cdot & \cdot & + \\
\cdot & \cdot & \cdot & \cdot & \cdot & + & \cdot & - & \cdot & + & \cdot & \cdot & - & \cdot & \cdot & - \\
\cdot & \cdot & \cdot & \cdot & \cdot & \cdot & + & - & \cdot & \cdot & + & \cdot & \cdot & - & \cdot & + \\
\cdot & \cdot & \cdot & \cdot & \cdot & \cdot & \cdot & \cdot & + & - & + & \cdot & \cdot & \cdot & - & - \\
\cdot & \cdot & \cdot & \cdot & \cdot & \cdot & \cdot & \cdot & \cdot & \cdot & \cdot & + & - & + & - & +
\end{pmatrix}$$

$D(16,5)$

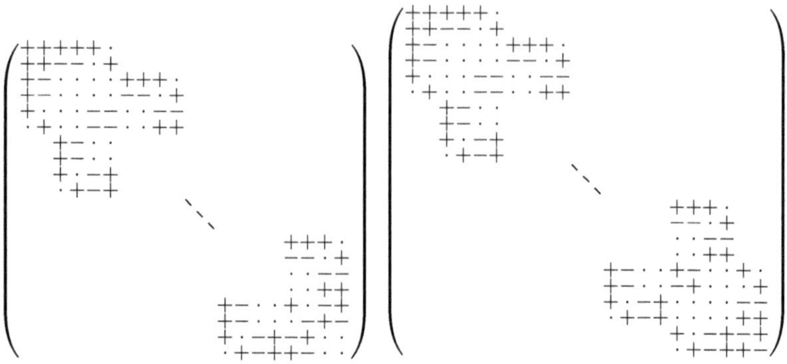

$E(4t+2,5)$ $F(4t+4,5)$

In the schematic representations we write $+$ *for* 1, $-$ *for* -1 *and dot for* 0. *Every unspecified entry is* 0.

The graph notation used in the preceding theorem is not uniform, as it is adopted from a broader context that is not fully elaborated here.

It is evident that every weighing matrix W of weight 5 gives rise to a bipartite signed graph $\dot{W}$ whose eigenvalues are $\sqrt{5}$ and $-\sqrt{5}$. More precisely, the adjacency matrix of $\dot{W}$ is a 2×2 block matrix whose diagonal blocks are zero matrices and whose off-diagonal blocks are W and $W^{\intercal}$, respectively. Consequently, the central challenge in classifying signed graphs with this spectrum lies in addressing the non-bipartite case. The corresponding result is transferred from [226].

Theorem 6.6 *A connected non-complete signed graph with degree* 5 *has two eigenvalues if and only if it is switching isomorphic to*

(i) one of the bipartite signed graphs $\dot{W}(6,5)$, $\dot{W}(8,5)$, $\dot{W}_{12,5}$, $\dot{W}_{14,5}$, $\dot{D}(16,5)$, $\dot{E}(4t+2,5)$ *or* $\dot{F}(4t+4,5)$, *or*

(ii) one of the non-bipartite signed graphs illustrated in Figure 6.1 *or a non-bipartite signed graph* $\dot{G}$ *such that either* $\mathrm{bd}(\dot{G}) \cong \dot{E}(4t+2,5)$ *or* $\mathrm{bd}(\dot{G}) \cong \dot{F}(4t+4,5)$,

where, wherever it occurs, the parameter t *satisfies* $t \geq 2$.

The complete signed graphs with the required spectral property are homogeneous of order 6. Then $\mathrm{bd}(\dot{W}_6) \cong \dot{W}(6,5)$, $\mathrm{bd}(\dot{W}_8) \cong \dot{W}(8,5)$ and $\mathrm{bd}(\dot{W}_{16}) \cong \dot{D}(16,5)$, that is, $\dot{W}_6$ is extracted from its bipartite double $\dot{W}(6,5)$, and similarly in the remaining cases. It has been shown in [226] that $\dot{E}(4t+2,5)$ and $\dot{F}(4t+4,5)$ feature as bipartite doubles.

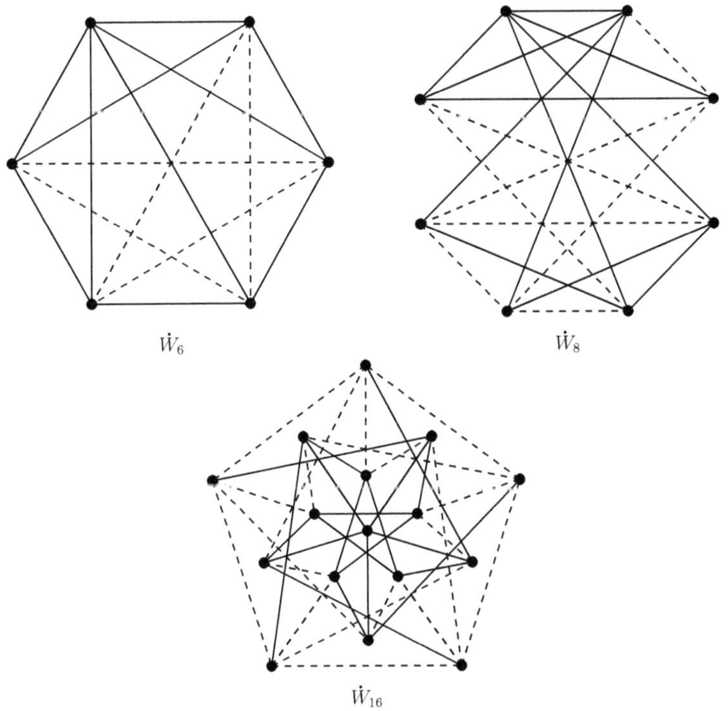

Figure 6.1 Signed graphs of Theorem 6.6(ii). Reproduced from [226], licensed under CC BY 4.0.

6.1.2 Least Eigenvalue ≥ -2

As established in the previous chapter, the least eigenvalue of a connected signed graph is greater than or equal to -2 if and only if the graph is either a signed line graph or an exceptional signed graph. We begin by considering the case of signed line graphs.

Theorem 6.7 ([217]) *A connected signed line graph $\mathscr{L}(\dot{G})$ has two eigenvalues if and only if $\dot{G}$ is switching isomorphic to*

(i) the star $K_{1,n-1}$,
(ii) a negative quadrangle or the simple signed graph obtained by inserting the (parallel) negative edge between the vertices of degree 2 in the path P_4,
(iii) the complete graph K_n or
(iv) a regular signed doubled graph with n vertices,

in all cases for $n \geq 3$.

Proof Assume that $\mathcal{L}(\dot{G})$ is connected, has m edges and possesses two eigenvalues. If $n \geq m$, then $\dot{G}$ must be one of the following:

- a signed tree, specifically the star $K_{1,n-1}$ for $n \geq 3$, since its signed line graph must be regular;
- a unicyclic signed graph, which in this context must be a negative quadrangle;
- a tree with a doubled edge, yielding the previously described simply signed graph.

These cases correspond to parts (i) and (ii) of the classification.

Now consider the case $n < m$. By Theorem 5.3, the shifted Laplacian characteristic polynomial $\Psi(x+2)$ must either have exactly one root or two distinct roots, one of which is -2.

In the first scenario, the diagonal matrix of vertex degrees D has constant diagonal entries and the adjacency matrix $A_{\dot{G}}$ is the zero matrix. This leads directly to case (iv).

In the second scenario, the Laplacian matrix (being positive semidefinite) has two eigenvalues, one of which is zero. Analyzing the minimal polynomial of the Laplacian matrix (that is, replacing A with $D - A_{\dot{G}}$ and μ with 0 in (6.1)) shows that $\dot{G}$ must be r-regular with eigenvalues of the adjacency matrix r and -1, corresponding to case (iii).

The converse implications in each case can be verified by direct computation; in particular, case (iv) follows by tracing the argument in reverse. □

Theorem 5.4 tells us that if G is r-regular with n vertices then the spectrum of $\mathcal{L}(\ddot{G})$ is

$$\left[2(r-1)^n,\ (-2)^{n(r-1)}\right]. \tag{6.2}$$

The spectrum of the remaining signed graphs obtained in Theorem 6.7 is computed easily.

In what follows, we provide several characterizations of exceptional signed graphs possessing two eigenvalues. Let m_λ and m_μ denote the multiplicity of λ and μ, respectively, and let t (resp. q) denote the difference between the number of positive and the number negative triangles (resp. quadrangles) contained.

Theorem 6.8 ([214]) *If $\dot{G}$ is an exceptional signed graph with two eigenvalues, then either $\dot{G}$ is the three-dimensional signed cube with negative quadrangles or its parameters $(n, r, \lambda, m_\lambda, \mu, m_\mu, t, q)$ have the form*

$$\left(\frac{k}{2}(\lambda+2),\ 2\lambda,\ \lambda,\ k,\ -2,\ \frac{k}{2}\lambda,\ \frac{k}{6}\lambda(\lambda^2-4),\ \frac{k}{8}\lambda(\lambda+2)(\lambda-1)(\lambda-5)\right),$$

where λ is a positive integer and either

- $k=6,\ 2\le\lambda\le 10$,
- $k=7,\ 2\le\lambda\le 16$ *and* λ *is even or*
- $k=8,\ 2\le\lambda\le 28$.

Proof Let $\mu>-2$. Then, by considering (6.1), either $\mu=-1$ or $\lambda=-\mu$. In the former case, the graph reduces to a complete signed graph, which is non-exceptional. In the latter case, the vertex degree r satisfies $r=-\lambda\mu<4$, and thus $r=3$, since any $r<3$ yields only non-exceptional solutions. Theorem 6.4 guarantees that the cube $\dot{Q}_3$ is the unique cubic signed graph with two eigenvalues; moreover, this graph is exceptional, a fact that can be verified by direct computation.

Now, assume that $\mu=-2$. Immediately, we have $r=2\lambda$. By virtue of Theorem 5.10, the multiplicity $m_\lambda=k$ satisfies $6\le k\le 8$. Using the relations $m_\lambda+m_{-2}=n$ and $\lambda m_\lambda-2m_{-2}=0$, we deduce that $n=\frac{1}{2}k(\lambda+2)$ and $m_{-2}=\frac{1}{2}k\lambda$.

Furthermore, the parameters t and q satisfy

$$\begin{aligned} m_\lambda\lambda^3-8m_{-2}&=6t,\\ m_\lambda\lambda^4+16m_{-2}&=(2r-1)rn+8q, \end{aligned}$$

where the right-hand sides represent the differences between the numbers of positive and negative closed walks of lengths 3 and 4, respectively.

Since $\dot{G}$ is an induced subgraph of $\dot{M}_k$ (defined in Subsection 5.3.2), we obtain upper bounds on λ. The proof is concluded by noting that all parameters are integers if and only if either k is even or k is odd and λ is even. □

Here is another result giving a specified decomposition of $\dot{M}_k$.

Theorem 6.9 ([214]) *Let $\dot{G}$ be a signed graph with a representation in E_k $(6\le k\le 8)$ having two eigenvalues, λ and -2, and $\frac{1}{2}k(\lambda+2)$ vertices. If n denotes the number of vertices in $\dot{M}_k$ then, unless $\dot{G}$ coincides with $\dot{M}_k$, there exists a signed graph $\dot{H}$ with eigenvalues $(2n/k)-\lambda-4$ (of multiplicity k) and -2 (of multiplicity $n-\frac{1}{2}k(\lambda+4)$), such that*

$$A_{\dot{M}_k}=\begin{pmatrix} A_{\dot{G}} & B^{\mathsf{T}}\\ B & A_{\dot{H}}\end{pmatrix}. \tag{6.3}$$

Proof Under the given assumptions, the multiplicity of the eigenvalue λ in $\dot{G}$ is k. The adjacency matrix $A_{\dot{M}_k}$ can be expressed in the block form (6.3), where $A_{\dot{H}}$ stands for the adjacency matrix of some signed graph $\dot{H}$. Using (6.1), we obtain the relation

$$A_{\dot{H}}B = -BA_{\dot{G}} + \left(\frac{2(n-k)}{k} - 2\right)B,$$

since the positive eigenvalue of $\dot{M}_k$ is $2(n-k)/k$.

Let $\mathbf{x}$ be an eigenvector of $\dot{G}$ corresponding to an eigenvalue ν (which is either λ or -2). Then

$$A_{\dot{H}}B\mathbf{x} = -BA_{\dot{G}}\mathbf{x} + \left(\frac{2(n-k)}{k} - 2\right)B\mathbf{x},$$

which yields

$$A_{\dot{H}}B\mathbf{x} = \left(-\nu + \frac{2(n-k)}{k} - 2\right)B\mathbf{x}. \tag{6.4}$$

If $B\mathbf{x} = \mathbf{0}$, then, again by (6.1),

$$A_{\dot{M}_k}\begin{pmatrix}\mathbf{x}\\ \mathbf{0}\end{pmatrix} = \begin{pmatrix}A_{\dot{G}}\mathbf{x}\\ \mathbf{0}\end{pmatrix} = \nu\begin{pmatrix}\mathbf{x}\\ \mathbf{0}\end{pmatrix}. \tag{6.5}$$

The case $\nu = \lambda$ in (6.5) cannot occur, since $\dot{M}_k$ and $\dot{G}$ do not share the same eigenvalues. Therefore, $(2n/k) - \lambda - 4$ is an eigenvalue of $\dot{H}$ with multiplicity at least k. Moreover, this multiplicity must be exactly k, as follows by interchanging the roles of $\dot{G}$ and $\dot{H}$ and considering (6.4); a larger multiplicity would imply either $\lambda = 2(n-k)/k$ or $\lambda = -2$, both of which are impossible.

Since $\dot{H}$ has $n - \frac{1}{2}k(\lambda+2)$ vertices, there must remain exactly $n - \frac{1}{2}k(\lambda+4)$ eigenvalues (counted with multiplicity). If $n - \frac{1}{2}k(\lambda+4) = 0$, then $\dot{H}$ is edgeless with k vertices, and the argument is complete; this corresponds to the degenerate case where $\dot{H}$ has only one eigenvalue. Otherwise, if $\overline{\lambda}$ denotes the average of the remaining eigenvalues, then from

$$k\left(\frac{2n}{k} - \lambda - 4\right) + \left(n - \frac{k}{2}(\lambda+4)\right)\overline{\lambda} = 0,$$

we deduce that $\overline{\lambda} = -2$, which implies that -2 is the unique remaining eigenvalue with the required multiplicity. □

We are now in position to give the following characterization of exceptional signed graphs with two eigenvalues.

Theorem 6.10 *If $\dot{G}$ is an exceptional signed graph with two eigenvalues, then either*

(i) $\dot{G}$ is the three-dimensional cube with negative quadrangles, $\dot{M}_6$, $\dot{M}_7$, $\dot{M}_8$, or

(ii) *the parameters of $\dot{G}$ are established in Theorem* 6.8 *and $\dot{G}$ is obtained from a decomposition of $\dot{M}_k$ according to the construction described in Theorem* 6.9.

The discussion concludes with an illustrative example.

Example 6.11 If G is an r-regular graph with k vertices, then the spectrum of $\mathscr{L}(\ddot{G})$ is obtained by replacing n with k in (6.2). For $k = 8$, the signed graph $\mathscr{L}(\ddot{G})$ is paired with an exceptional signed graph whose spectrum is

$$\left[2(14-r)^8,\ (-2)^{8(14-r)}\right].$$

The existence of this exceptional graph accounts for many specific cases in Theorem 6.8.

6.2 Regular with Three Eigenvalues

We start with the following consideration concerning integral eigenvalues. The first two statements do not require the regularity assumption.

Theorem 6.12 ([13]) *Let $\dot{G}$ be a signed graph whose adjacency matrix has three eigenvalues. If the multiplicities of these eigenvalues are not all equal, then one of the following holds:*

- *exactly one eigenvalue is integral and the remaining two are irrational algebraic conjugates with equal multiplicities;*
- *all three eigenvalues are integers.*

Proof Let the eigenvalues be denoted by λ, μ and ν, and assume that their multiplicities satisfy $m_\lambda \leq m_\mu \leq m_\nu$. Consider the polynomial

$$\frac{\Phi(x)}{\phi(x)^{m_\lambda}} = (x-\mu)^{m_\mu - m_\lambda}(x-\nu)^{m_\nu - m_\lambda}. \tag{6.6}$$

This polynomial has integral coefficients, as follows by polynomial long division, since both $\Phi(x)$ and $\phi(x)$ have integral coefficients and $\phi(x)$ is monic.

Suppose $m_\lambda < m_\mu$. By comparing the coefficients of the polynomial (6.6) with those of the binomials it factors into, we immediately deduce the following: if $m_\mu < m_\nu$, then $\mu, \nu \in \mathbb{Z}$, whereas if $m_\mu = m_\nu$, then either $\mu, \nu \in \mathbb{Z}$ or they are irrational algebraic conjugates. In all cases, the quadratic polynomial $(x-\mu)(x-\nu)$ has integral coefficients. Consequently, the quotient polynomial $\phi(x)/\big((x-\mu)(x-\nu)\big)$ also has integral coefficients, which implies that λ is integral.

If instead $m_\lambda = m_\mu$, then (6.6) reduces to $(x-\nu)^{m_\nu - m_\lambda}$, which, having integral coefficients, forces ν to be integral. Examining $\phi(x)/(x-\nu)$, it follows that λ and μ are either integral or irrational algebraic conjugates. □

We can say more under the assumption that one eigenvalue is simple.

Lemma 6.13 ([171]) *If $\dot{G}$ is connected with three eigenvalues, one of which is simple and non-integral, then $\dot{G}$ switches to a complete bipartite graph.*

Proof Denote by ν a simple non-integral eigenvalue and by μ, λ the remaining two eigenvalues. Without loss of generality, we may assume that $\nu \geq 0$, since otherwise we may consider the signed graph $-\dot{G}$. At least one of the remaining eigenvalues, say λ, is also simple and non-integral. If μ is simple, then $\dot{G} \cong K_{1,2}$; hence, we assume that μ is not simple. In this setting, we have

$$\nu, \lambda = \frac{1}{2}\left(a \pm \sqrt{b}\right),$$

where $a, b \in \mathbb{Z}$ and $b \neq 0$.

If $a = 0$, then $\mu = 0$ and, by Lemma 6.1, $\dot{G}$ is switching equivalent to a complete multipartite graph. Moreover, since λ is simple, it must be switching equivalent to a complete bipartite graph.

In what follows, we eliminate the possibility that $a \neq 0$. It suffices to consider $a > 0$, since the case $a < 0$ is analogous. For $a > 0$, using the fact that $\mathrm{trace}(A_{\dot{G}}) = 0$, we have $a = (2-n)\mu$, where n is the order of $\dot{G}$. It follows that μ is a negative integer.

Consider first $\mu = -1$. Then $\lambda = \frac{1}{2}(n-2-\sqrt{b})$. If $\lambda < -1$, it follows that $n < \sqrt{b}$, implying $\nu > n-1$, a contradiction. If $\lambda > -1$, then $\dot{G}$ must be complete; otherwise, it would contain a path on three vertices as an induced subgraph, which is impossible since the least eigenvalue of $\dot{G}$ would then be at most $-\sqrt{2}$. On the other hand, since μ has multiplicity $n-2$, $\dot{G}$ contains an induced subgraph on two vertices whose spectrum avoids -1, necessarily $2K_1$, which is a contradiction.

For $\mu = -2$, from the inequality (4.3) we deduce that $\dot{G}$ has at most four vertices, and this case can be resolved by direct inspection. Finally, for $\mu \leq -3$, $a \geq 3(n-2)$, which implies $\nu > n$, yielding the final contradiction. □

It is not difficult to see that if $\dot{G}$ is non-integral with spectrum $[\nu, \mu^m, \lambda^l]$, where $1+m+l=n$, but not switching equivalent to a complete bipartite graph, then $m=l$ and one of the following holds:

- $\nu = 0$, $\mu, \lambda = \pm\sqrt{b}/2$ for $b \equiv 0 \pmod 4$;
- $\nu = \frac{1}{2}(1-n)$, $\mu, \lambda = \frac{1}{2}(1 \pm \sqrt{b})$ for $b \equiv 1 \pmod 4$;
- $\nu = \frac{1}{2}(n-1)$, $\mu, \lambda = \frac{1}{2}(-1 \pm \sqrt{b})$ for $b \equiv 1 \pmod 4$.

6.2.1 Signed Line Graphs and Cubic Signed Graphs

The following theorem provides a complete characterization of signed line graphs with three eigenvalues arising from regular signed graphs.

Theorem 6.14 *The signed line graph $\mathcal{L}(\dot{G})$ of a connected r-regular signed graph has three eigenvalues if and only if one of the following holds:*

(i) $\dot{G}$ *is a positive quadrangle, any pentagon or a negative hexagon;*
(ii) $\dot{G}$ *has two eigenvalues, but is not switching equivalent to* K_n*;*
(iii) $\dot{G}$ *is switching equivalent to a strongly regular unsigned graph.*

Proof First, if $r = 2$ then $\mathcal{L}(\dot{G})$ is also a cycle, and the result follows from Example 1.11, giving (i).

For $r \geq 3$, Theorem 1.26 yields that the vertex degree occurs in the spectrum of a connected signed graph if and only if it is balanced. Now, if $\mathcal{L}(\dot{G})$ has three eigenvalues then, by Theorem 5.4, either $\dot{G}$ has two eigenvalues where r is not one of them, or $\dot{G}$ has three eigenvalues where r is one of them.

In the former case, we immediately arrive at (ii). In the latter case, the underlying graph must be strongly regular, and we are done.

The opposite implication is verified directly. □

We proceed with vertex degree 3.

Theorem 6.15 *Every connected cubic signed graph with three eigenvalues is switching equivalent to a homogeneous signed graph.*

Proof Let the signed graph under consideration be denoted by $\dot{G}$.

Suppose first that $\dot{G}$ has no integral eigenvalues. Then, by Theorem 6.12, all eigenvalues occur with the same multiplicity, namely $n/3$. Computing the fourth spectral moment of the adjacency matrix yields

$$\lambda^4 + \mu^4 + \nu^4 = 45 + 6q_i,$$

where q_i denotes the difference between the number of positive and negative quadrangles containing the vertex i. Given the identities $\lambda + \mu + \nu = 0$ and $\lambda^2 + \mu^2 + \nu^2 = 9$, it follows that $\lambda^4 + \mu^4 + \nu^4 = 81/2$, which in turn implies $q_i = -3/4$ which is a contradiction, as q_i must be an integer.

Assume that at least one eigenvalue of $\dot{G}$, say λ, is integral. If an eigenvalue equals 3 or -3, then $\dot{G}$ is switching equivalent to G or to its negation. Thus, we may assume that the spectrum of $\dot{G}$ lies entirely within the open interval $(-3, 3)$. Since λ is integral, the sum and product of the remaining two eigenvalues, denoted by $s = \mu + \nu$ and $z = \mu\nu$, must also be integral; this follows from the integrality of the coefficients of the minimal polynomial. Moreover, the bounds $z \in [-8,\ 8]$ and $s \in [-5,\ 5]$ apply. Substituting into the quadratic equation satisfied by μ (or ν), we get $\mu^2 - s\mu + z = 0$. Since μ and ν are real and distinct, the discriminant must be positive: $s^2 - 4z > 0$.

A complete enumeration of the integer pairs (s, z) belonging to the prescribed segments and satisfying the previous inequality yields no admissible solutions. Hence, no such configuration is possible. □

6.3 Common and Computational Results

Particular examples of signed graphs with two eigenvalues and non-regular signed graphs with three eigenvalues where at least one eigenvalue is simple, are constructed in [100] and [173, 181]. Numerous constructions of signed graphs with at most three eigenvalues are known, many of which are derived from association schemes or follow from Theorems 2.1 and 2.2; see [171, 173] for details. We illustrate one such construction involving positive cones over net-regular signed graphs. For the sake of simplicity, we write $K_1\nabla\dot{H}$ instead of $K_1\nabla^+\dot{H}$ to denote such a cone.

Theorem 6.16 ([171]) *Let $\dot{H}$ be a net-regular signed graph with spectrum $[\varrho, \mu^m, \lambda^l]$ such that $m, l \geq 2$, $1+m+l = n$, ϱ is the net degree and $\mu > \lambda$.*

(i) If $\varrho \notin (\lambda, \mu)$, then $K_1\nabla\dot{H}$ has at least three eigenvalues; if it has exactly three, then either $\mu(\mu - \varrho) = n$ and $\mu > 0 > \lambda > \varrho$ or $\lambda(\lambda - \varrho) = n$ and $\varrho > \mu > 0 > \lambda$.

(ii) If $\varrho \in (\lambda, \mu)$ and $K_1\nabla\dot{H}$ has at most three eigenvalues, then $\mu(\mu - \lambda) = n$ or $\lambda(\lambda - \mu) = n$, with two eigenvalues precisely if both equalities are satisfied, equivalently $\varrho = \mu + \lambda$.

Conversely, if $\mu(\mu - \varrho) = n$ or $\lambda(\lambda - \varrho) = n$, then $K_1\nabla\dot{H}$ has two eigenvalues if $\varrho = \mu + \lambda$, and three eigenvalues otherwise.

Proof Since ϱ is the unique main eigenvalue of $\dot{H}$, by Corollary 2.3 we have

$$P_{K_1\nabla\dot{H}}(x) = (x - \mu)^m (x - \lambda)^l (x^2 - \varrho x - n).$$

(i): Suppose that $K_1\nabla\dot{H}$ has two eigenvalues. Then its characteristic polynomial is of the form

$$x^2 - \varrho x - n = (x - \mu)(x - \lambda),$$

which implies $\varrho = \mu + \lambda$. However, this contradicts the assumed location of ϱ.

Now suppose that $K_1\nabla\dot{H}$ has three eigenvalues. Then the characteristic polynomial takes one of the following forms:

$$x^2-\varrho x-n=(x-\theta)(x-\mu)\quad\text{or}\quad x^2-\varrho x-n=(x-\theta)(x-\lambda),$$

for some $\theta\notin\{\mu,\lambda\}$. We examine the first possibility in detail. The equalities $\theta+\mu=\varrho$ and $\theta\mu=-n$ imply $\mu(\mu-\varrho)=n$. It remains to verify a particular chain of inequalities, for which we consider two cases: $\varrho<\lambda$ and $\varrho>\mu$.

Assume that $\varrho<\lambda$. Then $\theta=\varrho-\mu<0$ and, since $\theta\mu=-n$, it follows that $\mu>0$. If, in addition, $\lambda\geq 0$, then $K_1\nabla\dot{H}$ has exactly one negative eigenvalue (namely θ). By Lemma 6.1, this implies that the graph is switching equivalent to the negation of a complete multipartite graph with three eigenvalues. This is only possible if the cone is a star, in which case $m=1$.

Now consider the case $\varrho>\mu$. Similarly, $\theta=\varrho-\mu>0$, so $\theta\mu=-n$ implies $\mu<0$. However, this contradicts the second part of Lemma 6.1, completing the proof.

(ii): As before, starting from the identity

$$x^2-\varrho x-n=(x-\theta)(x-\mu),$$

we deduce that $\theta=\varrho-\mu$, and consequently $\mu(\mu-\varrho)=n$. If $K_1\nabla\dot{H}$ has two eigenvalues, then θ must equal either μ or λ.

In the first case, $\theta=\mu$ implies $\varrho=2\mu$, which contradicts the assumed location of ϱ. In the second case, $\theta=\lambda$ yields $\varrho=\mu+\lambda$. The alternative form

$$x^2-\varrho x-n=(x-\theta)(x-\lambda)$$

is entirely analogous and, in the case of exactly two eigenvalues, again leads to the conclusion $\varrho=\mu+\lambda$.

Conversely, suppose that $\mu(\mu-\varrho)=n$. Then the polynomial factorizes as

$$x^2-\varrho x-n=(x-(\varrho-\mu))(x-\mu),$$

implying that $K_1\nabla\dot{H}$ has at most three eigenvalues, and two if $\varrho-\mu=\lambda$. The argument is symmetric for the condition $\lambda(\lambda-\varrho)=n$. □

Evidently, $K_1\nabla\dot{H}$ is regular if and only if $\dot{H}$ is complete. As $\dot{H}$ is net-regular, $K_1\nabla\dot{H}$ is never net-regular.

Example 6.17 Examples of signed graphs for Theorem 6.16 with $\varrho=\mu+\lambda$ can easily be found. If $\dot{H}$ is a complete signed graph of order n whose negative edges induce a strongly regular graph with parameters $(n,2f,e,f)$, then the eigenvalues of $\dot{H}$ are $\varrho=n-2f-1$ and $\mu,\lambda=f-e\pm\sqrt{(e-f)^2+4f}-1$. Using the relation between the parameters of a strongly regular graph, which

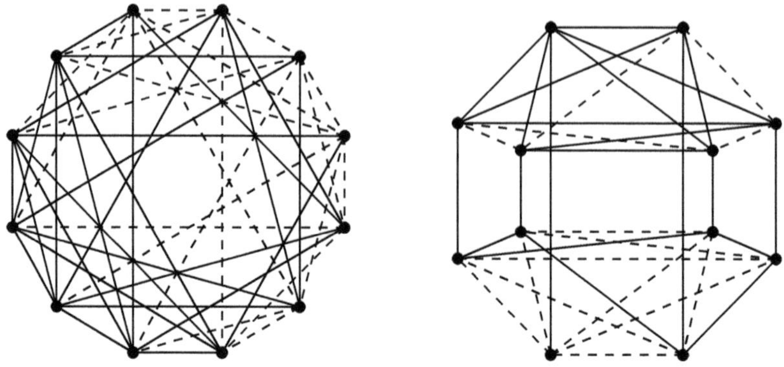

Figure 6.2 Signed graphs of Theorem 6.18. Reproduced from [226], licensed under CC BY 4.0.

can be found, for example, in [222, Theorem 3.4.6], we confirm that $\varrho = \mu + \lambda$ holds. Thus, by Theorem 6.16(ii), $K_1 \nabla \dot{H}$ has two eigenvalues.

We conclude the section with the results of a computer search for signed graphs with two eigenvalues and for regular signed graphs with three eigenvalues. These computational findings help illustrate the scope of the theoretical results presented above and provide concrete examples of the structures under study. The data also suggest patterns and constraints that may guide investigations in this area.

Theorem 6.18 ([233, 226]) *There are exactly five connected non-complete signed graphs with at most* 13 *vertices, vertex degree at least* 6 *and two eigenvalues:* $\mathscr{L}(K_5)$*, its negation, two signed graphs illustrated in Figure* 6.2 *and the negation of the first of them.*

Vertex degrees less than or equal to 5 are excluded by Theorems 6.4 and 6.6. Complete signed graphs are also excluded, as they are either homogeneous or correspond to regular two-graphs. Similar constraints will be imposed in the statement of the next theorem, allowing us to focus on non-trivial cases that illustrate the main phenomena.

The graphs $\mathscr{L}(K_5)$ and its negation represent the smallest examples with asymmetric spectra. The first signed graph depicted in Figure 6.2 is $\mathscr{L}(\ddot{K}_4)$, whose spectrum is given by $[4^4, (-2)^8]$. The second graph is switching equivalent to its negation and has spectrum $\left[\sqrt{6}^6, (-\sqrt{6})^6\right]$.

Theorem 6.19 ([13]) *Apart from those that are switching isomorphic to the homogeneous ones, there are exactly* 15 *connected non-complete regular signed graphs with at most* 10 *vertices, three eigenvalues and vertex degree at* *least* 4. *Data on* 11 *of these graphs is presented below; the remaining* 4 *are the negations of those with asymmetric spectra.*

```
· · + + + +
· · + − + −
+ + · · + +
+ − · · − +
+ + + − · ·
+ − + + · ·
```

(i) $2.73^2, -0.73^2, -2^2$

```
· · · · + + + +
· · · · + + + +
· · · · + + − −
· · · · + + − −
+ + + + · · · ·
+ + + + · · · ·
+ + − − · · · ·
+ + − − · · · ·
```

(ii) $2.83^2, 0^4, -2.83^2$

```
· · + · + + + +
· · · − + − + −
+ · · · + + + +
· − · · − + − +
+ + + − · · + ·
+ − + + · · · +
+ + + − + · · ·
+ − + + · + · ·
```

(iii) $3.83^2, -1^4, -1.83^2$

```
· · · − + · · − +
· · · + · − + · −
· · · · − + − + ·
− + · · · · + − ·
+ · − · · · − · +
· − + · · · · + −
· + − + − · · · ·
− · + − · + · · ·
+ − · · + − · · ·
```

(iv) $3^2, 0^5, -3^2$

```
· · · + − − + − −
· · · − + − − + −
· · · − − + − − +
+ − − · · · + − −
− + − · · · − + −
− − + · · · − − +
+ − − + − − · · ·
− + − − + − · · ·
− − + − − + · · ·
```

(v) $4^2, 1^2, -2^5$

```
· · · · · + + + + ·
· · · · · + + − · +
· · · · · + − · + −
· · · · · + · − − −
· · · · · · + + − −
+ + + + · · · · · ·
+ + − · + · · · · ·
+ − · − + · · · · ·
+ · + − − · · · · ·
· + − − − · · · · ·
```

(vi) $2.24^4, 0^2, -2.24^4$

```
· · · · + · + + + ·
· · · · + · − − · +
· · · · · + + · − +
· · · · · − · + − +
+ + · · · · + − · ·
· · + − · · · · − −
+ − + · + · · · · ·
+ − · + − · · · · ·
+ · − − · − · · · ·
· + + + · − · · · ·
```

(vii) $2.24^4, 0^2, -2.24^4$

```
· · · + · + · + · +
· · · · + · + · − +
· · · · · + + − + ·
+ · · · · + · + · −
· + · · · · − · + +
+ · + + · · · − · ·
· + + · − · · · − ·
+ · − + · − · · · ·
· − + · + · − · · ·
+ + · − + · · · · ·
```

(viii) $2.24^4, 0^2, -2.24^4$

```
· · + + + + + + + +
· · + − + − − + + −
+ + · · + + − − + +
+ − · · + − + − + −
+ + + + · · + + − −
+ − + − · · − + − +
+ − − + + − · · − +
+ + − − + + · · − −
+ + + + − − − − · ·
+ − + − − + + − · ·
```

(ix) $3.16^4, 0^2, -3.16^4$

```
· · + + + + + + + +
· · + + − − + + − −
+ + · · + + − + − +
+ + · · − − + − − +
+ − + − · · + + − +
+ − + − · · − − + −
+ + − + + − · · − −
+ + + − + − · · + +
+ − − − − + − + · ·
+ − + + + − − + · ·
```

(x) $4^2, 1.24^4, -3.24^4$

```
· · − − − − + + + +
· · − − − − + + + +
− − · · + + − − + +
− − · · + + − − + +
− − + + · · + + − −
− − + + · · + + − −
+ + − − + + · · − −
+ + − − + + · · − −
+ + + + − − − − · ·
+ + + + − − − − · ·
```

(xi) $4.47^2, 0^6, -4.47^2$

We have adopted the $(+, \cdot, -)$ notation to denote $(1, 0, -1)$-matrices, following the convention established in Theorem 6.5. The associated spectra have also been recorded.

6.4 Two or Three Net Laplacian Eigenvalues

Following on from the previous sections, we present characterizations of signed graphs with two net Laplacian eigenvalues, one of which is simple, as well as characterizations of regular signed graphs with three net Laplacian eigenvalues, where 0 is a simple eigenvalue. These results are drawn from [16].

If $\dot{G}$ is net-regular with net degree ϱ, then $D^{\pm}_{\dot{G}} = \varrho I$, which implies that ν is an eigenvalue of the net Laplacian $N_{\dot{G}} = D^{\pm}_{\dot{G}} - A_{\dot{G}}$ if and only if $\varrho - \nu$ is an eigenvalue of the adjacency matrix $A_{\dot{G}}$. Consequently, for net-regular signed graphs, the net Laplacian spectrum is determined directly from the spectrum of the adjacency matrix, and the results of the preceding sections apply accordingly.

A more interesting case arises when we drop the net regularity condition but retain the regularity of the underlying signed graph.

Suppose that $\dot{G}$ is a regular signed graph of degree r with two net Laplacian eigenvalues, 0 and ν. From the minimal polynomial, it follows that $N_{\dot{G}}(N_{\dot{G}} - \nu I) = O$, so $N_{\dot{G}}^2 - \nu N_{\dot{G}} = O$. To analyze this relation, we consider the diagonal entries of the left-hand side. Expanding $N_{\dot{G}} = D_{\dot{G}}^{\pm} - A_{\dot{G}}$, we obtain

$$N_{\dot{G}}^2 = (D_{\dot{G}}^{\pm})^2 - D_{\dot{G}}^{\pm}A_{\dot{G}} - A_{\dot{G}}D_{\dot{G}}^{\pm} + A_{\dot{G}}^2.$$

Since $D_{\dot{G}}^{\pm}$ is diagonal, the matrices $D_{\dot{G}}^{\pm}A_{\dot{G}}$ and $A_{\dot{G}}D_{\dot{G}}^{\pm}$ have zero diagonal entries. Moreover, because $\dot{G}$ is r-regular, the diagonal entries of $A_{\dot{G}}^2$ are all equal to r.

Therefore, for each vertex i of $\dot{G}$, its net degree $d_i^{\pm}$ satisfies the quadratic $x^2 - \nu x + r = 0$. This leads to the following result.

Theorem 6.20 *Let $\dot{G}$ be a regular signed graph with degree r. If $\dot{G}$ has two net Laplacian eigenvalues, then net degrees in $\dot{G}$ take two values which are solutions of $x^2 - \nu x + r = 0$, where ν is a non-zero net Laplacian eigenvalue.*

Examples of connected signed graphs for which 0 is not a simple net Laplacian eigenvalue can be found at the end of this section. Here we focus on two eigenvalues such that 0 has either minimum or maximum multiplicity.

Theorem 6.21 *Let $\dot{G}$ be a signed graph with two net Laplacian eigenvalues, 0 and ν. The following statements hold:*

(i) If 0 is simple, then $\dot{G}$ is complete and homogeneous.
(ii) If ν is simple, then $\dot{G}$ is the disjoint union of a complete signed graph $\dot{K}$ of even order and a set of isolated vertices, where the vertices of $\dot{K}$ are partitioned into two cells of equal size in such a way that the edges joining vertices within the each cell have a fixed sign, while the edges joining vertices belonging to different cells have the opposite sign.

Proof (i): If the eigenvalue ν has multiplicity $n-1$, then the matrix $N_{\dot{G}} - \nu I$ has rank one and a single non-zero eigenvalue, namely $-\nu$, with associated eigenvector $\mathbf{j}$. Therefore, $N_{\dot{G}} - \nu I = -(\nu/n)J$. Comparing the off-diagonal entries on both sides, we deduce that ν must be either n or $-n$.

In the case $\nu = n$, we obtain

$$N_{\dot{G}} = \nu I - \frac{\nu}{n}J = nI - J = (n-1)I - (J - I).$$

Similarly, for $\nu = -n$, we have

$$N_{\dot{G}} = -nI + J = (1-n)I - (I-J).$$

In both instances, the desired conclusion follows immediately.

(ii): Suppose that ν is simple, and let $\mathbf{x}$ be a corresponding unit eigenvector. Then the net Laplacian admits the spectral decomposition $N_{\dot{G}} = \nu\mathbf{x}\mathbf{x}^\intercal$. Note that if any entry of $\mathbf{x}$ is zero, the corresponding vertex is isolated in $\dot{G}$. Hence, we may assume that $\dot{G}$ contains no isolated vertices.

For vertices i and j with corresponding entries x_i and x_j in $\mathbf{x}$, we have $\nu x_i x_j = \pm 1$. In particular, this implies that $\nu x_i^2 = 1$. Since $\nu x_i^2 = d_i^{\pm}$ (the net degree of vertex i) and similarly $\nu x_j^2 = d_j^{\pm}$, the relation $\nu^2 x_i^2 x_j^2 = 1$ implies $d_i^{\pm} d_j^{\pm} = 1$. Therefore, $d_i^{\pm} = d_j^{\pm} = \pm 1$ for all vertices i, j, and consequently $\dot{G}$ is net-regular with net degree in $\{1, -1\}$.

From $\nu x_i^2 = 1$ for all i, the trace of $N_{\dot{G}}$ equals either n or $-n$, so $\nu = n$ or $\nu = -n$, and hence $x_i = \pm 1/\sqrt{n}$. Since $\mathbf{x}$ is orthogonal to $\mathbf{j}$, it follows that n is even; say $n = 2k$.

In this case, the adjacency matrix $A_{\dot{G}}$ takes one of the forms

$$\begin{pmatrix} J_k - I_k & -J_k \\ -J_k & J_k - I_k \end{pmatrix} \text{ or } \begin{pmatrix} I_k - J_k & J_k \\ J_k & I_k - J_k \end{pmatrix},$$

which yields the desired classification. □

Here is a specific case with three net Laplacian eigenvalues.

Theorem 6.22 *Let $\dot{G}$ be a regular signed graph with order n, degree r and three net Laplacian eigenvalues: the simple one 0, ν and μ. Then n divides $\nu\mu$ and the net degrees are solutions of $x^2 - (\nu+\mu)x + r + \nu\mu - \nu\mu/n = 0$.*

Proof The matrix $(N_{\dot{G}} - \nu I)(N_{\dot{G}} - \mu I)$ is of rank one with a single non-zero eigenvalue $\nu\mu$ and corresponding eigenvector $\mathbf{j}$. Consequently, we have

$$(N_{\dot{G}} - \nu I)(N_{\dot{G}} - \mu I) = \frac{\nu\mu}{n}J.$$

By examining the minimal polynomial, it follows that $\nu + \mu$ is an integer. This integrality ensures that the off-diagonal entries of the left-hand side are integers, which further implies that $\nu\mu$ is divisible by n.

Considering the diagonal entries yields the equation satisfied by the net degree of every vertex, as stated in the theorem. □

We conclude this section by presenting data obtained from a computer search of connected signed graphs on at most eight vertices possessing two eigenvalues of the net Laplacian matrix. One eigenvalue is zero, while the other is an integer whose absolute value does not exceed the number of vertices.

For each n, the complete unsigned graph K_n exhibits net Laplacian spectrum $[n^{n-1}, 0]$. When n is even, the signed graph $K_{n/2}\nabla^{+}(-K_{n/2})$ has $[n^{n/2}, 0^{n/2}]$. Below, we list the net Laplacian matrices and corresponding spectra of the remaining eight connected signed graphs with non-negative eigenvalues discovered by the search. In the matrices, the symbol $\times$ appears on the main diagonal; its precise value can be determined from the remaining entries in the corresponding row or column.

```
× − − +
− × + −
− + × −
+ − − ×
```
(i) $4,0^3$

```
× · − · − −
· × · − − +
− · × · − −
· − · × + −
− − − + × ·
− + − − · ×
```
(ii) $4^3,0^3$

```
× · − · − +
· × · + − −
− · × · + −
· + · × − −
− − + − × ·
+ − − − · ×
```
(iii) $4^2,0^4$

```
× − − + − +
− × + − + −
− + × − + −
+ − − × − +
− + + − × −
+ − − + − ×
```
(iv) $6,0^5$

```
× · · − · · − +
· × · · − + − −
· · × · + − − −
− · · × · · + −
· − + · × − · ·
· + − · − × · ·
− − − + · · × ·
+ − − − · · · ×
```
(v) $4^4,0^4$

```
× · · − · · − −
· × · · − − − +
· · × · − − + −
− · · × · · − −
· − − · × + · ·
· − − · + × · ·
− − + − · · × ·
− + − − · · · ×
```
(vi) $4^4,0^4$

```
× · · − · · − −
· × · · − − − +
· · × · − − + −
− · · × · · − −
· − − · × − · ·
· − − · − × · ·
− − + − · · × ·
− + − − · · · ×
```
(vii) $4^5,0^3$

```
× − − + − + + −
− × + − + − − +
− + × − + − − +
+ − − × − + + −
− + + − × − − +
+ − − + − × + −
+ − − + − + × −
− + + − + − − ×
```
(viii) $8,0^7$

Problems

6.1 ([92]) Prove that the signed cube with negative quadrangles $\dot{Q}_3$ is the unique connected bipartite cubic signed graph with two eigenvalues.

6.2 Prove Theorem 6.5.

6.3 ([160]) Determine all signed graphs that simultaneously appear as star complements for both 2 and -2 in signed graphs with spectrum $[2^{n/2}, (-2)^{n/2}]$.

6.4 By setting $(k,\lambda) = (7,14)$ and $(k,\lambda) = (8,26)$ in Theorem 6.8, show that the graph $\dot{G}$ is paired (in the sense of Theorem 6.9) with an edgeless graph on seven and eight vertices, respectively. Conclude that $\dot{G}$ is exceptional.

6.5 ([214]) In light of Theorem 6.8, show that there exists no exceptional signed graph for $k = 6$ with $\lambda \notin \{4, 5, 10\}$, nor for $(k,\lambda) = (8,27)$.

6.6 Characterize all exceptional signed graphs with two eigenvalues. (Difficult task.)

6.7 ([13]) Prove that every regular signed graph with at most three eigenvalues is walk-regular.

6.8 ([226]) Give an example demonstrating that $\dot{E}(4t+2,5)$ and $\dot{F}(4t+4,5)$ (as defined in Theorem 6.6) are bipartite doubles.

6.9 ([13]) Let a signed graph have three eigenvalues λ, μ and ν. Prove that their multiplicities are given by

$$m_\lambda = \frac{(r+\mu\nu)n}{(\lambda-\mu)(\lambda-\nu)}, \quad m_\mu = \frac{(r+\lambda\nu)n}{(\mu-\lambda)(\mu-\nu)}, \quad m_\nu = \frac{(r+\lambda\mu)n}{(\nu-\lambda)(\nu-\mu)}.$$

6.10 ([173]) Let $\dot{G}$ and $\dot{H}$ be connected net-regular signed graphs with spectra $[\rho^r, \mu^m]$ and $[\nu^\nu, \lambda^l]$, respectively. Assuming that ρ and ν are the main eigenvalues, determine whether the positive join $\dot{G}\nabla\dot{H}$ has two or three eigenvalues.

6.11 ([171]) Let $\dot{G}$ be a connected signed graph with spectrum $[\rho, \mu^m, \lambda^l]$, and let $\dot{H} = \dot{G} - v$. Let $\mathbf{r}$ be the characteristic $(1,0,-1)$-vector that determines the neighbourhood of v in $\dot{H}$. Show that if $\dot{H}$ has three eigenvalues, then $\mathbf{r}$ is an eigenvector of $\dot{H}$ associated with an eigenvalue distinct from μ and λ. Compute the spectrum of $\dot{H}$.

6.12 Give a characterization of signed graphs with two net Laplacian eigenvalues.

6.13 ([107]) Determine all signed graphs in which all but at most two eigenvalues lie in $\{1,-1\}$. (Prove and use the fact that two vertices sharing the same set of positive and negative neighbours contribute the eigenvalue 1 (resp. -1) if joined by a negative (resp. positive) edge.)

6.14 ([249]) Determine all signed graphs with all but at most three eigenvalues equal to -1.

6.15 Determine all signed graphs of order 14 with two eigenvalues, and those of order 11 or 12 with three eigenvalues. Similarly, identify all signed graphs on nine vertices having two net Laplacian eigenvalues.

7

Integrality and Cospectrality

We adapt two concepts from the theory of ordinary graphs to the setting of signed graphs. A signed graph is called *integral* if its spectrum consists entirely of integers. Two signed graphs are said to be *cospectral* if they are switching non-isomorphic yet share the same spectrum. In both cases, the term 'spectrum' refers to the multiset of the eigenvalues of a prescribed matrix associated with the signed graph.

7.1 Integral Signed Graphs

Integrality is a rare phenomenon in the sense that the proportion of integral signed graphs of fixed (and sufficiently large) order is comparatively small. Nevertheless, constructing integral signed graphs is not difficult. Many iterative constructions rely on NEPS products, as discussed in Subsection 2.3.2. Association schemes and related matrix transformations also provide promising tools for this purpose.

In the preceding chapters, we encountered numerous examples of integral signed graphs, most of which possess a very limited number of eigenvalues. Informally, a smaller number of eigenvalues often facilitates the emergence of integral signed graphs.

7.1.1 Small Vertex Degree or Small Size

Here we consider integral or Laplacian integral signed graphs with vertex degree at most 3 and k-cyclic Laplacian integral signed graphs with $k \leq 3$.

We first determine all cubic net-regular signed graphs with the desired spectral property.

Theorem 7.1 ([210]) *Let $\dot{G}$ be a connected cubic net-regular signed graph with a non-negative net degree ϱ. The following hold:*

- *If at least one of the numbers 3 or -3 is its eigenvalue, then $\dot{G}$ is determined in the following way:*
  - *(i) If 3 is an eigenvalue of $\dot{G}$, then*
    - *(a) for $\varrho = 3$, $\dot{G}$ is one of the 13 cubic underlying graphs G_1–G_{13} of signed graphs of Figure 7.1;*
    - *(b) for $\varrho = 1$, $\dot{G}$ switches to G_2, G_4, G_5, G_7, G_{10} or G_{12}.*
  - *(ii) If 3 is not an eigenvalue of $\dot{G}$, then*
    - *(a) the case $\varrho = 3$ cannot occur;*
    - *(b) for $\varrho = 1$, $\dot{G}$ switches to G_9 or G_{13}.*
- *If $\dot{G}$ avoids ± 3 in the spectrum, then it is the bipartite signed graph with six vertices whose negative edges form a perfect matching.*

Proof Suppose that at least one of ± 3 belongs to the spectrum of $\dot{G}$.

(i)(a): It is established in [185] that there exist exactly 13 connected integral cubic graphs. These graphs serve as the underlying graphs of the signed graphs depicted in Figure 7.1. By Theorem 1.26, the signed graph $\dot{G}$ is balanced, implying that it is isomorphic to one of these graphs.

(i)(b): Here, $\dot{G}$ is obtained by selecting a perfect matching in one of the graphs G_1–G_{13} and reversing the sign of all edges in that matching. Moreover, this sign reversal must yield a graph that is switching equivalent to the original. This condition implies that the vertex set can be partitioned into two subsets such that an edge is negative if and only if it connects vertices belonging to different subsets, as characterized in Theorem 1.3.

By examining the 13 graphs, we conclude that such a construction is possible precisely for the six graphs listed in the statement.

(ii)(a): This case follows directly.

(ii)(b): Here, $\dot{G}$ is non-bipartite and 3 is an eigenvalue of $-\dot{G}$. Considering G_9–G_{13} (the remaining ones are bipartite) as candidates for a graph which is switching equivalent to $-\dot{G}$, we arrive at exactly two solutions: G_9 and G_{13}. This concludes this part of the proof.

Suppose now that $\dot{G}$ avoids ± 3 in the spectrum, and assume that it is bipartite with $2n$ vertices and spectrum $\left[2^{m_2}, 1^{m1}, 0^{2m_0}, (-1)^{m_1}, (-2)^{m_2}\right]$.

By counting the difference between the numbers of positive and negative closed walks and considering the spectral moments, we obtain the system of equations

$$m_2 + m_1 + m_0 = n, \qquad 4m_2 + m_1 = 3n, \qquad 16m_2 + m_1 = 15n + 4q,$$

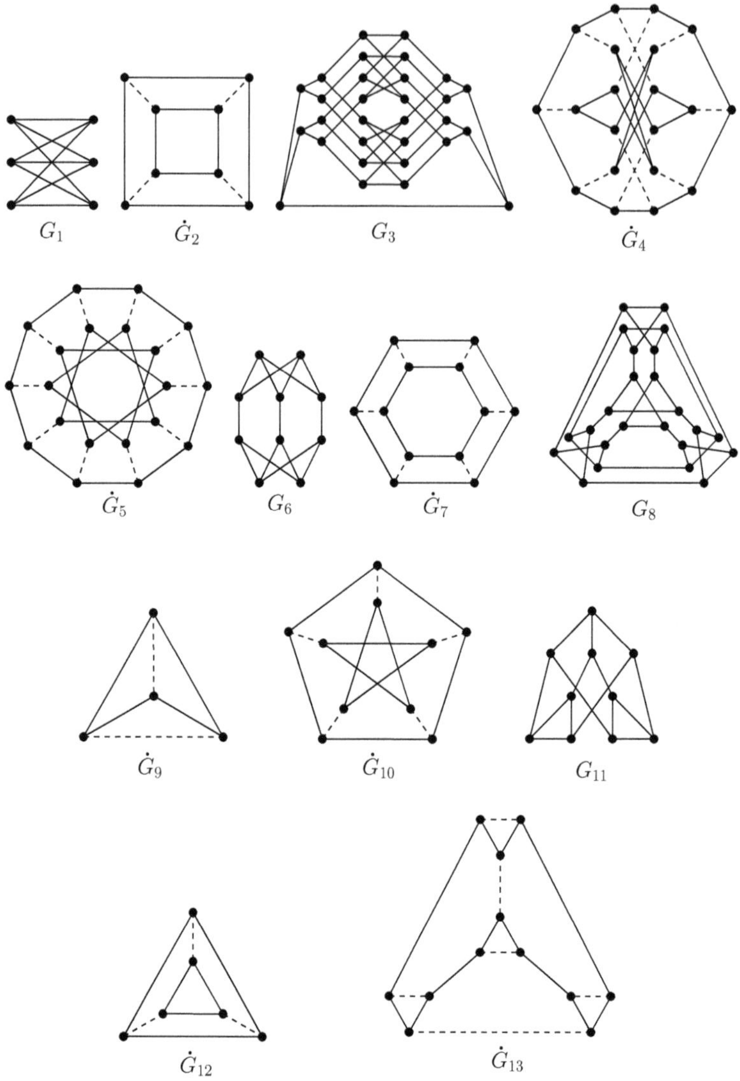

Figure 7.1 Cubic signed graphs for Theorem 7.1. Reproduced from [210], licensed under CC BY 4.0.

where q denotes the difference between the numbers of positive and negative quadrangles contained in the graph. Solving this system yields $m_1 = -(n + \frac{4}{3}q)$, along with certain parametrizations for the other multiplicities that are not essential here.

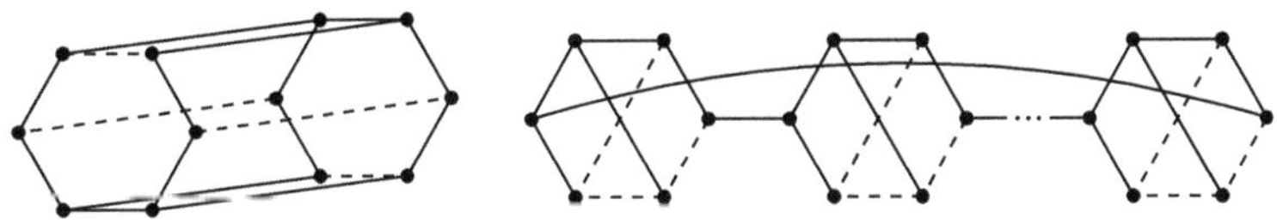

Figure 7.2 Cubic signed graphs for Theorem 7.2.

The last equality implies that q must be negative, which in turn means that $\dot{G}$ contains at least one negative quadrangle. If every vertex lies within distance 1 of such a quadrangle, then $\dot{G}$ has at most eight vertices. Otherwise, if there exists a vertex at distance 2 from a fixed negative quadrangle, then the vertex on the path between them is adjacent to exactly one vertex of the quadrangle. Otherwise, the largest eigenvalue of that induced subgraph would exceed 2 and, by the interlacing theorem, the same would hold for $\dot{G}$.

Hence, $\dot{G}$ contains a negative quadrangle with a pendant path of length 2. By analyzing the possible neighbourhoods of vertices in such a subgraph and taking into account the conditions of 3-regularity and net regularity, we conclude that the largest eigenvalue of $\dot{G}$ must be greater than 2.

Therefore, it suffices to consider connected signed graphs on at most eight vertices. This enumeration can be carried out straightforwardly, yielding the solution stated.

If $\dot{G}$ is non-bipartite, then the bipartite solution arises as its bipartite double. It can be easily verified that such a situation is impossible. □

It is worth noting that additional net-regular signed graphs can be obtained by assigning appropriate signatures to the underlying graphs. For instance, in the case of G_1, this can be achieved by reversing the signs of the edges belonging to a perfect matching. However, these net-regular signed graphs are not integral.

If the assumption of net regularity is dropped, then the following result holds.

Theorem 7.2 ([247]) *If $\dot{G}$ is a connected integral cubic, but not net-regular, signed graph, then $\dot{G}$ switches to one of signed graphs of Figure 7.2.*

This result is established in the context of Laplacian eigenvalues. The proof hinges on two key observations regarding an unbalanced signed graph with integral Laplacian spectrum and maximum vertex degree at most three: (i) the Laplacian spectrum does not contain the eigenvalues 0 or 6; (ii) the matrix $L_{\dot{G}} - I$ is either positive semidefinite (if 1 is an eigenvalue) or positive def-

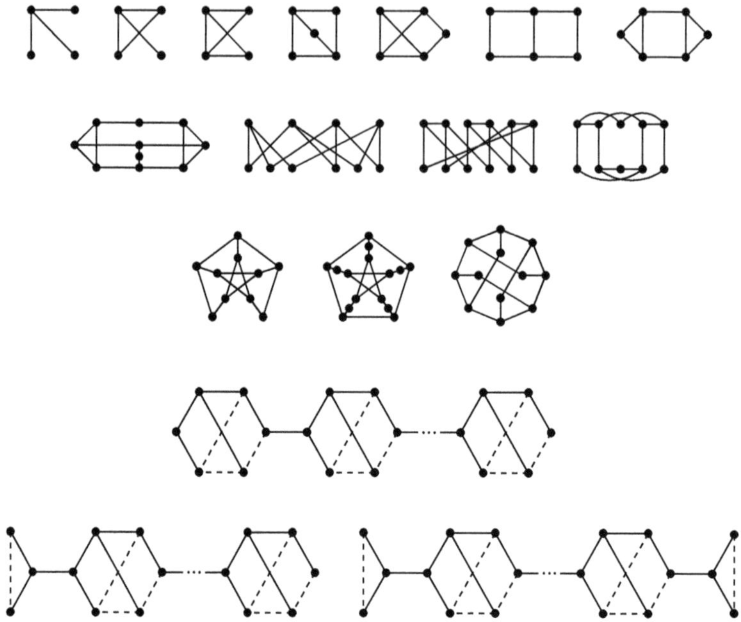

Figure 7.3 Subcubic signed graphs for Theorem 7.3.

inite (otherwise). Statement (i) follows from the graph's unbalanced nature, while (ii) is derived by analyzing the corresponding matrix.

Theorem 7.2 is extended to connected *subcubic* signed graphs, where the term 'subcubic' refers to non-cubic signed graphs whose vertex degrees do not exceed 3. This extension builds upon a related result in the unsigned graph domain, which is formulated in the following theorem.

Theorem 7.3 ([133, 247]) *If $\dot{G}$ is a connected Laplacian integral subcubic signed graph, then $\dot{G}$ is one of the signed graphs of Figure* 7.3.

Ordinary graphs of Figure 7.3 are determined in [133], whereas the remaining ones (three infinite families and an individual signed graph) are reported in [247]. The method is essentially the same, and relies on facts (i) and (ii) above.

Gathering the previous results, we arrive at the following.

Theorem 7.4 ([247]) *If $\dot{G}$ is a connected Laplacian integral signed graph with vertex degree at most* 3*, then $\dot{G}$ switches to one of (i) signed graphs of Figures* 7.1 *and* 7.2*, (ii) signed graphs of Figure* 7.3 *or (iii) negations of the non-bipartite signed graphs listed above.*

We remark that all connected integral subcubic signed graphs (with respect to the adjacency matrix) have been classified in [246].

It is worth noting that ordinary quartic integral graphs remain largely unknown; however, there exist numerous partial results on this topic [238]. Some particular results pertaining to signed graphs are provided in [210].

Finally, Laplacian integral k-cyclic signed graphs for $k \leq 3$ are characterized in [254]. In the case of trees, the unique solutions are the n-vertex stars, as follows from the well-known classification of Laplacian integral trees [145]. For unicyclic graphs, the only unbalanced solution is the negation of C_3, whereas the balanced solutions, as shown in [145], are C_3, C_4, C_6 and the graph obtained by adding an edge to an n-vertex star. The remaining cases – bicyclic and tricyclic – include analogous results for ordinary graphs, as well as several infinite families and individual unbalanced signed graphs.

7.1.2 Net Laplacian Spectrum of ∇-Cographs

At the end of Section 2.1, we noted that an unsigned cograph is defined as a P_4-free graph, while a threshold graph is considered as a cograph that is additionally $\{C_4, 2K_2\}$-free. Within the framework of signed graphs, these notions are extended by defining a *signed cograph* (resp. *signed threshold graph*) as a signed graph whose underlying graph is a cograph (resp. threshold graph).

In this subsection, we employ Theorem 1.20 to demonstrate that the net Laplacian spectrum of certain signed cographs, and consequently of certain signed threshold graphs, is integral. Recall that the positive and negative join operations are defined before the same theorem.

Following [157], an unsigned cograph is defined iteratively as follows. The single-vertex graph K_1 is a cograph; if G_1 and G_2 are cographs, then their disjoint union $G_1 \cup G_2$ and their join $G_1 \nabla G_2$ are also cographs. Similarly, a threshold graph $G \cong G(\mathbf{b})$ is established by its binary generating sequence $\mathbf{b} = (b_1, b_2, \dots, b_n)$ via the recursive construction $G_1 \cong G_1(b_1) \cong K_1$, and, assuming that $G_{i-1}(b_1, b_2, \dots, b_{i-1})$ has been constructed, the graph $G_i(b_1, b_2, \dots, b_i)$ is formed by adding a vertex that is non-adjacent to all vertices of G_{i-1} if $b_i = 0$ or by adding a vertex adjacent to all vertices of G_{i-1} if $b_i = 1$.

We follow the previous constructions to introduce particular signed cographs. A ∇-*cograph* is a signed graph defined in the following way:

- K_1 is a ∇-cograph.
- If $\dot{G}_1$ and $\dot{G}_2$ are ∇-cographs, then $\dot{G}_1 \cup \dot{G}_2$, $\dot{G}_1 \nabla^+ \dot{G}_2$ and $G_1 \nabla^- G_2$ are ∇-cographs.

A ∇-*threshold graph* is defined on the basis of a $(1,0,-1)$-generating sequence $\mathbf{b} = (b_1, b_2, \ldots, b_n)$ as follows:

- $\dot{G}_1 \cong \dot{G}_1(b_1) \cong K_1$ is a ∇-threshold graph.
- If $\dot{G}_{i-1}(b_1, b_2, \ldots, b_{i-1})$ is constructed, then $\dot{G}_i(b_1, b_2, \ldots, b_i)$ is formed by adding a vertex non-adjacent to the vertices of $\dot{G}_{i-1}$ if $b_i = 0$, or adding a vertex joined by a positive edge to the vertices of $\dot{G}_{i-1}$ if $b_i = 1$, or adding a vertex joined by a negative edge to the vertices of $\dot{G}_{i-1}$ if $b_i = -1$.

It follows from Theorem 1.20 that every ∇-cograph and every ∇-threshold graph is net Laplacian integral. An analogous result holds for every unsigned cograph, as established by the unsigned variant of Theorem 1.20 (cf. [157]). However, this theorem does not generalize to the Laplacian spectrum of signed graphs.

7.2 Cospectral Signed Graphs

Many structural properties of a signed graph can be studied through the lens of spectral graph theory. A natural question arising in this context is whether a signed graph is uniquely *determined by its spectrum*, that is, whether it is the only signed graph (up to switching isomorphism) having the given spectrum. Otherwise, there exists at least one other signed graph with the same spectrum, in which case they are said to be mutually *cospectral* with respect to the spectrum of a prescribed matrix. It is important to emphasize that, in our approach, switching-isomorphic signed graphs are not regarded as cospectral.

In what follows, we focus on the spectrum of the adjacency matrix. The smallest pair of cospectral unsigned graphs occurs on five vertices; there exists exactly one such pair, depicted in Figure 7.4(a), along with their common spectrum. In the setting of signed graphs, there exists another cospectral pair of the same order, shown in Figure 7.4(b). In both examples, the pair consists of one connected and one disconnected graph. The smallest order for which connected cospectral mates exist is six. In fact, there are exactly 121 signed graphs on six vertices that are not determined by their spectrum. Further details are available online at [224].

It is clear that if two signed graphs are cospectral, then they must have the same order and size. Many additional shared structural invariants can be derived from Theorem 1.9 in the context of the adjacency matrix spectrum, as well as from analogous results for other prescribed matrices.

We have met some cospectral pairs. For example, $\dot{T}_{14}$ and $\dot{S}_{14}$, as well as $\dot{T}_{16}$

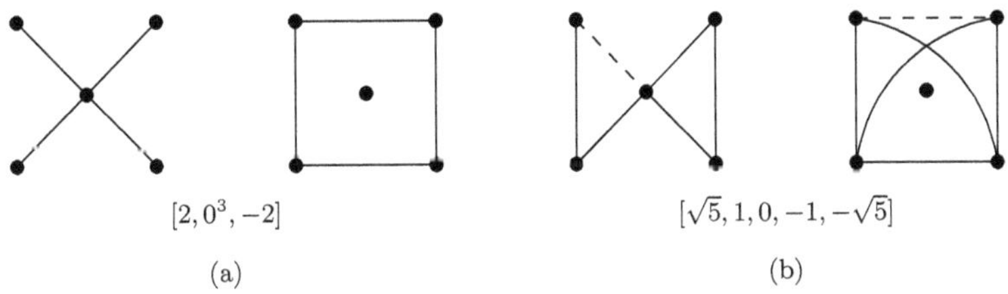

Figure 7.4 The smallest pairs of cospectral signed graphs.

and $\dot{Q}_4$ of Figure 3.4, are such pairs. It is not difficult to construct more examples on the basis of association schemes or Theorems 2.1 and 2.2.

In the following, we present two techniques – both adapted from the theory of unsigned graphs – for constructing cospectral signed graphs. Additionally, we introduce a construction method that is specific to signed graphs. The results in the next two subsections are drawn from [28], while the subsequent subsection is based on the findings of [201, 202, 216]. Finally, in Subsection 7.2.4, we examine signed graphs that are determined by the spectrum of a prescribed matrix.

7.2.1 GM-Switching

In [97], Godsil and McKay introduced a graph construction, which is nowadays called Godsil–McKay switching or *GM-switching*, allowing us to produce pairs of cospectral graphs.

Given a graph G with vertex set V, let $\pi = \{C_1, C_2, \dots, C_t, D\}$ be a partition of V, with $|C_i| = n_i$ and $|D| = d$. For a vertex $v \in D$, assume that the following hold:

- any two vertices in C_i have the same number of neighbours in C_j;
- v has either 0, $n_i/2$ or n_i neighbours in C_i, for $1 \leq i, j \leq t$.

Then the graph G^π constructed from G by local switching with respect to the partition π is obtained as follows. For every $v \in D$ and $1 \leq i \leq t$ such that v has $n_i/2$ neighbours in C_i, we delete these $n_i/2$ edges and join v instead to the remaining $n_i/2$ vertices of C_i. It turns out that the graphs G and G^π share the same spectrum. Note that they may or may not be isomorphic.

We proceed with the signed variant of the previous construction, extending the results from the setting of $(0,1)$-matrices to the set of $(1,0,-1)$-matrices. We need the following lemma.

Lemma 7.5 *For a positive integer m, let $Q_m = (2/m)J_m - I_m$. Also let* $\mathbf{x} =$

$(x_1, x_2, \ldots, x_m)^\intercal$ *be a vector with entries in* $\{1, 0, -1\}$. *The following statements hold:*

(i) *If* $\mathbf{x}$ *has sum* 0, *then* $Q_m\mathbf{x} = -\mathbf{x}$.
(ii) *If* m *is even and* $\mathbf{x}$ *has half entries* 0 *and half* 1, *then* $Q_m\mathbf{x} = \mathbf{j}_m - \mathbf{x}$.
(iii) *If* m *is even and* $\mathbf{x}$ *has half entries* 0 *and half* -1, *then* $Q_m\mathbf{x} = -\mathbf{j}_m - \mathbf{x}$.
(iv) *If* $\mathbf{x} = \mathbf{j}_m$, *then* $Q_m\mathbf{x} = \mathbf{x}$.
(v) *If* $\mathbf{x} = -\mathbf{j}_m$, *then* $Q_m\mathbf{x} = \mathbf{x}$.

Proof Parts (ii), (iii) and (v) follow by inspecting the matrix Q_m, so we focus on the remaining two parts. Note that

$$(Q_m)_{ij} = \begin{cases} (2-m)/m & \text{if } i = j, \\ 2/m & \text{if } i \neq j. \end{cases}$$

For (i), we have

$$(Q_m x)_i = \sum_{k=1}^{m} q_{ik}x_k = \frac{2}{m}(0 - x_i) + \frac{2-m}{m}x_i = -x_i.$$

Part (iv) follows since, by definition, Q_m has row sums and column sums equal to 1. □

Now let $\dot{G} = (G, \sigma)$ be a signed graph with the vertex set V. Assume that $\pi = \{C_1, C_2, \ldots, C_t, D\}$ is a partition of V_G, with $|C_i| = n_i$ and $|D| = d$; so, the same as before. For each $v \in V$, define $d_i^+(v)$ and $d_i^-(v)$ as the number of positive and negative edges from v to C_i, respectively. Define the ith net degree: $d_i^\pm(v) = d_i^+(v) - d_i^-(v)$.

Suppose the following:

- For $1 \leq i, j \leq t$, any two vertices in C_i have the same jth net degree.
- For $1 \leq i \leq t$ and $v \in D$:
  - either $d_i^\pm(v) = 0$,
  - or $d_i^+(v) = n_i/2$ and $d_i^-(v) = 0$,
  - or $d_i^-(v) = n_i/2$ and $d_i^+(v) = 0$,
  - or $d_i^+(v) = n_i$,
  - or $d_i^-(v) = n_i$.

Then the signed graph $\dot{G}^\pi$ constructed from $\dot{G}$ by local switching with respect to π is obtained as follows. For every $v \in D$ and $1 \leq i \leq t$:

- If $d_i^\pm(v) = 0$, change the sign of every edge from v to C_i.
- If $d_i^+(v) = n_i/2$ and $d_i^-(v) = 0$, delete the existing positive edges from v to C_i and add positive edges to the remaining $n_i/2$ vertices in C_i.

- If $d_i^-(v) = n_i/2$ and $d_i^+(v) = 0$, perform the same operation for negative edges.

We call this *GM-switching for signed graphs*. It is supported by the following result.

Theorem 7.6 *The signed graphs $\dot{G}$ and $\dot{G}^\pi$ share the same spectrum.*

Proof With an appropriate vertex labelling, the adjacency matrix of $\dot{G}$ has the form

$$A_{\dot{G}} = \begin{pmatrix} C_1 & C_{12} & \cdots & C_{1t} & D_1 \\ C_{12}^{\mathsf{T}} & C_2 & \cdots & C_{2t} & D_2 \\ \vdots & \vdots & \ddots & \vdots & \vdots \\ C_{1t}^{\mathsf{T}} & C_{2t}^{\mathsf{T}} & \cdots & C_t & D_t \\ D_1^{\mathsf{T}} & D_2^{\mathsf{T}} & \cdots & D_t^{\mathsf{T}} & D \end{pmatrix},$$

where:

- C_i is symmetric with constant row/column sums;
- C_{ij} is the $n_i \times n_j$ matrix with constant row and column sums;
- each column of D_i satisfies one of the five conditions of Lemma 7.5;
- D is symmetric of order d.

Define $Q = \mathrm{diag}(Q_{n_1}, Q_{n_2}, \ldots, Q_{n_t}, I_d)$, where Q_{n_i} is defined in Lemma 7.5. Then

$$Q^{-1}A_{\dot{G}}Q = \begin{pmatrix} C_1 & C_{12} & \cdots & C_{1t} & \widetilde{D}_1 \\ C_{12}^{\mathsf{T}} & C_2 & \cdots & C_{2t} & \widetilde{D}_2 \\ \vdots & \vdots & \ddots & \vdots & \vdots \\ C_{1t}^{\mathsf{T}} & C_{2t}^{\mathsf{T}} & \cdots & C_t & \widetilde{D}_t \\ \widetilde{D}_1^{\mathsf{T}} & \widetilde{D}_2^{\mathsf{T}} & \cdots & \widetilde{D}_t^{\mathsf{T}} & D \end{pmatrix},$$

where the columns of D_i transform to $\widetilde{D}_i$ by Lemma 7.5. So, $Q^{-1}A_{\dot{G}}Q = A_{\dot{G}^\pi}$, and we are done. □

7.2.2 WQH-Switching

WQH-switching is a generalization of GM-switching proposed by Wang, Qiu and Hu [252]. Observe that GM-switching conjugates the adjacency matrix A_G by a rational orthogonal matrix Q. The new construction, called *generalized GM-switching* (or *WQH-switching*), uses a different rational matrix and yields new cospectral graphs not obtainable via classical GM-switching.

Here is a brief outline. Let G be a graph with vertex set $V(G)$, and let p be an odd prime. Assume that $V(G) = V_1 \sqcup V_2 \sqcup V$, with $|V_1| = |V_2| = p$ and $|V| = d$. We write $d_i(v)$ to denote the number of edges connecting v to some vertex of V_i. For each $v \in V_1$ and $u \in V_2$, we assume the following:

- $d_1(v) - d_2(v) = d_2(u) - d_1(u) = l$.
- Each vertex in V is either:
  - adjacent to all vertices of V_1 and none in V_2,
  - or adjacent to all vertices of V_2 and none in V_1,
  - or has an equal number of neighbours in V_1 and V_2.

WQH-switching constructs a new graph G' by relocating edges in the following way. For $v \in V$ adjacent to all vertices of V_1 (resp. V_2), reconnect v to all vertices of V_2 (resp. V_1) instead. Then G and G' share the same spectrum.

In [252], the condition that p is an odd prime is used to prove some results concerning the possibility of a graph admitting a WQH-switching. However, the matrix machinery still works whenever the size of the parts V_1 and V_2 is any positive integer. In the extension to signed graphs, the common cardinality of the vertex subsets V_1 and V_2 is a positive integer m; we still write d for the cardinality of V. The ith net degree of a vertex is defined as in the previous subsection.

Suppose that the following hold:

- $d_1^{\pm}(v) - d_2^{\pm}(v) = d_2^{\pm}(u) - d_1^{\pm}(u) = l$, for every $v \in V_1$ and $u \in V_2$.
- For each $v \in V$:
  - either $d_1^{+}(v) = m$ and $d_1^{-}(v) = d_2^{+}(v) = d_2^{-}(v) = 0$ (and similarly with V_1 and V_2 exchanged),
  - or $d_1^{-}(v) = m$ and $d_1^{+}(v) = d_2^{+}(v) = d_2^{-}(v) = 0$ (and similarly with V_1 and V_2 exchanged),
  - or $d_1^{\pm}(v) = d_2^{\pm}(v)$,
  - or $d_1^{+}(v) = d_2^{-}(v) = m$ (and similarly with V_1 and V_2 exchanged).

We construct $\dot{G}'$ from $\dot{G}$ by applying the following switching rules to each $v \in V$:

- If $d_1^{+}(v) = m$ and $d_1^{-}(v) = d_2^{+}(v) = d_2^{-}(v) = 0$, delete positive edges from v to V_1 and connect v to all vertices in V_2 by positive edges (and similarly if V_1 and V_2 are exchanged).
- If $d_1^{-}(v) = m$ and $d_1^{+}(v) = d_2^{+}(v) = d_2^{-}(v) = 0$, delete negative edges from v to V_1 and connect v to all vertices in V_2 by negative edges (and similarly if V_1 and V_2 are exchanged).

- If $d_1^+(v) = d_2^-(v) = m$, switch signs of edges from v to V_1 (positive to negative) and to V_2 (negative to positive). Again, similarly if V_1 and V_2 are exchanged.

This is *WQH-switching for signed graphs*, and here is the supporting theorem.

Theorem 7.7 *For a signed graph $\dot{G}$, let $\dot{G}'$ be constructed by WQH-switching. Then $\dot{G}$ and $\dot{G}'$ share the same spectrum.*

Proof The proof is based on the fact that the adjacency matrices $A_{\dot{G}}$ and $A_{\dot{G}'}$ are conjugated by the rational orthogonal matrix

$$Q = \operatorname{diag}(U_{2m}, I_d), \quad \text{where} \quad U_{2m} = I_{2m} + \frac{1}{m}\begin{pmatrix} -J_m & J_m \\ J_m & -J_m \end{pmatrix}.$$

Let the adjacency matrix of $\dot{G}$ be written as

$$A_{\dot{G}} = \begin{pmatrix} M & C \\ C^\intercal & D \end{pmatrix}, \quad \text{where} \quad M = \begin{pmatrix} A_1 & B \\ B^\intercal & A_2 \end{pmatrix}.$$

Given the structure of $\dot{G}$, the following hold:

- A_1 and A_2 are symmetric matrices of order m.
- For each $1 \leq i, i' \leq m$,

$$\sum_{j=1}^{m} ((A_1)_{ij} - B_{ij}) = \sum_{j=1}^{m} ((A_2)_{i'j} - B_{ji'}) = l.$$

- Each column $\mathbf{c}_j$ of C is of length $2m$, partitioned into two m-length vectors $\mathbf{c}_{j,1}$ and $\mathbf{c}_{j,2}$ such that:
  - $\mathbf{c}_{j,1} = \mathbf{j}_m$, $\mathbf{c}_{j,2} = \mathbf{0}_m$, or vice versa,
  - or $\mathbf{c}_{j,1} = -\mathbf{j}_m$, $\mathbf{c}_{j,2} = \mathbf{0}_m$, or vice versa,
  - or $\sum_{h=1}^{m} \mathbf{c}_{j,1}^h = \sum_{h=1}^{m} \mathbf{c}_{j,2}^h$, where $\mathbf{c}_{j,i}^h$ is the hth entry of $\mathbf{c}_{j,i}$ for $i \in \{1,2\}$,
  - or $\mathbf{c}_{j,1} = \mathbf{j}_m$, $\mathbf{c}_{j,2} = -\mathbf{j}_m$, or vice versa.

Now,

$$QA_{\dot{G}}Q = \begin{pmatrix} U_{2m} & 0 \\ 0 & I_d \end{pmatrix}\begin{pmatrix} M & C \\ C^\intercal & D \end{pmatrix}\begin{pmatrix} U_{2m} & 0 \\ 0 & I_d \end{pmatrix} = \begin{pmatrix} U_{2m}MU_{2m} & U_{2m}C \\ C^\intercal U_{2m} & D \end{pmatrix}.$$

We claim that $U_{2m}MU_{2m} = M$. A direct computation gives

$$U_{2m}MU_{2m} = M + N,$$

where N is computed as follows:

$$N = \frac{1}{m}\begin{pmatrix} A_1 & B \\ B^\intercal & A_2 \end{pmatrix}\begin{pmatrix} -J_m & J_m \\ J_m & -J_m \end{pmatrix} + \frac{1}{m}\begin{pmatrix} -J_m & J_m \\ J_m & -J_m \end{pmatrix}\begin{pmatrix} A_1 & B \\ B^\intercal & A_2 \end{pmatrix}$$
$$+ \frac{1}{m^2}\begin{pmatrix} -J_m & J_m \\ J_m & -J_m \end{pmatrix}\begin{pmatrix} A_1 & B \\ B^\intercal & A_2 \end{pmatrix}\begin{pmatrix} -J_m & J_m \\ J_m & -J_m \end{pmatrix}.$$

By computing each block of N, it follows that N is the zero matrix. Hence, $U_{2m}MU_{2m} = M$, as claimed.

Next, consider how U_{2m} acts on columns of the form $\mathbf{c}_j = \begin{pmatrix} \mathbf{c}_{j,1} \\ \mathbf{c}_{j,2} \end{pmatrix}$. We obtain the following:

$$U_{2m}\begin{pmatrix} \mathbf{c}_{j,1} \\ \mathbf{c}_{j,2} \end{pmatrix} = \begin{pmatrix} (I_m - \frac{1}{m}J_m)\mathbf{c}_{j,1} + \frac{1}{m}J_m\mathbf{c}_{j,2} \\ \frac{1}{m}J_m\mathbf{c}_{j,1} + (I_m - \frac{1}{m}J_m)\mathbf{c}_{j,2} \end{pmatrix},$$

and so, for $1 \leq i \leq m$,

$$(U_{2m}C)^i_{j,1} = c^i_{j,1} + \frac{1}{m}\sum_{h=1}^{m}(c^h_{j,2} - c^h_{j,1}),$$
$$(U_{2m}C)^i_{j,2} = c^i_{j,2} + \frac{1}{m}\sum_{h=1}^{m}(c^h_{j,1} - c^h_{j,2}).$$

Therefore, U_{2m} either switches, or negates, or preserves the semicolumns as required by the switching rules. This concludes the proof. □

7.2.3 A Seminal Construction

As established in (6.2), the spectrum of the signed line graph $\mathscr{L}(\ddot{G})$ of a signed doubled graph $\ddot{G}$, where G is a regular graph, depends solely on the order n and degree r of G. In other words, if G and H are non-isomorphic regular graphs of the same order and degree, then $\mathscr{L}(\ddot{G})$ and $\mathscr{L}(\ddot{H})$ will have identical spectra. This simple yet powerful observation provides a construction method for generating many cospectral regular signed graphs. Notably, this technique is exclusive to the domain of signed graphs: according to our current knowledge, it has no analogue in the unsigned setting.

Furthermore, in what follows we extend this construction to signed line graphs of signed doubled graphs derived from graphs that share the same vertex degree sequence. This represents an original and, at the first glance, surprising construction developed by the author through a series of publications culminating in [201].

The following analysis determines the spectrum of $\mathscr{L}(\ddot{G})$; the corresponding eigenspaces are also computed. The argument begins with a few auxiliary lemmas that rely on the already used fact that the matrices $B_\eta^\mathsf{T} B_\eta$ and $B_\eta B_\eta^\mathsf{T}$ share the same non-zero eigenvalues. Moreover, the identities given in (5.3) remain valid when the signed graph $\dot{G}$ is replaced with the signed doubled graph $\ddot{G}$; that is,

$$B_\eta B_\eta^\mathsf{T} = L_{\ddot{G}} \quad \text{and} \quad B_\eta^\mathsf{T} B_\eta = A_{\mathscr{L}(\ddot{G})} + 2I. \tag{7.1}$$

Lemma 7.8 *Given a graph G without isolated vertices, let $\ddot{G}$ be the corresponding signed doubled graph. The following statements hold:*

(i) The Laplacian matrix of $\ddot{G}$ is positive semidefinite.
(ii) If B_η is a vertex–edge incidence matrix of $\ddot{G}$, then $\ker(B_\eta^\mathsf{T}) = \mathbf{0}$.

Proof (i): The Laplacian matrix $L_{\ddot{G}}$ has vertex degrees on the main diagonal and zeros outside this diagonal. Since G has no isolated vertices, the main diagonal has no zero entries, which gives the desired result.

(ii): If $\mathbf{x} \in \ker(B_\eta^\mathsf{T})$, then $\mathbf{x} \in \ker(B_\eta B_\eta^\mathsf{T})$. The first identity of (7.1), in conjunction with part (i), leads to $\mathbf{x} = \mathbf{0}$. □

Lemma 7.9 *Let G be a graph without isolated vertices. The dimension of the eigenspace $\mathscr{E}(-2)$ for $\mathscr{L}(\ddot{G})$ is $2m-n$, where n and m are the order and the size of G, respectively.*

Proof By Lemma 7.8(i), the matrix $B_\eta B_\eta^\mathsf{T}$ has n non-zero eigenvalues, which means that the multiplicity of zero in the $2m \times 2m$ matrix $B_\eta^\mathsf{T} B_\eta$ is $2m-n$. The desired result follows from the second identity of (7.1). □

We now compute the eigenvalues distinct from -2 and the corresponding eigenvectors.

Theorem 7.10 *Given a graph G without isolated vertices, let $\ddot{G}$ be the corresponding signed doubled graph. Suppose that the vertices $1, 2, \ldots, n$ of G are arranged in a non-increasing order according to their degrees $d_1, d_2, \ldots, d_n$. The eigenvalues distinct from -2 of $\mathscr{L}(\ddot{G})$ are $2(d_i - 1)$, for $1 \leq i \leq n$.*

An eigenvector associated with $2(d_i - 1)$ is $B_\eta^\mathsf{T} \mathbf{e}_i$, where B_η is a vertex–edge incidence matrix of $\ddot{G}$ and $\mathbf{e}_i$ is the ith vector of the canonical basis of $\mathbb{R}^n$.

Proof The Laplacian matrix $L_{\ddot{G}}$ is the diagonal matrix $\mathrm{diag}(2d_1, 2d_2, \ldots, 2d_n)$, where d_i denotes the degree of vertex i in the underlying graph. Consequently, the corresponding eigenvalues are $2d_1, 2d_2, \ldots, 2d_n$; they are associated with the standard basis vectors $\mathbf{e}_1, \mathbf{e}_2, \ldots, \mathbf{e}_n$, respectively.

Combining this with (7.1), we conclude that the eigenvalues of the signed line graph $\mathscr{L}(\ddot{G})$ are $2(d_1-1), 2(d_2-1), \ldots, 2(d_n-1)$. The proof is completed by observing that $A_{\mathscr{L}(\ddot{G})}$ shares its eigenvectors with $B_\eta^\intercal B_\eta$. □

Here is an application of the previous results – the main result in this subsection: a pivotal and distinguishing construction.

Theorem 7.11 *Let G and H be the graphs of order n that share the same vertex degree sequence $d_1, d_2, \ldots, d_n$. The signed line graphs $\mathscr{L}(\ddot{G})$ and $\mathscr{L}(\ddot{H})$ share the same spectrum.*

Proof Without loss of generality, assume that the degrees are ordered as $d_1 \geq d_2 \geq \cdots \geq d_n$, and let k denote the largest index for which $d_k > 0$. Then, by Lemma 7.9 and Theorem 7.10, the signed line graphs $\mathscr{L}(\ddot{G})$ and $\mathscr{L}(\ddot{H})$ have identical spectra, consisting of the eigenvalue -2 with multiplicity $\sum_{i=1}^{k} d_i - k$ (where this sum equals twice the size of G) and $2(d_i-1)$, for $1 \leq i \leq k$. □

Although there exist non-isomorphic signed graphs that produce switching-isomorphic signed line graphs, it is established in [221] that if G and H of Theorem 7.11 are non-isomorphic, then $\mathscr{L}(\ddot{G})$ and $\mathscr{L}(\ddot{H})$ are not switching-isomorphic, that is, cospectral. It is also worth noting that the corresponding construction primarily pertains to unsigned graphs, because once a pair of graphs with identical vertex degrees is fixed, the associated signed doubled graphs and their line graphs are uniquely determined.

Example 7.12 The smallest pair of non-isomorphic graphs that share the same vertex degrees are the graphs G and H of order 5, illustrated in Figure 7.5. The corresponding signed doubled graphs and their signed line graphs are also depicted in the same figure. This example additionally demonstrates that the vertex degrees of the corresponding signed line graphs may differ.

According to Lemma 7.9 and Theorem 7.10, the common spectrum of the signed line graphs $\mathscr{L}(\ddot{G})$ and $\mathscr{L}(\ddot{H})$ is $[4, 2^3, 0, (-2)^5]$.

Finally, we compute the eigenspace for -2; this result is not related to cospectrality, but extends the considerations of Section 5.5.

Theorem 7.13 *Given a graph G without isolated vertices, let $\ddot{G}$ be the corresponding signed doubled graph. Then $\mathscr{E}(-2) = \ker(B_\eta)$, where $\mathscr{E}(-2)$ is the eigenspace for -2 in $A_{\mathscr{L}(\ddot{G})}$ and B_η defines the vertex–edge orientation of $\ddot{G}$.*

Proof Let $\mathbf{x} \in \mathscr{E}(-2)$. From the second identity of (7.1), we have $B_\eta^\intercal B_\eta \mathbf{x} = \mathbf{0}$. By virtue of Lemma 7.8(ii), $B_\eta^\intercal$ has a trivial kernel, which implies $B_\eta \mathbf{x} = \mathbf{0}$, that is, $\mathbf{x} \in \ker(B_\eta)$.

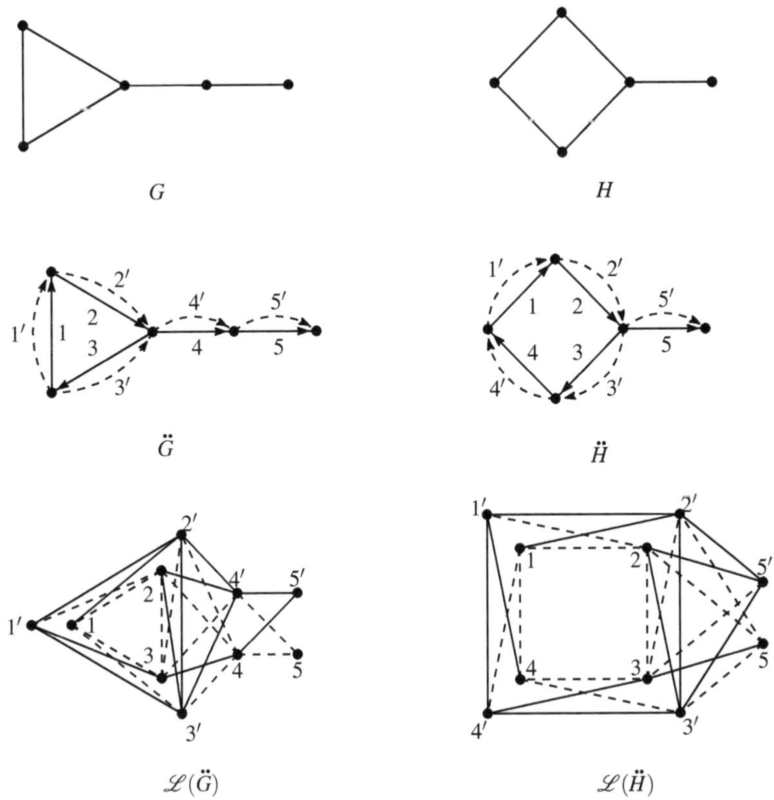

Figure 7.5 Graphs G and H sharing the same degree sequence, signed doubled graphs $\ddot{G}$ and $\ddot{H}$, and signed line graphs $\mathscr{L}(\ddot{G})$ and $\mathscr{L}(\ddot{H})$. For $\ddot{G}$ and $\ddot{H}$, an arrow indicates a vertex–edge orientation; for the sake of simplicity, only one orientation of each edge is given. Reproduced from [201], licensed under CC BY 4.0.

Supposing that $\mathbf{x} \in \ker(B_\eta)$, we compute

$$A_{\mathscr{L}(\ddot{G})}\mathbf{x} = \left(B_\eta^\intercal B_\eta - 2I\right)\mathbf{x} = B_\eta^\intercal B_\eta \mathbf{x} - 2I\mathbf{x} = \mathbf{0} - 2\mathbf{x},$$

giving $\mathbf{x} \in \mathscr{E}(-2)$. □

7.2.4 Spectral Determination

As previously mentioned, every signed graph with at most four vertices is determined by the spectrum of its adjacency matrix. The same holds for all but two pairs of signed graphs on five vertices.

Many results on spectral determination are well established for ordinary graphs, yet most of them do not extend directly to the setting of signed graphs. For instance, although signed cycles have a seemingly simple structure, it turns out that, with the exception of the negative four-vertex cycle $\dot{C}_4$, every even signed cycle has a cospectral mate and is thus not determined by its spectrum. In contrast, every odd signed cycle is determined by its spectrum, as shown in [6]. According to the same reference, all negative cycles and all positive odd cycles are determined by their Laplacian spectra, while positive even cycles possess Laplacian cospectral mates.

Other results in this area primarily focus on sparse signed graphs. In [33], it is shown that a *signed lollipop graph*, that is, a signed graph formed by attaching a hanging path to a vertex of a signed cycle, is determined by its Laplacian spectrum. On the other hand, [123] demonstrates that a *signed infinity-graph*, a signed graph obtained by identifying one vertex of a cycle with a vertex of another cycle, is not determined by the Laplacian spectrum.

In this subsection, we show that ∇-threshold graphs, as defined in Subsection 7.1.2, do not share the same net Laplacian spectrum. This observation provides a relaxed version of spectral determination: while it confirms that no two ∇-threshold graphs are net Laplacian cospectral, it does not exclude the possibility that a ∇-threshold graph shares its spectrum with a non-∇-threshold signed graph.

Theorem 7.14 ([16]) *No two ∇-threshold graphs have the same net Laplacian spectrum.*

Proof Throughout the proof, we abbreviate net Laplacian eigenvalues and net Laplacian spectrum to eigenvalues and spectrum, respectively. Suppose that $\dot{G}_1$ and $\dot{G}_2$ form the smallest pair (with respect to the order) of ∇-threshold graphs that share the same spectrum. Let n denote their common order.

First, suppose that $\dot{G}_1$ is disconnected. We have $\dot{G}_1 \cong \dot{G}_1' \cup c_1K_1$, where $\dot{G}_1'$ is a connected ∇-threshold graph of order $k \geq 2$. Without loss of generality, assume that $\dot{G}_1'$ is a positive join. By Theorem 1.20, the largest eigenvalue of $\dot{G}_1'$ is k and the least eigenvalue is greater than $-k$. Since $\dot{G}_1$ and $\dot{G}_2$ are cospectral, the same must hold for $\dot{G}_2$: its spectrum is bounded below by $-k+1$, and it also contains k as an eigenvalue. Because $k < n$, this implies that $\dot{G}_2$ is also disconnected; otherwise, either n or $-n$ would appear in its spectrum, contradicting the bounds. Thus, we have $\dot{G}_2 \cong \dot{G}_2' \cup c_2K_1$, where $\dot{G}_2'$ is a connected ∇-threshold graph. As before, $\dot{G}_2'$ must also be a positive join with largest eigenvalue k and least eigenvalue greater than $-k$, which implies that $\dot{G}_1'$ and $\dot{G}_2'$ are cospectral. Consequently, $c_1 = c_2$, and $\dot{G}_1$ and $\dot{G}_2$ are not a minimal counterexample.

Now suppose that $\dot{G}_1$ is connected, and assume without loss of generality that it is a positive join. Then n is the largest eigenvalue of $\dot{G}_1$, and also of $\dot{G}_2$. It follows that $\dot{G}_2$ must also be a positive join. In other words, we can write $\dot{G}_1 \cong K_1\nabla^+\dot{G}_1'$ and $\dot{G}_2 \cong K_1\nabla^+\dot{G}_2'$ for some ∇-threshold graphs $\dot{G}_1'$ and $\dot{G}_2'$. Let $\nu_1 = n, \nu_2, \dots, \nu_n = 0$ be the common eigenvalues of $\dot{G}_1$ and $\dot{G}_2$. Then, by Theorem 1.20, the eigenvalues of $\dot{G}_1'$ and $\dot{G}_2'$ are $\nu_2 - 1, \nu_3 - 1, \dots, \nu_{n-1} - 1$ and 0. This shows that $\dot{G}_1'$ and $\dot{G}_2'$ are also cospectral, which contradicts the minimality assumption. □

For the record, every unsigned threshold graph is determined by the Laplacian spectrum [111].

Problems

7.1 Determine whether the NEPS of a signed graph is integral.

7.2 Re-prove the result of [185] identifying the 13 connected cubic integral unsigned graphs.

7.3 Prove Theorem 7.2 using spectral moments.

7.4 Prove that the definitions of an unsigned cograph given at the beginning of Subsection 7.1.2 are equivalent.

7.5 ([210]) Show that a connected integral bipartite quartic net-regular signed graph, whose net degree is 2 and appears as a simple eigenvalue, has at most 120 vertices. Determine feasible spectra of such signed graphs.

7.6 Use Theorem 1.20 to construct an infinite family of signed graphs whose net Laplacian spectrum is symmetric and consists entirely of integers.

7.7 Determine whether the spectrum of a ∇-cograph is non-negative.

7.8 Prove that GM-switching and WQH-switching for unsigned graphs necessarily produce graphs with the same spectrum.

7.9 Demonstrate an example of cospectral signed graphs constructed via GM-switching or WQH-switching.

7.10 Derive necessary and sufficient conditions under which GM-switching and WQH-switching produce signed graphs that are not switching-isomorphic. (Difficult task.)

7.11 Show that an r-dimensional signed cube with negative quadrangles $\dot{Q}_r$ is determined by its spectrum whenever $r \neq 4$. Show that the minimal spectral radius of all signed r-dimensional cubes is attained by $\dot{Q}_r$.

7.12 Use the idea from the previous problem to solve the following. Let an unsigned graph be the underlying graph of a signed graph $\dot{G}$ with exactly two symmetric eigenvalues. Prove that, among all possible signatures on this underlying graph, the minimal spectral radius is attained by $\dot{G}$.

7.13 Under what conditions do cospectral signed graphs necessarily produce cospectral signed line graphs?

7.14 Examine whether a signed graph whose spectrum lies in $[-2, 2]$ is determined by its spectrum.

7.15 Determine whether a ∇-threshold graph is determined by its net Laplacian spectrum.

7.16 Investigate signed graphs of relatively small order and determine, among the prescribed spectra (the spectrum of the adjacency matrix, the Laplacian spectrum and the net Laplacian spectrum) for which spectrum the phenomenon of cospectrality occurs with the lowest frequency. The objective is to identify which spectral representation most effectively distinguishes non-isomorphic signed graphs, thereby providing insight into the discriminative power of different spectral invariants in the context of signed graph theory.

8

Polynomial Reconstruction

The reconstruction of a graph from partial information is one of the most intriguing problems in graph theory. Among the oldest and most famous open conjectures in this area is the question of whether every graph with at least three vertices can be uniquely reconstructed from the collection of its vertex-deleted subgraphs. This conjecture first appeared in 1942 [132] and is commonly attributed to Ulam, who is said to have posed it to his Ph.D. student Kelly [240]. Accordingly, it is broadly known as the *Ulam conjecture*. Over the years, significant progress has been made and many special cases have been resolved; however, the conjecture remains open and no counterexamples are known.

In 1973, Cvetković formulated a spectral variant of this conjecture (see [72, p. 69]), which asks whether the characteristic polynomial of a graph G with at least three vertices can be reconstructed from the collection of characteristic polynomials of its vertex-deleted subgraphs. This problem was independently considered by Schwenk [183] and has since attracted substantial attention. Despite numerous positive partial results, the general problem remains open, since no counterexample has been discovered to date; see [70, 74, 105, 183] for foundational contributions, [186, 187, 193, 197, 198] for results on specific graph classes and [200, 229] for comprehensive treatments. A survey covering the first 50 years of research on polynomial reconstruction is provided in [188]. The enduring appeal of the problem lies in its fundamental simplicity and its deep connections with various graph invariants.

More recently, Simić and Stanić [195] extended the polynomial reconstruction problem to the domain of signed graphs. Given a signed graph $\dot{G}$, let

$$\mathscr{P}(\dot{G}) = \{\Phi_{\dot{G}_1}, \Phi_{\dot{G}_2}, \dots, \Phi_{\dot{G}_n}\},$$

where $\dot{G}_i = \dot{G} - v_i$ $(1 \leq i \leq n)$, be the collection of characteristic polynomials

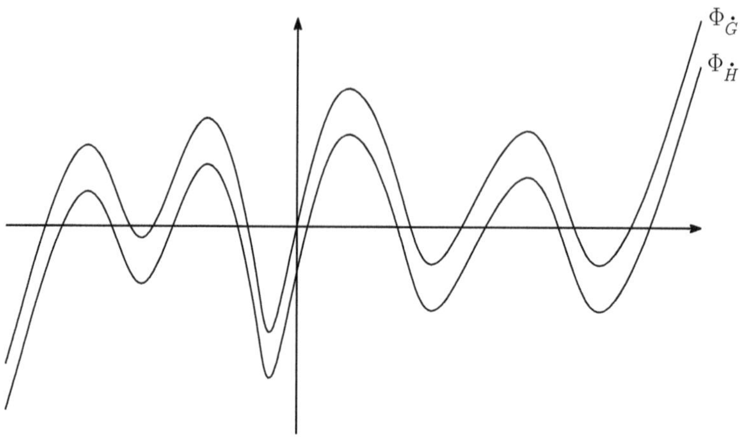

Figure 8.1 A sketch of the distinct characteristic polynomials of $\dot{G}$ and $\dot{H}$ when they share the same polynomial deck. Reproduced from [188].

of vertex-deleted subgraphs of $\dot{G}$. $\mathscr{P}(\dot{G})$ is referred to as the *polynomial deck* of $\dot{G}$. Since

$$\Phi'_{\dot{G}}(x) = \sum_{i=1}^{n} \Phi_{\dot{G}_i}(x) \tag{8.1}$$

(see, for example, [73, p. 60]), we can readily determine the characteristic polynomial of a signed graph from its polynomial deck except for the constant term. In other words, if $\dot{G}$ and $\dot{H}$ have distinct characteristic polynomials but share the same polynomial deck, then we have $\Phi_{\dot{G}}(x) = \Phi_{\dot{H}}(x) + a$ for $a \in \mathbb{Z} \setminus \{0\}$, which implies that spectra of $\dot{G}$ and $\dot{H}$ are disjoint, and characteristic polynomials of these graphs can be sketched as in Figure 8.1.

Taking into account the identity (8.1), we deduce that if any eigenvalue of the signed graph $\dot{G}$ is known, then the constant term of its characteristic polynomial is uniquely determined (see [105]). In particular, this applies if some polynomial from the deck possesses a multiple root; by eigenvalue interlacing, this root must also appear in the spectrum of $\dot{G}$. More generally, if the value of the characteristic polynomial at any given point is deduced from the polynomial deck, the reconstruction problem is resolved.

A natural question arises: does a counterexample to this reconstruction problem exist? Indeed, there is one. Consider two signed cycles of the same order, one balanced and the other unbalanced. These two signed graphs share the same polynomial deck, yet, as shown in Example 1.11, their characteristic polynomials differ. At present, no other counterexamples with at least three vertices are known. Consequently, the following problem remains open.

Problem 8.1 (Polynomial reconstruction for signed graphs) Given two signed graphs $\dot{G}$ and $\dot{H}$ with at least three vertices such that none of them is a signed cycle, is it true that $\mathscr{P}(\dot{G}) = \mathscr{P}(\dot{H}) \Rightarrow \Phi_{\dot{G}} = \Phi_{\dot{H}}$?

This problem has been positively resolved by a computer search for signed graphs with at most eight vertices [215]. We show that switching does not affect the polynomial deck.

Theorem 8.2 ([195]) *Switching-equivalent signed graphs share the same polynomial deck.*

Proof Let $\dot{G}$ and $\dot{H}$ be two switching-equivalent signed graphs on the same set of vertices $v_1, v_2, \dots, v_n$. The signed graphs $\dot{G} - v_i$ and $\dot{H} - v_i$, $1 \leq i \leq n$, are switching equivalent as well (where the corresponding switching is the restriction of that between $\dot{G}$ and $\dot{H}$). Thus, they have the same characteristic polynomials. □

In other words, the polynomial reconstruction of a signed graph is unique if and only if the polynomial reconstruction of any switching-equivalent signed graph is unique. For instance, since every signed tree is switching equivalent to its underlying unsigned tree, and since the polynomial reconstruction of trees is known to be unique [74], it follows that the polynomial reconstruction of signed trees is also unique.

We proceed with the following convention. In the context of polynomial reconstruction (therefore, the whole of this chapter), it is convenient to write the characteristic polynomial as

$$\Phi_{\dot{G}}(x) = \det(xI - A(\dot{G})) = x^n + a_{n-1}x^{n-1} + \cdots + a_1 x + a_0.$$

So, in contrast to Theorem 1.9, the coefficients a_i, $0 \leq i \leq n-1$, are ordered decreasingly, and therefore the constant term is $a_0 = a_0(\dot{G})$. By the same theorem, we have

$$a_{n-i}(\dot{G}) = \sum_{\dot{H} \in \dot{\mathscr{H}}_i} \sigma(\dot{H}_{\mathrm{c}})\big((-1)^{p(\dot{H})} 2^{q(\dot{H})}\big), \quad 1 \leq i \leq n, \tag{8.2}$$

where $\dot{H}$ is a subgraph of $\dot{G}$, spanned over i vertices, whose components are edges or cycles (of length at least 3), $p(\dot{H})$ is the number of components of $\dot{H}$, $q(\dot{H})$ is the number of cycles in $\dot{H}$ and $\dot{H}_{\mathrm{c}}$ is a subgraph of $\dot{H}$ containing only its cycles. Clearly, $\dot{\mathscr{H}}_i$ is the set of all subgraphs of $\dot{H}$ of order i.

We list some invariants that are reconstructible from the polynomial deck.

Lemma 8.3 ([195]) *The following invariants of a signed graph $\dot{G}$ with at least three vertices are recoverable from its polynomial deck:*

(i) *the number of vertices n;*
(ii) *the number of edges m;*
(iii) *the vertex degree sequence* $d_{v_1} \geq d_{v_2} \geq \cdots \geq d_{v_n}$;
(iv) *the length of the shortest odd cycle and the number of such cycles;*
(v) *the spectral moments* $M_0, M_1, \ldots, M_{n-1}$ *of* $A_{\dot{G}}$.

Proof The order of $\dot{G}$ is equal to one plus the degree of any polynomial in the polynomial deck, which gives (i). By (8.2), $m = -a_{n-2}(\dot{G})$, and this coefficient can be determined from $\mathscr{P}(\dot{G})$ provided $n > 2$, giving (ii).

Similarly, the number m_i of edges of each vertex-deleted subgraph $\dot{G}_i$, $1 \leq i \leq n$, is deduced from its characteristic polynomial. Thus, $m - m_i$ are the required vertex degrees. Hence, (iii).

Part (iv) follows directly from (8.2). For (v), we have $M_0 = n$, $M_1 = 0$ and $M_2 = 2m$. The remaining spectral moments are obtained from the coefficients $a_{n-3}(\dot{G}), a_{n-4}(\dot{G}), \ldots, a_1(\dot{G})$ by making use of Newton's formulas. □

In the following sections, we consider (i.e., resolve or partially resolve) Problem 8.1 for certain classes of signed graphs. Throughout these sections, $\dot{G}$ denotes a signed graph whose polynomial deck is under consideration, and $\dot{H}$ denotes a signed graph (if any) that, together with $\dot{G}$, forms a counterexample to the reconstruction problem. In this context, the pair $(\dot{G}, \dot{H})$ is referred to as a *counterexample pair*.

We also employ the notion of partners between vertices of $\dot{G}$ and $\dot{H}$. The vertices $v \in V(\dot{G})$ and $v' \in V(\dot{H})$ are called *partners* if their corresponding vertex-deleted subgraphs have identical characteristic polynomials, that is,

$$\Phi_{\dot{G}-v}(x) = \Phi_{\dot{H}-v'}(x).$$

By examining the proof of Lemma 8.3(iii), it follows that partners necessarily share the same vertex degree.

8.1 Disconnected Signed Graphs

Observe that at least one of the signed graphs $\dot{G}$ and $\dot{H}$ from a counterexample pair must be connected. Otherwise, the largest eigenvalue of each graph appears as a root of some polynomials in their common polynomial deck, implying that these signed graphs share a common eigenvalue. However, this is impossible in the light of the discussion visualized in Figure 8.1.

Many results on the polynomial reconstruction of disconnected graphs rely solely on the eigenvalue interlacing property, meaning that all such results can be directly transferred to the signed domain. In particular, the following two theorems hold.

Theorem 8.4 ([74, 187]) *Let $\dot{H}$ be a disconnected signed graph whose components are $\dot{H}_1, \dot{H}_2, \ldots, \dot{H}_k$. Then $\Phi_{\dot{H}}$ is reconstructible if either $k > 2$ or $k = 2$ and $n(\dot{H}_1) \neq n(\dot{H}_2)$. Otherwise, if $k = 2$ and $n(\dot{H}_1) = n(\dot{H}_2)$ then reconstructibility holds in the following situations:*

(i) $m(\dot{H}_1) - m(\dot{H}_2) \geq \delta(\dot{H}_1)$, where $\delta(\dot{H}_1)$ is the minimum vertex degree in $\dot{H}_1$;
(ii) $m(\dot{H}_1) > m(\dot{H}_2)$ and $\dot{H}_1$ has a pendant vertex.

There is a corollary.

Corollary 8.5 *Let $\dot{H}$ be a disconnected signed graph with two components, $\dot{H}_1$ and $\dot{H}_2$, each of order n. Then the polynomial reconstruction is unique whenever*

$$m(\dot{H}_1) - m(\dot{H}_2) \geq n - 1 \geq \lfloor \bar{d}(\dot{H}_1) \rfloor,$$

where $\bar{d}$ stands for the average vertex degree.

Proof Since $\lfloor \bar{d}(H_1) \rfloor \geq \delta(H_1)$, the proof follows from Theorem 8.4(i). □

In the previous corollary let $\Delta(m) = m(\dot{H}_1) - m(\dot{H}_2)$. Then, if $\Delta(m) \in \{0, 1\}$, we get no benefits. If $\Delta(m) = 2$, then the corollary ensures uniqueness whenever $\bar{d}(\dot{H}_1) < 3$.

We proceed with the following characterization.

Theorem 8.6 ([74, 187]) *Let $\dot{H}$ be a disconnected signed graph with two components, $\dot{H}_1$ and $\dot{H}_2$, each of order n. Let $\lambda_1 > \lambda_2 > \cdots > \lambda_n$ be the eigenvalues of $\dot{H}_1$, and $\mu_1 > \mu_2 > \cdots > \mu_{n-1}$ the eigenvalues of $\dot{H}_2 - u$, where u is an arbitrary vertex of $\dot{H}_2$. Then the polynomial reconstruction is unique whenever the following chain of inequalities does not hold:*

$$\lambda_1 > \mu_1 > \lambda_2 > \mu_2 > \cdots > \mu_{n-1} > \lambda_n. \tag{8.3}$$

If (8.3) holds, we say that the eigenvalues of $\dot{H}_2 - u$ *strictly interlace* the eigenvalues of $\dot{H}_1$. The latter two theorems give very strong restrictions on a putative (disconnected) signed graph having a (connected) partner. More restrictions are given below.

Theorem 8.7 ([195]) *If $(\dot{G}, \dot{H})$ is a counterexample pair and if $\dot{H}$ is disconnected with two components, $\dot{H}_1$ and $\dot{H}_2$, then:*

(i) $\lambda_1(\dot{G}) > \lambda_1(\dot{H})$.

(ii) $\lambda_1(\dot{G}) > \lambda_2(\dot{G})$.
(iii) *The largest roots of polynomials in the common polynomial deck take at most two distinct values, namely* $\lambda_1(\dot{H}_1)$ *and* $\lambda_1(\dot{H}_2)$.
(iv) $\dot{H}_1$ *and* $\dot{H}_2 - u$ *(u being any vertex of* $\dot{H}_2$*) have no common eigenvalues.*

Proof (i): Since $\dot{H}$ is disconnected, there exists a vertex u of $\dot{G}$ such that $\lambda_1(G-u) = \lambda_1(\dot{H})$. But then $\lambda_1(\dot{G}) \geq \lambda_1(\dot{G}-u)$ (by the interlacing), and the rest follows immediately.

(ii): Otherwise, $\lambda_1(\dot{G}) = \lambda_2(\dot{G})$ by the interlacing and, moreover, the equality $\lambda_1(\dot{G}-u) = \lambda_1(\dot{G})$ holds for any vertex $u \in V(\dot{G})$. Consequently, by Theorem 8.6, we would have $\lambda_1(H_1) = \lambda_1(H_2) = \lambda_1(H) = \lambda_1(G)$, which is not allowed.

Part (iii) follows from Theorem 8.6. If (iv) fails to hold, then $\dot{H} - u$ would have repeated eigenvalues. □

8.2 Signed Graphs with Pendant Vertices

Additional restrictions arise in the case of signed graphs containing pendant vertices.

Theorem 8.8 ([186, 195]) *If* $(\dot{G}, \dot{H})$ *is a counterexample pair and* $\dot{G}$ *contains a pendant vertex u whose unique neighbour is v, then the following statements hold:*

(i) $a_0(\dot{G}) = -a_0(\dot{G}-u-v) = -a_1(\dot{G}-v)$.
(ii) *The partner of v in* $\dot{H}$ *(i.e.,* v'*) is not adjacent to a pendant vertex.*
(iii) *Neither* $\dot{G}$ *nor* $\dot{H}$ *is singular (has 0 as an eigenvalue).*
(iv) *If* $\dot{H}$ *is disconnected with components* $\dot{H}_1$ *and* $\dot{H}_2$*, such that* $\lambda_1(\dot{H}_1) \geq \lambda_1(\dot{H}_2)$*, then* $\dot{H}_1$ *has no pendant vertices, and if* $\lambda_1(\dot{H}_1) = \lambda_1(\dot{H}_2)$ *then* $\dot{H}$ *has no pendant vertices.*
(v) *If* w' *is a neighbour of* u' *in* $\dot{H}$*, then* $d_{\dot{H}}(w') = d_{\dot{G}}(v)$*, and if* $\dot{H}$ *is disconnected then* v' *is not in the same component as* u' *(or* w'*).*
(vi) $\mathscr{P}(\dot{G})$ *contains two polynomials (one is* $\Phi_{\dot{G}-v}(x)$*, the other is* $\Phi_{\dot{G}-w}(x)$*, where w is a partner of* w'*) which differ only in the coefficient of x.*

Proof Parts (i), (ii) and (iii) follow by a simple computation based on (1.2).

(iv): Assume that $\dot{H}_1$ has a pendant vertex. Let u be its partner in $\dot{G}$, with v as its (unique) neighbour. Let us denote the partners of u and v in $\dot{H}$ by u' and v'. From (8.3), we in turn have that $\lambda_1(\dot{H}_2) = \lambda_1(\dot{H} - u') = \lambda_1(\dot{G} - u)$. On the contrary, since u is an isolated vertex in $\dot{G} - v$, we have

$$\lambda_1(\dot{G}-u-v) = \lambda_1(\dot{G}-v) = \lambda_1(\dot{H}-v') \geq \min\{\lambda_1(H_1), \lambda_1(\dot{H}_2)\} = \lambda_1(\dot{H}_2).$$

Therefore, $\lambda_1(\dot{G}-u-v) \geq \lambda_1(\dot{G}-u)$ and $\lambda_1(\dot{G}-u) = \lambda_1(\dot{G}-u-v)$ (by the interlacing). However, by (1.2), $\lambda_1(\dot{G}-u)$ is an eigenvalue of $\dot{G}$ and of $\dot{H}$ (via $\dot{H}_2$), which is impossible. The last claim of (iv) follows from the previous one.

(v): From (1.2),

$$\Phi_{\dot{G}}(x) - \Phi_{\dot{H}}(x) = -\left(\Phi_{\dot{G}-u-v}(x) - \Phi_{\dot{H}-u'-w'}(x)\right).$$

Hence, $\Phi_{\dot{G}-u-v}(x)$ and $\Phi_{\dot{H}-u'-w'}(x)$ differ only up to an additive constant. Consequently, if $n \geq 4$ then $\dot{G}-v$ and $\dot{H}-w'$ have the same size, whence v and w' have the same degree. For $n = 3$, the same conclusion holds immediately.

To prove the second claim, observe that $\lambda_1(\dot{G}-u) \geq \lambda_1(\dot{G}-v)$, and assume that $\lambda_1(\dot{H}_1) > \lambda_1(\dot{H}_2)$ (by part (iv), equality cannot hold since otherwise $\dot{H}$ would not have pendant vertices). Hence, u' is in $\dot{H}_2$. If v' is also in $\dot{H}_2$, then by (8.3),

$$\lambda_1(\dot{H}-u') = \lambda_1(\dot{H}-v') = \lambda_1(\dot{H}_1) = \lambda_1(\dot{H}),$$

which is impossible.

(vi): Observe that

$$x\left(\Phi_{\dot{G}}(x) - \Phi_{\dot{H}}(x)\right) = -\left(\Phi_{\dot{G}-v}(x) - \Phi_{\dot{H}-w'}(x)\right).$$

Given that $\Phi_{\dot{G}}(x) - \Phi_{\dot{H}}(x)$ is constant and $\Phi_{\dot{H}-w'}(x) = \Phi_{\dot{G}-w}(x)$, the claim follows. □

Recall that a hanging path in a (signed) graph is a path whose end vertex is identified with a fixed vertex in the remainder of the graph. The following theorem contributes to the polynomial reconstruction problem for disconnected graphs containing hanging paths.

Theorem 8.9 ([195]) *If $\dot{G}$ is a signed graph which is connected and contains a hanging path of length at least two, then its partner $\dot{H}$ cannot be disconnected.*

The proof is based on (1.2) and Theorems 8.7 and 8.8.

We now employ the results of Section 2.2; in particular, the equalities (2.3) and (2.4) are used. Moreover, both are simplified since every eigenvalue of every vertex-deleted subgraph is simple; otherwise, the polynomial reconstruction would be unique. Recall that the determinant of an n-vertex signed graph $\dot{G}$ is related to the constant term of the characteristic polynomial by $\det(\dot{G}) = (-1)^n a_0(\dot{G})$. Furthermore, if $\dot{G}-v$ is singular, then we denote by Π_v the product of the non-zero eigenvalues of $\dot{G}-v$. For convenience, we assume that zero takes both signs.

Theorem 8.10 ([215]) *Let $(\dot{G}, \dot{H})$ be a counterexample pair (if any) for the polynomial reconstruction.*

(i) If at least one vertex-deleted subgraph of $\dot{G}$ is singular, then we have $\mathrm{sign}(a_0(\dot{G})) = \mathrm{sign}(a_0(\dot{H}))$.
(ii) If $\dot{G}$ has a pendant vertex, then $\mathrm{sign}(a_0(\dot{G})) = \mathrm{sign}(a_0(\dot{H}))$ *and both* $a_0(\dot{G}), a_0(\dot{H}) \neq 0$.

Proof (i): Let $v \in V(\dot{G})$ and $v' \in V(\dot{H})$ be partners such that $\dot{G} - v$ and $\dot{H} - v'$ are singular. In this case, 0 is a simple eigenvalue in both signed graphs, which means that $a_0(\dot{G})$ and $a_0(\dot{H})$ can be computed by (2.4). Let $\mathbf{r}_v$ and $\mathbf{r}_{v'}$ be the characteristic vectors corresponding to v and v' in $\dot{G}$ and $\dot{H}$, respectively, and suppose that the unit vectors $\mathbf{x}$ and $\mathbf{y}$ span the nullspaces of $\dot{G} - v$ and $\dot{H} - v'$, respectively.

Taking into account that $(\mathbf{r}_v^\intercal \mathbf{x})^2, (\mathbf{r}_{v'}^\intercal \mathbf{y})^2 \geq 0$, we compute

$$\begin{aligned}\mathrm{sign}(a_0(\dot{G})) &= \mathrm{sign}\left((-1)^{n-1}(\mathbf{r}_v^\intercal \mathbf{x})^2 \Pi_v\right) = (-1)^{n-1}(\mathbf{r}_v^\intercal \mathbf{x})^2 \,\mathrm{sign}(\Pi_v)\\ &= (-1)^{n-1}(\mathbf{r}_{v'}^\intercal \mathbf{y})^2 \,\mathrm{sign}(\Pi_{v'}) = \mathrm{sign}\left((-1)^{n-1}(\mathbf{r}_{v'}^\intercal \mathbf{y})^2 \Pi_{v'}\right)\\ &= \mathrm{sign}(a_0(\dot{H})).\end{aligned}$$

(ii): If $\dot{G}$ has a pendant vertex, say v, then by deleting its neighbour we obtain a singular vertex-deleted subgraph, and the desired equality follows from the previous part.

Moreover, considering this same vertex-deleted subgraph, the unit eigenvector $\mathbf{x}$ associated with the simple eigenvalue 0 takes the value 1 at v and 0 at all other vertices. Together with (2.3), this implies that $a_0(\dot{G}) \neq 0$. Since $\dot{G}$ has a pendant vertex, $\dot{H}$ also has a pendant vertex (by Lemma 8.3(iii)). Repeating the procedure, this time for the neighbour of a pendant vertex of $\dot{H}$, we similarly obtain $a_0(\dot{H}) \neq 0$. □

If $\dot{G}$ is singular, then $a_0(\dot{G}) = 0$. On the other hand, if $\dot{G}$ has a pendant vertex and the polynomial reconstruction is not unique, Theorem 8.10(ii) guarantees $a_0(\dot{G}) \neq 0$. Thus, we have established the following result.

Corollary 8.11 *The polynomial reconstruction is unique for singular signed graphs with pendant vertices.*

The same result for unsigned graphs has been proved in [186]. It is worth mentioning that the singularity of $\dot{G}$ is not reconstructible from its polynomial deck.

Theorem 8.12 ([215]) *Suppose that $\dot{G}$ has a pendant vertex and $|\Pi_v| < 2$ holds for all singular vertex-deleted subgraphs. The polynomial reconstruction is unique.*

Proof Let $\dot{H}$ be as before. Let $v \in V(\dot{G})$ and $w' \in V(\dot{H})$ be vertices adjacent to pendant vertices. Considering the unit vectors spanning the nullspaces of $\dot{G} - v$ and $\dot{H} - w'$, respectively, we observe that for both signed graphs the unique non-zero coordinate is 1, corresponding to the isolated vertex in each case. Consequently, for $\dot{G}$ and $\dot{H}$, the term $(\mathbf{r}^\intercal\mathbf{x})^2$ in (2.3) equals 1, which yields $a_0(\dot{G}) = (-1)^{n-1}\Pi_v$ and $a_0(\dot{H}) = (-1)^{n-1}\Pi_{w'}$.

Since $a_0(\dot{G})$ and $a_0(\dot{H})$ are coefficients of integral polynomials, we have $a_0(\dot{G}), a_0(\dot{H}) \in \mathbb{Z}$, and thus $\Pi_v, \Pi_{w'} \in \mathbb{Z}$. By the assumptions of the theorem, $|\Pi_v|, |\Pi_{w'}| < 2$, which implies $\Pi_v, \Pi_{w'} \in \{1, 0, -1\}$. Hence,

$$a_0(\dot{G}), a_0(\dot{H}) \in \{1, 0, -1\}.$$

Applying Theorem 8.10(ii), we conclude that $a_0(\dot{G}) = a_0(\dot{H})$, which leads to a contradiction. □

8.3 Further Results

It is worth mentioning that Problem 8.1 remains unresolved for regular signed graphs, despite having a straightforward solution in the unsigned case [70]. The question of unique polynomial reconstruction for disconnected or bipartite graphs remains open, and consequently it is still not settled for signed graphs either, except in the case of signed cycles. Nonetheless, progress has been made for signed graphs in certain areas: Problem 8.1 has been affirmatively resolved for some classes of disconnected signed graphs [195, 215], various classes of bipartite signed graphs [195, 215] and particular signed graphs possessing pendant vertices [215].

We list more relevant results. In many cases, a proof is omitted.

8.3.1 Least Eigenvalue Close to -2

The main result of [196] reads as follows.

Theorem 8.13 *Let $\dot{G}$ be a signed graph of order $n \geq 3$ whose vertex-deleted subgraphs have the least eigenvalue greater than -2. Then the polynomial reconstruction is unique, unless the underlying graph of $\dot{G}$ is a cycle.*

The proof is based on the classification given in Theorem 5.18. Accordingly, each vertex-deleted subgraph belongs to either (L1), (L2) or (E1) of this theorem. Further consideration examines these possibilities by means of more sophisticated methods. There is an unsigned version of Theorem 8.13 allowing vertex-deleted subgraphs to have the least eigenvalue that is greater than or equal to -2 (so, the equality is allowed) [197].

8.3.2 Unicyclic Signed Graphs

The main result of [195] is the following theorem.

Theorem 8.14 *Let $\dot{G}$ be a unicyclic signed graph. The polynomial reconstruction is unique, unless the underlying graph of $\dot{G}$ is a cycle.*

The proof is based on examining the constant term $a_0(\dot{G})$ computed in the following lemma.

Lemma 8.15 *Let $\dot{U}$ be a unicyclic signed graph of order n, $\dot{C}$ the unique signed cycle of $\dot{U}$ with length g, and let $s = \sigma(\dot{C})$. Then:*

(i) If n and g are odd then

$$a_0(\dot{U}) = \begin{cases} 2s(-1)^{\frac{1}{2}(n-g)+1} & \text{if } \dot{U} - V(\dot{C}) \text{ has a perfect matching,} \\ 0 & \text{otherwise.} \end{cases}$$

(ii) If n is odd and g even, then $a_0(\dot{U}) = 0$.

(iii) If n is even and g odd, then

$$a_0(\dot{U}) = \begin{cases} (-1)^{\frac{1}{2}n} & \text{if } \dot{U} \text{ has a perfect matching,} \\ 0 & \text{otherwise.} \end{cases}$$

(iv) If n and g are even, then

$$a_0(\dot{U}) = \begin{cases} 2(-1)^{\frac{1}{2}n}(1 - s(-1)^{\frac{1}{2}g}) & \text{if } \dot{U} - V(\dot{C}) \text{ has a perfect matching,} \\ (-1)^{\frac{1}{2}n} & \text{if } \dot{U} \text{ has a perfect matching,} \\ 0 & \text{otherwise.} \end{cases}$$

8.3.3 Signed Bundles

A *signed bundle* is a signed graph composed of an arbitrary number of signed cycles sharing exactly one common vertex (see Figure 8.2). A particular case of a signed bundle is a single cycle, for which the polynomial reconstruction is known to be non-unique. In what follows, we show that this is the only exception. Non-regular signed bundles, that is, signed bundles that are not signed cycles, can be characterized as signed graphs in which precisely one vertex has degree different from 2; moreover, this degree is always even.

We need the following lemma.

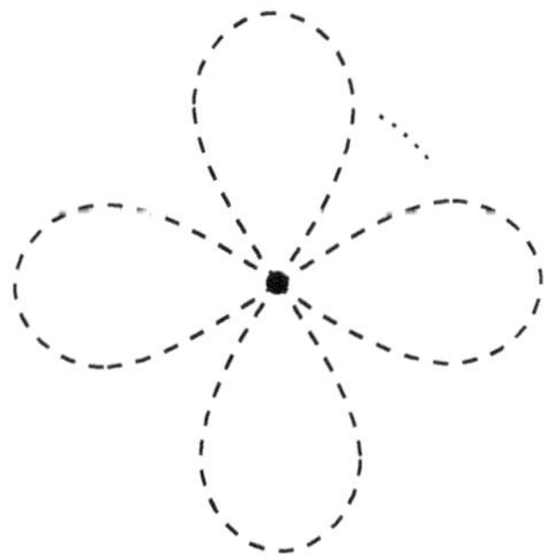

Figure 8.2 A bundle.

Lemma 8.16 ([215]) *Let $\dot{G}$ and $\dot{H}$ be non-regular signed bundles that share the same underlying graph, in which no two cycles are of the same length. If $a_0(\dot{G}) \neq a_0(\dot{H})$, then $\Phi_{\dot{G}}(x) - \Phi_{\dot{H}}(x) \neq a_0(\dot{G}) - a_0(\dot{H})$.*

Proof Assume that $a_0(\dot{G}) \neq a_0(\dot{H})$ and suppose that

$$\Phi_{\dot{G}}(x) - \Phi_{\dot{H}}(x) = a_0(\dot{G}) - a_0(\dot{H}). \tag{8.4}$$

Let v be the vertex of largest degree in the common underlying graph G of $\dot{G}$ and $\dot{H}$. Denote by $C_1, C_2, \dots, C_k$ the set of cycles in G containing v, and by $P_1, P_2, \dots, P_k$ the corresponding paths obtained by deleting v from each cycle. Without loss of generality, assume that the cycles are indexed so that their lengths satisfy $|V(C_1)| \leq |V(C_2)| \leq \cdots \leq |V(C_k)|$. For each $1 \leq i \leq k$, let $\dot{C}_i$ and $\dot{C}'_i$ denote the respective signed cycles in $\dot{G}$ and $\dot{H}$ underlain by C_i.

On the basis of (1.1), we compute

$$\begin{aligned}\Phi_{\dot{G}}(x) - \Phi_{\dot{H}}(x) &= x\big(\Phi_{\dot{G}-v}(x) - \Phi_{\dot{H}-v}(x)\big) - \sum_{w \sim v} \big(\Phi_{\dot{G}-v-w}(x) - \Phi_{\dot{H}-v-w}(x)\big) \\ &\quad - 2\sum_{i=1}^{k} \big(\sigma(\dot{C}_i)\Phi_{\dot{G}-\dot{C}_i}(x) - \sigma(\dot{C}'_i)\Phi_{\dot{H}-\dot{C}'_i}(x)\big).\end{aligned}$$

Since every signed path switches to its underlying path, the first two terms of the right-hand side are zero. Considering the third one, we get

$$\Phi_{\dot{G}}(x) - \Phi_{\dot{H}}(x) = 2\sum_{i=1}^{k} \big(\sigma(\dot{C}'_i) - \sigma(\dot{C}_i)\big) \prod_{j \neq i} \Phi_{P_j}(x).$$

The polynomials $\prod_{j\neq 1} \Phi_{P_j}(x), \prod_{j\neq 2} \Phi_{P_j}(x), \dots, \prod_{j\neq k} \Phi_{P_j}(x)$ are ordered decreasingly by their degree. Let s_i denote the degree of $\prod_{j\neq i} \Phi_{P_j}(x)$. From (8.4), we obtain $\sigma(\dot{C}'_1) - \sigma(\dot{C}_1) = 0$, otherwise the term x^{s_1} would appear in the difference. By iterating the same argument, we deduce that $\sigma(\dot{C}_2') - \sigma(\dot{C}_2) = 0$ to avoid the presence of x^{s_2}, and so forth for all i.

Hence, $\sigma(\dot{C}_i') - \sigma(\dot{C}_i) = 0$ for every i, giving $\Phi_{\dot{G}}(x) - \Phi_{\dot{H}}(x) = 0$. □

Theorem 8.17 ([215]) *The polynomial reconstruction of non-regular signed bundles is unique.*

Proof Let $\dot{G}$ be a non-regular signed bundle and suppose that $(\dot{G}, \dot{H})$ forms a counterexample to Problem 8.1. Since exactly one vertex in $\dot{G}$ has degree different from 2, by Lemma 8.3(iii) and Theorem 8.4, there are exactly two possibilities for $\dot{H}$: either it is a non-regular signed bundle or a disjoint union of a non-regular signed bundle and a signed cycle. The latter is immediately excluded since every signed cycle has a repeated eigenvalue (see Example 1.11).

Now, let v and v' denote the vertices of largest degree in $\dot{G}$ and $\dot{H}$, respectively. As these are the unique vertices whose degrees differ from 2, they must be partners, meaning they have the same degree and satisfy $\Phi_{\dot{G}-v}(x) = \Phi_{\dot{H}-v'}(x)$. Note that both $\dot{G}-v$ and $\dot{H}-v'$ are disjoint unions of paths and are switching isomorphic. If two paths in $\dot{G}-v$ have the same length, then $\dot{G}-v$ would have a repeated eigenvalue, which is impossible. Therefore, $\dot{G}$ and $\dot{H}$ are non-regular signed bundles sharing the same underlying graph in which no two cycles have the same length; that is, they satisfy the assumptions of Lemma 8.16. The conclusion of that lemma contradicts the initial assumption of this proof. □

8.3.4 Bicyclic Signed Graphs without Pendant Vertices

Since Problem 8.1 has been resolved affirmatively for signed trees and unicyclic signed graphs, the next natural step is to consider bicyclic signed graphs. We address the problem under the additional assumption that the signed graph in question has no pendant vertices. This restriction simplifies certain structural considerations while still encompassing a broad and non-trivial class of examples.

Theorem 8.18 ([215]) *The polynomial reconstruction of bicyclic signed graphs without pendant vertices is unique.*

Proof There are three types of such signed graphs: signed theta-graphs, signed dumbbells, and signed infinity-graphs. The first two are illustrated in Figure 5.4, while signed infinity-graphs arise as a special case of signed dumbbells in Subsection 7.2.4.

First, observe that a signed infinity-graph is a particular case of a signed bundle, which has already been resolved in Proposition 8.17.

Next, note that the polynomial reconstruction of a signed dumbbell is unique since it has a vertex-deleted subgraph consisting of a signed cycle and another component, and hence has a repeated eigenvalue.

Let $\dot{G}$ be a signed theta-graph, and let $(\dot{G},\dot{H})$ be a counterexample pair. Then $\dot{H}$ is either a signed theta-graph or a disjoint union of a signed theta-graph and a signed cycle. The latter possibility is immediately eliminated, so we only consider the former.

Let $\dot{C}_1,\dot{C}_2,\dot{C}_3$ (resp. $\dot{C}'_1,\dot{C}'_2,\dot{C}'_3$) be the induced cycles of $\dot{G}$ (resp. $\dot{H}$), ordered non-decreasingly by length. Let v,w be the vertices of degree 3 in $\dot{G}$. Their partners v',w' are the vertices of degree 3 in $\dot{H}$. Now, $\dot{G}-v$ and $\dot{H}-v'$ share the same spectrum and both are switching isomorphic to T-shape trees. It is established in [140] that two T-shape trees sharing the same spectrum are necessarily isomorphic. This implies that $\dot{C}_i$ is equal in length to $\dot{C}'_i$ for $1<i<3$.

Suppose now that v and w are non-adjacent. Observe that $\dot{G}-\dot{C}_i$ and $\dot{H}-\dot{C}'_i$ switch to a fixed path, say P_i. Proceeding as in the proof of Lemma 8.16, we compute

$$0\neq a_0(\dot{G})-a_0(\dot{H})=\Phi_{\dot{G}}(x)-\Phi_{\dot{H}}(x)=2\sum_{i=1}^{3}\big(\sigma(\dot{C}'_i)-\sigma(\dot{C}_i)\big)\Phi_{P_i}(x). \quad (8.5)$$

Denote by l_i the length of P_i. Evidently, $l_1\geq l_2\geq l_3\geq 0$.

If $l_1>l_2$, then $\sigma(\dot{C}'_1)-\sigma(\dot{C}_1)=0$, since otherwise the term x^{l_1+1} would appear on the right-hand side of (8.5). Similarly, $l_2>l_3$ yields $\sigma(\dot{C}'_i)-\sigma(\dot{C}_i)=0$ for $i\in\{1,2\}$, making the right-hand side zero, which is impossible. If $l_2=l_3$, the term x^{l_2+1} appears unless $\sigma(\dot{C}'_2)-\sigma(\dot{C}_2)=\sigma(\dot{C}_3)-\sigma(\dot{C}'_3)$, but then all remaining terms vanish, again making the right-hand side zero.

Next, if $l_1=l_2$ then $\sigma(\dot{C}'_1)-\sigma(\dot{C}_1)=\sigma(\dot{C}_2)-\sigma(\dot{C}'_2)$, which reduces the difference to $\Phi_{\dot{G}}(x)-\Phi_{\dot{H}}(x)=2\big(\sigma(\dot{C}'_3)-\sigma(\dot{C}_3)\big)\Phi_{P_3}(x)$, again leading to a contradiction as above.

Finally, consider the case where $v\sim w$ and $v'\sim w'$. In this scenario, there are only paths P_1 and P_2, so the sum in (8.5) has only two summands. This case can be addressed by slight modifications of the previous argument. □

Problems

8.1 So far, we have encountered the Ulam conjecture and its spectral counterpart concerning unsigned graphs. Two additional problems arise naturally, and all four are illustrated in the following 'what is given – what to reconstruct' diagram, where $\mathscr{D}(G)$ denotes the collection of vertex-deleted subgraphs:

	$\mathscr{D}(G)$	$\mathscr{P}(G)$
G	A	B
Φ_G	C	D

Clearly, the Ulam conjecture corresponds to problem A and the spectral counterpart corresponds to C, whereas the remaining two are evident from the diagram. If G is reconstructible from $\mathscr{D}(G)$, then Φ_G is reconstructible from $\mathscr{P}(G)$, recorded as $A \Rightarrow C$. Deduce the other implications.

8.2 Show that the identity $\Phi'_{\dot{G}}(x) = \sum_{i=1}^{n} \Phi_{\dot{G}_i}(x)$ transfers from the theory of unsigned graphs to signed graphs without significant modification.

8.3 Show that polynomial reconstruction for unsigned graphs is unique whenever at least one vertex-deleted subgraph is integral. (Prove that, for sufficiently large order, such a subgraph must have a repeated eigenvalue.)

8.4 Show that polynomial reconstruction is unique for disconnected or bipartite unsigned graphs. (Difficult problems.)

8.5 Prove Theorems 8.4 and 8.6.

8.6 Analyse the chain of inequalities in Theorem 8.6 and identify a sufficient condition under which it fails to hold. Find a pair of signed graphs whose eigenvalues satisfy the same chain.

8.7 Show that polynomial reconstruction is unique for signed graphs of order n having at least $n/2$ pendant vertices.

8.8 Extend the results of Subsections 8.3.3 and 8.3.4 by allowing the existence of pendant vertices attached to the considered signed graphs.

8.9 Show that polynomial reconstruction is unique for tricyclic signed graphs without pendant vertices.

8.10 Show that polynomial reconstruction is unique for unsigned regular graphs and investigate the same problem in the setting of signed graphs with vertex degree at least 3.

9

Related Structures

In this chapter, the focus is on certain mathematical structures closely related to signed graphs. Specifically, we explore constructions of linear ternary codes generated by strongly regular signed graphs, investigate connections between signed graphs and systems of lines in a Euclidean space that are either pairwise orthogonal or subtend fixed angles, and examine the interplay between the spectrum of signed graphs and the skew spectrum of associated oriented graphs.

9.1 Linear Ternary Codes

The results in this section are drawn from [211]. The codewords of a linear ternary code are vectors over the finite field $\mathbb{F}_3$ consisting of integers modulo 3. For convenience, we represent the elements of $\mathbb{F}_3$ as $\{1,0,-1\}$, where the element 2 is identified with -1. Our focus is on codes generated by matrices of the form $A+\beta I$, where A is the adjacency matrix of a signed graph, I is the identity matrix and $\beta \in \mathbb{F}_3$. Every linear ternary code admitting at least one generator matrix of this form falls within the scope of our approach.

In the following two subsections we provide a concise introduction to linear ternary codes and examine the impact of switching operations on codes associated with signed graphs. Further discussions on such codes are presented in later subsections.

9.1.1 Linear Codes

We begin with formal definitions. A *linear q-ary code* of *length n* and *dimension k* is a k-dimensional linear subspace $\mathscr{C}$ of the vector space $\mathbb{F}_q^n$, where $\mathbb{F}_q$

is the field of prime order q. The elements of $\mathscr{C}$ are called *codewords*. The *size* of $\mathscr{C}$ is the number of codewords and equals q^k.

The *weight* $w(\mathbf{x})$ of a codeword $\mathbf{x}$ is the number of its non-zero entries and the *distance* between two codewords is the Hamming distance between them, that is, the number of entries in which they differ. The *distance* d of $\mathscr{C}$ is the minimum distance between distinct codewords and equals the minimum weight of its non-zero codewords [54, 144].

A linear code $\mathscr{C}$ with these parameters is called an $[n,k,d]$ code (or $[n,k,d]_q$ code). Simultaneously with the introduced notation, we use $|\mathscr{C}|$ and $\operatorname{dist}(\mathscr{C})$ to denote the dimension and the distance of $\mathscr{C}$, respectively.

All codewords of $\mathscr{C}$ are linear combinations of a selected set of codewords that form a basis of $\mathscr{C}$. The elements of this set are usually seen as the rows of a matrix A called the *generator* or the *generating matrix* for $\mathscr{C}$. Evidently, over $\mathbb{F}_q$, $\operatorname{rank}(A) = |\mathscr{C}|$.

The *dual code* $\mathscr{C}^\perp$ is defined by

$$\mathscr{C}^\perp = \{\mathbf{c} \in \mathbb{F}_q^n : \mathbf{x} \cdot \mathbf{c} = 0 \pmod{q} \text{ for all } \mathbf{x} \in \mathscr{C}\}.$$

It is known that $|\mathscr{C}| + |\mathscr{C}^\perp| = n$ [144]. We say that $\mathscr{C}$ is *self-orthogonal* if $\mathscr{C} \subseteq \mathscr{C}^\perp$ (i.e., if its codewords are pairwise orthogonal). The code $\mathscr{C}$ is called *self-dual* if $\mathscr{C} = \mathscr{C}^\perp$.

Clearly, $\mathbf{c} \in \mathscr{C}^\perp$ holds if and only if $A\mathbf{c} = \mathbf{0}$ (over $\mathbb{F}_q^n$), and A is also called the *parity-check matrix* for $\mathscr{C}^\perp$. To specify that $\mathscr{C}$ is generated by A, we write $\mathscr{C}_A$.

In what follows, we deal with linear ternary codes generated by symmetric matrices, so q is always 3 and A (or any other generator) is assumed to be a symmetric $n \times n$ matrix. Then n is the length of the corresponding code.

Note that the weight of every codeword of a linear self-orthogonal ternary code is a multiple of three. To see this, observe that

$$0 = \mathbf{x} \cdot \mathbf{x} = w(\mathbf{x}) \pmod{3},$$

where the first equality follows from self-orthogonality.

9.1.2 Switching Equivalence and the Corresponding Codes

If $\dot{G}$ and $\dot{H}$ are switching equivalent, then we have $A_{\dot{H}} = S^{-1}A_{\dot{G}}S$, where S is a diagonal switching matrix. Accordingly, the generating matrix of a code based on $\dot{H}$ has the form $S^{-1}A_{\dot{G}}S + \beta I$. In fact, this matrix is obtained from $A_{\dot{G}} + \beta I$ by negating all rows and all columns that correspond to -1 entries in S. In other words, switching-equivalent signed graphs generate equivalent codes. (We recall that codes are equivalent if their generating matrices can be obtained from one another by elementary operations on rows and columns; for a formal definition, the reader is referred to [144].)

We will deal with generating matrices of the form $\alpha A+\beta I+\gamma J$, where A is the adjacency matrix of a signed graph and $\alpha,\beta,\gamma\in\mathbb{F}_3$. Such a matrix can be rewritten as $A'+(\beta+\gamma)I$, where A' is the adjacency matrix of another signed graph, so this concept does not deviate from the one announced before. However, the switching equivalence of $\dot{G}$ and $\dot{H}$ does not imply the equivalence of codes generated by $\alpha A_{\dot{G}}+\beta I+\gamma J$ and $\alpha A_{\dot{H}}+\beta I+\gamma J$. Moreover, if $A_{\dot{H}}=S^{-1}A_{\dot{G}}S$, then $\mathscr{C}_{\alpha A_{\dot{G}}+\beta I+\gamma J}$ is equivalent to $\mathscr{C}_{\alpha A_{\dot{H}}+\beta I+\gamma S^{-1}JS}$, proving which is an easy exercise.

9.1.3 Codes of Signed Graphs with Two Eigenvalues

We demonstrate the technique in the case of connected signed graphs with two eigenvalues (which are necessarily strongly regular). The eigenvalues in question are denoted by λ and μ, and their multiplicities are m_λ and m_μ. Assume first that λ and μ are integral. We write 3-rank(A) to denote the rank of A over $\mathbb{F}_3$.

Theorem 9.1 *Let $\dot{G}$ be a signed graph with n vertices and two eigenvalues, λ of multiplicity m_λ and μ of multiplicity m_μ. If λ and μ are integral and A is the adjacency matrix of $\dot{G}$, then the following statements hold:*

(i) *If $\lambda\mu=0 \pmod 3$ and $\lambda+\mu=0 \pmod 3$, then $\mathscr{C}_A$ is self-orthogonal with $|\mathscr{C}_A|\leq\min\{m_\lambda,m_\mu\}$. Also, $\mathscr{C}_{A+I}=\mathscr{C}_{A-I}=\mathbb{F}_3^n$.*

(ii) *If $\lambda\mu=0 \pmod 3$ and $\lambda+\mu\neq 0 \pmod 3$, then $\mathscr{C}_{A-(\lambda+\mu)I}=\mathscr{C}_A^{\perp}$ and exactly one of the equalities $\lambda=0 \pmod 3$, $\mu=0 \pmod 3$ holds. If $\lambda=0 \pmod 3$, then $|\mathscr{C}_A|=m_\mu$ and $|\mathscr{C}_{A-(\lambda+\mu)I}|=m_\lambda$.*

(iii) *If $\lambda\mu\neq 0 \pmod 3$ and $\lambda+\mu=0 \pmod 3$, then $\mathscr{C}_A=\mathbb{F}_3^n$ and $\mathscr{C}_{A-I}=\mathscr{C}_{A+I}^{\perp}$, where $\{|\mathscr{C}_{A-I}|,|\mathscr{C}_{A+I}|\}=\{m_\lambda,m_\mu\}$.*

(iv) *If $\lambda\mu\neq 0 \pmod 3$ and $\lambda+\mu\neq 0 \pmod 3$, then $\mathscr{C}_A=\mathscr{C}_{A-(\lambda+\mu)I}=\mathbb{F}_3^n$ and $\mathscr{C}_{A-\lambda I}=\mathscr{C}_{A-\mu I}\subseteq\mathscr{C}_{A-\lambda I}^{\perp}$, with $|\mathscr{C}_{A-\lambda I}|\leq\min\{m_\lambda,m_\mu\}$.*

Proof We prove parts (i) and (ii), as the remaining ones are proved in a similar way. We have

$$A^2+bA-rI=O, \tag{9.1}$$

with $b=-(\lambda+\mu)$ and $r=-\lambda\mu$.

(i): The identity (9.1) reduces to $A^2=O \pmod 3$, which implies that $\mathscr{C}_A$ is self-orthogonal. Under the assumptions of this part of the statement, we have $\lambda=\mu=0 \pmod 3$. Over the real numbers, we have rank$(\lambda I-A)=m_\mu$, and thus 3-rank$(\lambda I - A) =$ 3-rank$(A) \leq m_\mu$, and similarly 3-rank$(A) \leq m_\lambda$, which gives the assertion.

Next, we have $(A - I)(A + I) = A^2 - I = -I \pmod 3$. In other words, the canonical basis of $\mathbb{F}_3^n$ lies in both $\mathscr{C}_{A-I}$ and $\mathscr{C}_{A+I}$.

(ii): Here we have $A(A + bI) = O \pmod 3$. This identity immediately gives $\mathscr{C}_{A+bI} \subseteq \mathscr{C}_A^\perp$. To prove the opposite inclusion, we take $\mathbf{c} \in \mathscr{C}_A^\perp$. From $A\mathbf{c} = 0 \pmod 3$, we obtain $(A + bI)\mathbf{c} = b\mathbf{c} \pmod 3$, and so $b\mathbf{c} \in \mathscr{C}_{A+bI}$, which means that $\mathbf{c} \in \mathscr{C}_{A+bI}$, that is, $\mathscr{C}_{A+bI} = \mathscr{C}_A^\perp$.

From $\lambda\mu = 0 \pmod 3$ we find that at least one of $\lambda = 0 \pmod 3$, $\mu = 0 \pmod 3$ must hold. From $\lambda + \mu \neq 0 \pmod 3$ we see that these equalities do not hold simultaneously, which leads to the desired assertion.

Assume that $\lambda = 0 \pmod 3$. Then $\mu = -b \pmod 3$. Now, over the real numbers, $|xI - A| = (x - \lambda)^{m_\lambda}(x - \mu)^{m_\mu}$, giving $|xI - A| = x^{m_\lambda}(x - \mu)^{m_\mu} \pmod 3$, which implies 3-rank$(A) \geq m_\mu$. Similarly, over the real numbers we have $|xI - (A + bI)| = |(x - b)I - A| = (x - b + \lambda)^{m_\lambda}(x - b - \mu)^{m_\mu}$, giving $|xI - (A + bI)| = (x + \mu)^{m_\lambda} x^{m_\mu} \pmod 3$, which exactly as before leads to 3-rank$(A + bI) \geq m_\lambda$. From $\mathscr{C}_{A+bI} = \mathscr{C}_A^\perp$,

$$m_\mu \leq \text{3-rank}(A) = |\mathscr{C}_A| = n - |\mathscr{C}_{A+bI}| = n - \text{3-rank}(A + bI) \leq n - m_\lambda = m_\mu,$$

and thus $|\mathscr{C}_A| = m_\mu$ and $|\mathscr{C}_{A+bI}| = m_\lambda$. □

We proceed with the case in which λ and μ are non-integral.

Theorem 9.2 *Let $\dot{G}$ be a signed graph with n vertices and two eigenvalues, λ and μ. If λ and μ are non-integral, then $\lambda + \mu = 0$ and λ^2 is an integer. If A denotes the adjacency matrix of $\dot{G}$, then the following statements hold:*

- *(i) If $\lambda^2 = 0 \pmod 3$, then $\mathscr{C}_A$ is self-orthogonal, $|\mathscr{C}_A| \leq n/2$ and $\mathscr{C}_{A-I} = \mathscr{C}_{A+I} = \mathbb{F}_3^n$.*
- *(ii) If $\lambda^2 \neq 0 \pmod 3$, then $\mathscr{C}_A = \mathbb{F}_3^n$.*
  - *(a) For $\lambda^2 = 1 \pmod 3$, we have $\mathscr{C}_{A-I} = \mathscr{C}_{A+I}^\perp$, $|\mathscr{C}_{A-I}| = |\mathscr{C}_{A+I}| = n/2$.*
  - *(b) For $\lambda^2 = -1 \pmod 3$, we have $\mathscr{C}_{A-I} = \mathscr{C}_{A+I} = \mathbb{F}_3^n$.*

Proof Except for calculating the dimension of $\mathscr{C}_A$ in (i) and the proof of (ii)(b), the argument proceeds exactly as in part (i) of Theorem 9.1.

For the dimension in (i), we have $|\mathscr{C}_A| + |\mathscr{C}_A^\perp| = n$ and $\mathscr{C}_A \subseteq \mathscr{C}_A^\perp$, which gives the desired result. The result of (ii)(b) follows from

$$(A - I)(A + I) = (\lambda^2 - 1)I = I,$$

and we are done. □

The exposition is concluded by presenting a class of self-dual codes arising from a class of related matrices.

Theorem 9.3 *Let M be an $n \times n$ $(1,0,-1)$-matrix with eigenvalues $\sqrt{r}$ and $-\sqrt{r}$, where r is a positive integer.*

(i) The spectrum of

$$A = \begin{pmatrix} M & I \\ I & -M \end{pmatrix}$$

is $[\sqrt{r+1}^{\,n}, (-\sqrt{r+1})^n]$.
(ii) For $r = -1 \pmod 3$, $\mathscr{C}_A$ is self-dual. Otherwise, $\mathscr{C}_A = \mathbb{F}_3^{2n}$.

Proof (i): Since the blocks of $xI - A$ mutually commute, we have

$$|xI - A| = (xI_n - M)(xI_n + M) - I_n = (x^2 - 1)I_n - M^2 = (x^2 - 1 - r)I_n,$$

which gives the desired spectrum.

(ii): $A^2 = (r+1)I$, and so for $r \neq -1 \pmod 3$, we have $\mathscr{C}_A = \mathbb{F}_3^{2n}$. For $r = -1 \pmod 3$, $\mathscr{C}_A \subseteq \mathscr{C}_A^{\perp}$. Clearly, the rows of the $n \times 2n$ matrix $(M \mid I)$ are linearly independent, which gives $|\mathscr{C}_A| \geq n$. The previous inclusion, together with $|\mathscr{C}_A| + |\mathscr{C}_A^{\perp}| = 2n$, gives $|\mathscr{C}_A| = n$, and then we necessarily have $|\mathscr{C}_A^{\perp}| = n$. Hence, $\mathscr{C}_A = \mathscr{C}_A^{\perp}$. □

9.1.4 Examples

Here one can find several examples that illustrate the previous results, followed by additional remarks at the end of the subsection.

Example 9.4 We consider signed graphs with two eigenvalues and maximum vertex degree at most 5, as determined in Subsection 6.1.1. There is exactly one connected non-complete such signed graph with vertex degree 3: $\dot{Q}_3$ illustrated in Figure 3.4. Its spectrum is $[\sqrt{3}^{\,4}, (-\sqrt{3})^4]$, so it falls under Theorem 9.2(i). Therefore, the corresponding code is self-orthogonal and its distance is 3.

For vertex degree 4, the signed graphs $\dot{S}_{14}$ and $\dot{Q}_4$ are covered by Theorem 9.1(iii). Regarding the remaining family, if A is the adjacency matrix of $\mathscr{L}(\ddot{C}_i)$ with $i \geq 3$, then by Theorem 9.1(iii) we have $\mathscr{C}_{A-I} = \mathscr{C}_{A+I}^{\perp}$ and $|\mathscr{C}_{A-I}| = |\mathscr{C}_{A+I}| = i$. Furthermore, the minimum distances satisfy

$$\mathrm{dist}(\mathscr{C}_{A-I}) \leq r + 1 = 5,$$

where r is the vertex degree, and similarly for $\mathrm{dist}(\mathscr{C}_{A+I})$.

Finally, non-complete signed graphs with two eigenvalues and vertex degree 5 are covered by Theorem 9.2(ii), and hence $\mathscr{C}_A = \mathscr{C}_{A+I} = \mathscr{C}_{A-I} = \mathbb{F}_3^n$.

Example 9.5 According to Theorem 4.18, there exists a unique r-dimensional signed cube containing only negative quadrangles. In the previous example, we considered the cases $\dot{Q}_3$ and $\dot{Q}_4$. What follows is a generalization to arbitrary r. The spectrum of $\dot{Q}_r$ is $[\sqrt{r}^r, (-\sqrt{r})^r]$. Now, if A is the adjacency matrix of $\dot{Q}_r$, then by Theorems 9.1(i), 9.1(iii) and 9.2, we obtain the following results:

- If $r = 0 \pmod 3$, then $\mathscr{C}_A$ is self-orthogonal and $|\mathscr{C}_A| \leq r$.
- If $r = 1 \pmod 3$, then $\mathscr{C}_{A-I} = \mathscr{C}_{A+I}^{\perp}$ and $|\mathscr{C}_{A-I}| = |\mathscr{C}_{A+I}| = r$.
- If $r = -1 \pmod 3$, then $\mathscr{C}_A = \mathscr{C}_{A-I} = \mathscr{C}_{A+I} = \mathbb{F}_3^{2r}$.

Example 9.6 Let A be the $rn \times rn$ adjacency matrix of the signed line graph $\mathscr{L}(\ddot{G})$, where $\ddot{G}$ is a regular signed doubled graph of order n and vertex degree $2r$ ($r \geq 2$). On the basis of the results in the previous subsection, we arrive at:

- If $r = 0 \pmod 3$, then $\mathscr{C}_{A-I}$ is self-orthogonal with $|\mathscr{C}_{A-I}| \leq n$.
- If $r = 1 \pmod 3$, then $\mathscr{C}_{A-I} = \mathscr{C}_A^{\perp}$ with $|\mathscr{C}_{A-I}| = n$ and $|\mathscr{C}_A| = (r-1)n$.
- If $r = -1 \pmod 3$, then $\mathscr{C}_{A-I} = \mathscr{C}_{A+I}^{\perp}$ with $|\mathscr{C}_{A-I}|, |\mathscr{C}_{A+I}| \in \{n, (r-1)n\}$.

Example 9.7 Let A be the $n \times n$ adjacency matrix of an exceptional signed graph $\dot{G}$ with eigenvalues λ and -2, where the multiplicity of λ is k with $6 \leq k \leq 8$. We have:

- If $\lambda = 1 \pmod 3$, then $\mathscr{C}_{A-I}$ is self-orthogonal with $|\mathscr{C}_{A-I}| \leq k$.
- If $\lambda = -1 \pmod 3$, then $\mathscr{C}_{A-I} = \mathscr{C}_{A+I}^{\perp}$ with $|\mathscr{C}_{A-I}|, |\mathscr{C}_{A+I}| \in \{k, \frac{1}{2}k\lambda\}$.

Example 9.8 Every $n \times n$ matrix of Theorem 9.3 gives rise to an infinite family of self-dual codes constructed in the following way. Denote this matrix by M_0 and, for $i \geq 1$, set

$$M_i = \begin{pmatrix} M_{i-1} & I \\ I & -M_{i-1} \end{pmatrix}.$$

By Theorem 9.3(i), the spectrum of M_i is $[\sqrt{r+i}^{2^{i-1}n}, (-\sqrt{r+i})^{2^{i-1}n}]$. By the second part of the same theorem, M_i generates a self-dual code whenever we have $r + i = 0 \pmod 3$.

Needless to add, Theorem 9.3 applies to the adjacency matrix of a signed graph with two symmetric eigenvalues in the role of M.

By taking M_{i-1} to be the weighing matrix $W(6,5)$ of Theorem 6.5, we obtain that M_i generates the famous Golay ternary code [98]. Its dimension and distance are both equal to 6.

The next example belongs to a broader context in which the corresponding signed graph has more than two eigenvalues, but is still strongly regular.

Example 9.9 The 13-*Paley graph* is a strongly regular graph of order 13 and degree 6. Its Seidel matrix A corresponds to the complete signed graph with spectrum $\left[\sqrt{13}^6, 0, (-\sqrt{13})^6\right]$. It is not difficult to compute that $\mathscr{C}_{A-I}$ has dimension 7 and distance 5. Similarly, both the dimension and distance of $\mathscr{C}_{A-I+J}$ equal 6.

An $[n,k,d]_q$ code is called *optimal* if n is the minimum number such that there exists at least one code with these parameters. In the previous examples, codes based on $\mathscr{L}(\ddot{C}_3)$, $\mathscr{L}(\ddot{C}_4)$, the Golay code and the code $\mathscr{C}_{A-I}$ of the latter example are optimal.

The length of every optimal code is bounded from below by the Griesmer bound [102]. We have checked that this bound is attained for the code based on $\mathscr{L}(\ddot{C}_3)$ and the Golay code.

9.2 Line Systems

A *line system* refers to a set of one-dimensional subspaces of $\mathbb{R}^k$, where the *distance* between a pair of lines is equal to the (smaller) angle between them. A system of n lines is uniquely determined by a set of unit vectors $\{\mathbf{y}_1, \mathbf{y}_2, \dots, \mathbf{y}_n\}$, where each vector spans one line. If the angle between the lines i and j is ϕ_{ij}, then we have $\mathbf{y}_i \cdot \mathbf{y}_j = \pm\cos\phi_{ij}$, and if we set

$$N = (\mathbf{y}_1 \mid \mathbf{y}_2 \mid \cdots \mid \mathbf{y}_n),$$

then $N^{\intercal}N$ is the $n \times n$ positive semidefinite Gram matrix of all possible inner products of the corresponding vectors.

Following [233], we consider lines that are pairwise orthogonal or at a fixed distance ϕ $(0 < \phi < \pi/2)$. By setting $\psi = \cos\phi$, we get that the corresponding Gram matrix has the form

$$N^{\intercal}N = I + \psi A, \tag{9.2}$$

where A is the adjacency matrix of some signed graph. Clearly, replacing $\mathbf{y}_i$ with $-\mathbf{y}_i$ does not affect the line system, but changes the corresponding signed graph into a switching-equivalent one.

This approach may be regarded as a generalization of the extensively studied concept of equidistant lines and its connection to Seidel matrices of unsigned graphs (see [139, 162, 199] for the foundational results, and [54] for a comprehensive treatment).

If $\mathbf{y}_1, \mathbf{y}_2, \dots, \mathbf{y}_n$ are linearly dependent, the matrix $N^{\intercal}N$ is singular and thus $-1/\psi$ is the least eigenvalue of A; if $k < n$, its multiplicity is at least $n - m_{-1/\psi}$.

Conversely, if $-1/\psi$ is the least eigenvalue of an $n\times n$ adjacency matrix A, then the matrix $(1/\psi)I+A$ is positive semidefinite of rank $n-m_{-1/\psi}$ and has the form $N^{\intercal}N$, and so it gives the system of lines embedded into $\mathbb{R}^{n-m_{-1/\psi}}$. Therefore, there is a natural relation between a signed graph and a line system represented by N.

By setting $\Lambda=(1/k)\sum_{i=1}^{k}\lambda_k$ and considering the spectral moments that arise from trace$(A)=0$ and trace$(A^2)=2m$, we compute

$$\begin{aligned}0\leq\sum_{i=1}^{k}(\lambda_i-\Lambda)^2&=\sum_{i=1}^{k}\lambda_i^k-2\Lambda\sum_{i=1}^{k}\lambda_i+k\Lambda^2\\&=2m-\frac{n-k}{\psi^2}-2\frac{(n-k)^2}{k\psi^2}+\frac{(n-k)^2}{k\psi^2}\\&=2m-n\frac{n-k}{k\psi^2},\end{aligned}$$

giving

$$n(n-k)\leq 2mk\psi^2, \tag{9.3}$$

which is the *relative bound* for the number of our lines in the k-dimensional Euclidean space. (It is 'relative' because it depends on m and ψ.) The equality in (9.3) is attained if and only if $\lambda_1=\lambda_k$, that is, if and only if A is the adjacency matrix of a signed graph with two eigenvalues ($(n-k)/k\psi$ and $-1/\psi$). Such a signed graph is regular, and if its vertex degree is r, then $2m=rn$ and the bound can be rewritten as

$$n-k\leq rk\psi^2.$$

The lines the relative bound is attained for are called *extremal*. There might exist a maximal set of described lines in $\mathbb{R}^k$ for which this bound is not attained. One may find a sequence of quadruples (n,r,k,ψ) attaining the bound, but the real question is whether the corresponding signed graph exists or not.

Example 9.10 The signed graph illustrated in Figure 1.3 has been met a few times on the previous pages, say in Theorem 3.34 and Figure 3.4 (where it was denoted by $\dot{T}_6$) or in Theorem 6.4 (after which it was recognized as the signed line graph $\mathscr{L}(\ddot{C}_3)$).

This signed graph corresponds to the quadruple $(6,4,3,1/2)$ that attains the relative bound. Thus, there are sic lines in $\mathbb{R}^3$ that are pairwise at angle $\pi/3$ or $\pi/2$. For example, they are determined by the (transposes of) vectors $(1,0,0)$, $(0,1,0)$, $(1/2,1/2,\sqrt{2}/2)$, $(-1/2,-1/2,\sqrt{2}/2)$, $(1/2,-1/2,\sqrt{2}/2)$ and $(1/2,1/2,-\sqrt{2}/2)$.

The discussion is concluded with the following result.

Theorem 9.11 ([233]) *Under the assumption that $N^{\intercal}N$ is singular, the matrix A of (9.2) has two eigenvalues, λ and μ, if and only if the matrix $NN^{\intercal}$ has exactly one non-zero eigenvalue $1-(\lambda/\mu)$.*

Proof If λ and μ are the eigenvalues of A, then by the identity (9.2), the eigenvalues of $N^{\intercal}N$ are $\lambda\psi+1$ and $\mu\psi+1$, where $\psi=\cos\phi$. Since this matrix is singular, $\psi=-1/\mu$ and thus its eigenvalues become $1-(\lambda/\mu)$ and 0. Because the matrices $N^{\intercal}N$ and $NN^{\intercal}$ share the same non-zero eigenvalues, the first implication follows. The opposite implication is obtained by reversing the argument. □

9.3 Oriented Graphs

We show that the theory based on the skew adjacency matrix of oriented graphs is strongly connected to the theory of spectra of signed graphs, and that many problems concerning oriented graphs can be transferred to the framework of signed graphs.

For a graph G of order n, an *oriented graph* G' is a pair (G,σ'), where σ' is the *edge orientation* satisfying $\sigma'(ij)\in\{i,j\}$ for every $ij\in E$. As for signed graphs, G is the underlying graph. If $\sigma'(ij)=j$, we say that the edge ij is *oriented* from i to j and designate this by $i\rightarrow j$; an oriented edge is also known as an *arc*.

The *skew adjacency matrix* $S_{G'}=(s_{ij})$ is the $n\times n$ matrix defined by

$$s_{ij}=\begin{cases}0 & \text{if } ij\notin E,\\ 1 & \text{if } \sigma'(ij)=j,\\ -1 & \text{if } \sigma'(ij)=i.\end{cases}$$

The eigenvalues of $S_{G'}$ are called the *skew eigenvalues* of G' and, together with their multiplicities, form the *skew spectrum* of G'. Observe that $S_{G'}$ is skew-symmetric, since it satisfies $S_{G'}=-S_{G'}^{\intercal}$, meaning that its spectrum consists of purely imaginary numbers. If $\mu\neq 0$ is an eigenvalue of $S_{G'}$, its conjugate $\overline{\mu}=-\mu$ is also an eigenvalue of $S_{G'}$, and thus the non-zero eigenvalues come in pairs μ and $-\mu$ (with equal multiplicity); hence, the rank of $S_{G'}$ is even.

An alternative matrix associated with oriented graphs, defined as $H_{G'}=\mathrm{i}S_{G'}$, where i is the imaginary unit, was introduced in a broader context by Guo and Mohar [104]. This approach is essentially equivalent to the one considered here. It is also related to the considerations presented in Section 10.2.

9.3.1 Association

The majority of this subsection is transferred from [223]. Clearly, if $S_{G'}$ is the skew adjacency matrix of an oriented graph $G' = (G, \sigma')$ and $A_{\dot{G}}$ is the adjacency matrix of a signed graph $\dot{G} = (G, \dot{\sigma})$, then $S^{\intercal}_{G'}S_{G'} = -S^2_{G'} = A^2_{\dot{G}}$ holds whenever

$$-\sum_{k=1}^{n} s_{ik}s_{kj} = \sum_{k=1}^{n} a_{ik}a_{kj} \quad \text{for } 1 \leq i \leq j \leq n.$$

The signature $\dot{\sigma}$ is *associated* with the orientation σ' (and also $\dot{G}$ is *associated* with G') if

$$\dot{\sigma}(ik)\dot{\sigma}(jk) = s_{ik}s_{jk} \text{ holds for every pair of edges } ik \text{ and } jk. \tag{9.4}$$

Being associated is a symmetric relation. We write $\dot{\sigma} \sim \sigma'$ to indicate that $\dot{\sigma}$ and σ' are mutually associated. Here is a useful lemma.

Lemma 9.12 *If $\dot{\sigma} \sim \sigma'$, then $-S^2_{G'} = A^2_{\dot{G}}$.*

Proof The (i, j)-entry of $-S^2_{G'}$, $-(s_{ij})^2$, is given by

$$-(s_{ij})^2 = -\sum_{k=1}^{n} s_{ik}s_{kj} = \sum_{k=1}^{n} s_{ik}s_{jk} = \sum_{k=1}^{n} \dot{\sigma}(ik)\dot{\sigma}(jk) = \sum_{k=1}^{n} a_{ik}a_{kj} = (a_{ij})^2,$$

where the last but one equality follows by the symmetry of $A_{\dot{G}}$. □

The identity $-S^2_{G'} = A^2_{\dot{G}}$ is not exclusive for associated graphs. We decide whether $\dot{\sigma} \sim \sigma'$ holds.

Theorem 9.13 *For a graph G and an orientation σ', there exists a signature $\dot{\sigma}$ associated with σ' if and only if G is bipartite.*

Proof Throughout the proof, let σ' denote an arbitrary orientation defining the skew adjacency matrix $S_{G'} = (s_{ij})$.

Suppose that there exists a signature $\dot{\sigma}$ associated with σ', while the underlying graph G is non-bipartite. Then G necessarily contains an odd cycle with vertices $i_1, i_2, \ldots, i_t$ arranged in their natural cyclic order, where t is odd.

By applying the relation in (9.4), we obtain $\dot{\sigma}(i_1i_2) = \dot{\sigma}(i_2i_3)\, s_{i_1i_2}\, s_{i_3i_2}$. Substituting for $\dot{\sigma}(i_2i_3)$ yields $\dot{\sigma}(i_1i_2) = \dot{\sigma}(i_3i_4)\, s_{i_2i_3}\, s_{i_4i_3}\, s_{i_1i_2}\, s_{i_3i_2}$. Continuing iteratively in this manner, we arrive at

$$\begin{aligned}\dot{\sigma}(i_1i_2) &= \dot{\sigma}(i_1i_2)s_{i_ti_1}s_{i_2i_1}s_{i_{t-1}i_t}s_{i_1i_t}\cdots s_{i_2i_3}s_{i_4i_3}s_{i_1i_2}s_{i_3i_2}\\ &= \dot{\sigma}(i_1i_2)(-1)^t = -\dot{\sigma}(i_1i_2),\end{aligned}$$

which is a contradiction as $\dot{\sigma}(i_1i_2) \neq 0$.

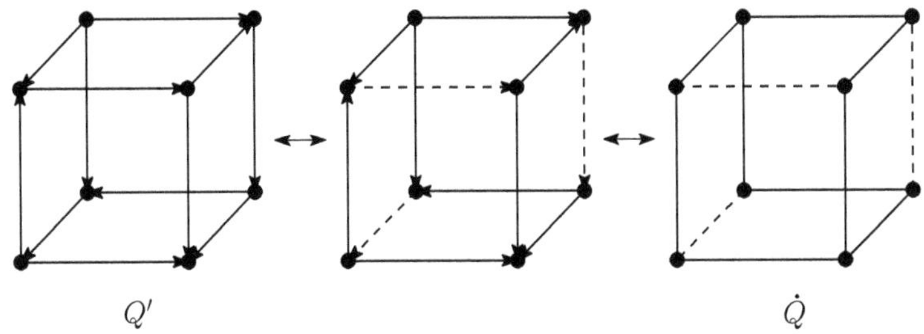

Figure 9.1 An oriented cube Q' and an associated signed cube $\dot{Q}$. Reproduced from [223].

Assume now that G is bipartite with colour classes V_1 and V_2. Without loss of generality, we also assume that G is connected. We claim that the signature

$$\dot{\sigma}(ij) = \begin{cases} 1 & \text{if } \sigma'(ij) \in V_1, \\ -1 & \text{if } \sigma'(ij) \in V_2 \end{cases} \tag{9.5}$$

is associated with σ'. To confirm this, let ik, jk be arbitrary edges of G. Since G is bipartite, i and j are in the same colour class, while k is in the other. If $s_{ik}s_{jk} = 1$, then either $\sigma'(ik) = \sigma'(jk) = k$ or $\sigma'(ik) = i$ and $\sigma'(jk) = j$. In each case, from (9.5) we obtain that $\dot{\sigma}(ik), \dot{\sigma}(jk)$ are of the same sign, which yields $\dot{\sigma}(ik)\dot{\sigma}(jk) = 1$. Similarly, $s_{ik}s_{jk} = -1$ occurs when exactly one of $\sigma'(ik), \sigma'(jk)$ is k, and then by (9.5), $\dot{\sigma}(ik), \dot{\sigma}(jk)$ differ in sign, which gives $\dot{\sigma}(ik)\dot{\sigma}(jk) = -1$. *Summa summarum*, we have $\dot{\sigma}(ik)\dot{\sigma}(jk) = s_{ik}s_{jk}$, which means that $\dot{\sigma}$ is associated with σ', and the proof is complete. □

An oriented graph G' is associated with a signed graph $\dot{G}$ if and only if it is also associated with its negation $-\dot{G}$. Consequently, G' admits exactly two associated signed graphs that share the same spectrum. Similarly, a signed graph $\dot{G}$ is associated with an oriented graph G' if and only if it is also associated with the oriented graph obtained by reversing the orientation of every arc in G'. Thus, each signed graph $\dot{G}$ corresponds to precisely two oriented graphs that possess identical skew spectra.

Henceforth, whenever we state 'let $\dot{G}$ be a signed graph associated with G'' (or similar expressions), we implicitly allow $\dot{G}$ to represent either of these two signed graphs differing by negation. To illustrate this phenomenon, we present the following example.

Example 9.14 Consider the oriented cube $Q' = (Q, \sigma')$ depicted in Figure 9.1. By Theorem 9.13, there exists a signed cube $\dot{Q}$ associated with Q', which can be constructed as follows.

Begin by assigning an arbitrary sign to a chosen edge. Subsequently, the signs of all remaining edges are uniquely determined by the rule given in (9.4). Practically, two adjacent edges of $\dot{Q}$ are assigned positive signs if and only if the corresponding edges in Q' are both oriented either towards or away from their common vertex. Applying this procedure yields the complete signature of $\dot{Q}$, as illustrated in the second part of Figure 9.1. By disregarding the orientations, we obtain the signed graph $\dot{Q}$, which is the three-dimensional signed cube with negative quadrangles, denoted previously by $\dot{Q}_3$.

Conversely, starting from $\dot{Q}$, assigning an arbitrary orientation to a fixed edge uniquely determines the oriented graph Q'.

9.3.2 Switching Classes

The concept of switching equivalence for oriented graphs is defined in a manner analogous to that for signed graphs. Two oriented graphs sharing the same underlying graph are said to be *switching equivalent* if there exists a subset $U \subseteq V$ of the vertex set such that one graph can be obtained from the other by reversing the orientation of every edge between U and its complement $V \setminus U$. More formally, oriented graphs G' and H' are switching equivalent if there exists a diagonal switching matrix S with entries ± 1 on the main diagonal such that $S_{H'} = S^{-1} S_{G'} S$. The notion of switching isomorphism is defined analogously, incorporating graph isomorphisms in addition to switching.

It follows immediately that if a signed graph $\dot{G}$ is associated with an oriented graph G', and if $\dot{G}$ (resp. G') is switched to $\dot{H}$ (resp. H'), then $\dot{H}$ is associated with H'. Furthermore, the same switching matrix effects the transformation from $\dot{G}$ to $\dot{H}$, as well as from G' to H'.

Theorem 9.15 ([216]) *For a graph G, the number of switching isomorphism classes of oriented graphs with underlying graph G is equal to the number of switching isomorphism classes of signed graphs with the same underlying graph.*

Proof Let the vertices of G be labelled by $1, 2, \dots, n$, and let G' be a representative of a switching isomorphism class of oriented graphs. Construct the signed graph $\dot{G}$ as follows. For each pair of vertices i, j with $i < j$ such that there is an arc between them in G', insert a positive edge between i and j in $\dot{G}$ if the arc is oriented from i to j; otherwise, insert a negative edge between these vertices. Note that this construction is symmetric in the sense that if G' induces $\dot{G}$, then $\dot{G}$ also induces G'. Perform this construction for every class representative.

We claim that the signed graphs thus obtained form a complete set of representatives for all switching isomorphism classes of signed graphs. To see this, suppose to the contrary that two such signed graphs, say $\dot{G}$ and $\dot{H}$ (induced by switching non-isomorphic oriented graphs G' and H', respectively), are switching isomorphic. Then there exist a permutation matrix P and a switching matrix S such that

$$A_{\dot{H}} = P^{-1}S^{-1}A_{\dot{G}}SP.$$

Applying the same transformation to the skew adjacency matrix $S_{G'}$ yields

$$S_{H'} = P^{-1}S^{-1}S_{G'}SP,$$

contradicting the assumption that G' and H' are switching non-isomorphic. Hence, the signed graphs obtained from the oriented class representatives are themselves switching non-isomorphic.

Finally, suppose that there exists another signed graph $\dot{G}$ representing a distinct switching isomorphism class, not among those constructed above. By the pigeonhole principle, the oriented graph G' induced by $\dot{G}$ must be switching isomorphic to an oriented graph induced by some other signed graph $\dot{H}$ in the constructed set. This implies, as before, that $\dot{G}$ and $\dot{H}$ are switching isomorphic, contradicting the distinctness of the classes. □

9.3.3 Fundamental Spectral Relations

The principal spectral connection is established in the following theorem. The definition of the *bipartite double* of an oriented graph G' parallels that of $\mathrm{bd}(\dot{G})$ for signed graphs. Suppose the vertices of G' are labelled $v_1, v_2, \ldots, v_n$. Then the vertex set of $\mathrm{bd}(G')$ consists of pairs $v_1 1, v_1 2, v_2 1, v_2 2, \ldots, v_n 1, v_n 2$. There is an arc from $v_p j$ to $v_q k$ in $\mathrm{bd}(G')$ if and only if there is an arc from v_p to v_q in G' and $j \neq k$.

As with signed graphs, the bipartite double $\mathrm{bd}(G')$ is bipartite, and it is connected if and only if G' is connected and non-bipartite. If G' itself is bipartite, then $\mathrm{bd}(G')$ decomposes into two disjoint copies of G'.

In matrix terms, the skew adjacency matrix of the bipartite double satisfies

$$S_{\mathrm{bd}(G')} = S_{G'} \otimes A_{K_2},$$

where $\otimes$ denotes the Kronecker product. Consequently, the skew eigenvalues of $S_{\mathrm{bd}(G')}$ are precisely the skew eigenvalues of G' multiplied by ± 1.

Theorem 9.16 ([223]) *For a graph G, let $G' = (G, \sigma')$ be an oriented graph with* $\mathrm{rank}(S_{G'}) = 2k$. *The following hold:*

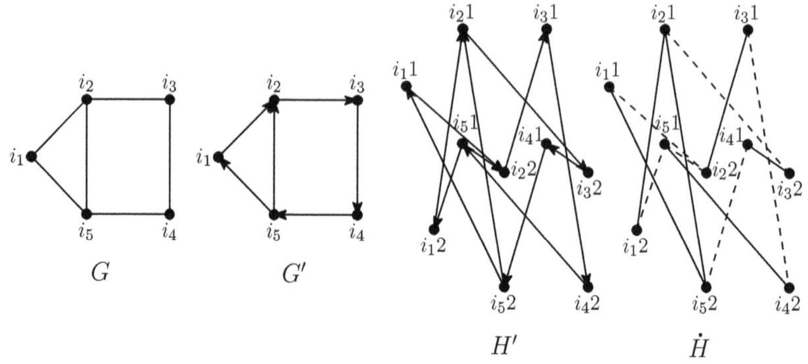

Figure 9.2 An example of G, G', $H' \cong \mathrm{bd}(G')$ and $\dot{H}$. Reproduced from [216].

(i) If G is bipartite and $\dot{\sigma} \sim \sigma'$, then $\pm \mathrm{i}\lambda_1, \pm \mathrm{i}\lambda_2, \ldots, \pm \mathrm{i}\lambda_k, 0^{n-2k}$ are the skew eigenvalues of G' if and only if $\pm\lambda_1, \pm\lambda_2, \ldots, \pm\lambda_k, 0^{n-2k}$ are the eigenvalues of $\dot{G} = (G, \dot{\sigma})$.

(ii) If G is non-bipartite, $H' \cong (\mathrm{bd}(G), \sigma'')$ is the bipartite double of G' and $\dot{\sigma} \sim \sigma''$, then $\pm \mathrm{i}\lambda_1, \pm \mathrm{i}\lambda_2, \ldots, \pm \mathrm{i}\lambda_k, 0^{n-2k}$ are the skew eigenvalues of G' if and only if

$$(\pm\lambda_1)^2, (\pm\lambda_2)^2, \ldots, (\pm\lambda_k)^2, 0^{2(n-2k)}$$

are the eigenvalues of $\dot{H} \cong (\mathrm{bd}(G), \dot{\sigma})$.

Proof (i): By Lemma 9.12 we have $-S_{G'}^2 = A_{\dot{G}}^2$, and thus $\pm \mathrm{i}\lambda_1, \pm \mathrm{i}\lambda_2, \ldots,$ $\pm \mathrm{i}\lambda_k, 0^{n-2k}$ are the skew eigenvalues of G'. Then

$$(\lambda_1^2)^2, (\lambda_2^2)^2, \ldots, (\lambda_k^2)^2 \tag{9.6}$$

are the non-zero eigenvalues of $A_{\dot{G}}^2$; we retain the use of exponential notation for representing multiplicities. Since $\dot{G}$ is bipartite, its spectrum is symmetric with respect to the origin, and so its eigenvalues are $\pm\lambda_1, \pm\lambda_2, \ldots, \pm\lambda_k, 0^{n-2k}$. Conversely, if these constitute the eigenvalues of $\dot{G}$, then the non-zero eigenvalues of $-S_{G'}^2$ are given by (9.6). Since every skew eigenvalue of G' is paired with its complex conjugate, we get the assertion.

(ii): The necessity follows from (i) since the skew eigenvalues of H' are $(\pm \mathrm{i}\lambda_1)^2, (\pm \mathrm{i}\lambda_2)^2, \ldots, (\pm \mathrm{i}\lambda_k)^2, 0^{2(n-2k)}$. The sufficiency follows from the same part and the fact that the non-zero skew eigenvalues come as complex conjugates. □

An example for part (ii) is given in Figure 9.2. In other words, if G is bipartite, then there is a signature $\dot{\sigma}$ (induced by σ') such that the skew spectrum of G' is obtained from the spectrum of $\dot{G}$, and vice versa. If G is non-bipartite, then the skew spectrum of G' is obtained from the spectrum of $\dot{H} \cong (\mathrm{bd}(G), \dot{\sigma})$.

9.3.4 Characteristic Polynomials

Here are relations between the coefficients of the characteristic polynomials.

Theorem 9.17 ([232]) *If $\sum_{i=0}^{n} s_i x^{n-i}$ is the characteristic polynomial of (the skew adjacency matrix of) an oriented graph G', then $s_i = 0$ for i odd and $s_i \geq 0$ for i even. If G' is bipartite and $\sum_{i=0}^{\lfloor n/2 \rfloor} a_{2i} x^{n-2i}$ is the characteristic polynomial of an associated signed graph, then*

$$a_{2i} = (-1)^i s_{2i}. \tag{9.7}$$

If G' is non-bipartite and $\sum_{i=0}^{n} a_{2i} x^{n-2i}$ is the characteristic polynomial of $\dot{H}$ (where $\dot{H}$ is as in the formulation of Theorem 9.16(ii)), then

$$a_{2i} = (-1)^i \sum_{\substack{j=-\min\{i,n-i\} \\ i \equiv j \ (\mathrm{mod}\, 2)}}^{\min\{i,n-i\}} s_{i-j} s_{i+j}. \tag{9.8}$$

Proof Under the assumptions of Theorem 9.16, the characteristic polynomial of the skew adjacency matrix of G' is

$$x^n + s_1 x^{n-1} + \cdots + s_{n-1} x + s_n = x^{n-2k}(x^2 + \lambda_1^2)(x^2 + \lambda_2^2) \cdots (x^2 + \lambda_k^2). \tag{9.9}$$

We immediately obtain $s_i = 0$ for i odd (since s_i is the ith elementary symmetric polynomial in the eigenvalues) and $s_i \geq 0$ for i even.

If G' is bipartite, then the characteristic polynomial of an associated signed graph is

$$\sum_{i=0}^{\lfloor n/2 \rfloor} a_{2i} x^{n-2i} = x^{n-2k}(x^2 - \lambda_1^2)(x^2 - \lambda_2^2) \cdots (x^2 - \lambda_k^2),$$

which yields that even coefficients alternate in sign, and $|a_{2i}| = s_{2i}$. This leads to (9.7).

If G' is non-bipartite, then the characteristic polynomial of $\dot{H}$ is

$$\sum_{i=0}^{n} a_{2i} x^{n-2i} = \left(x^{n-2k}(x^2 - \lambda_1^2)(x^2 - \lambda_2^2) \cdots (x^2 - \lambda_k^2)\right)^2.$$

Comparing this with (9.9), we arrive at

$$a_{2i} = (-1)^i \sum_{j=-\min\{i,n-i\}}^{\min\{i,n-i\}} s_{i-j} s_{i+j},$$

but we already know that odd coefficients under the sum are zero, so the summation reduces to even coefficients, which leads to (9.8). □

For $2i \leq n$, (9.8) is simplified to

$$a_{2i} = (-1)^i \sum_{j=0}^{i} s_{2j} s_{2(i-j)}.$$

On the basis of the previous theorem, one may derive the following coefficient theorem for the skew characteristic polynomial – an interpretation of the Sachs formula.

Theorem 9.18 ([59]) *Let $G' = (G, \sigma')$ and $\sum_{i=0}^{n} s_i x^{n-i}$ be an oriented graph and its skew characteristic polynomial, respectively. Then*

$$s_i = \begin{cases} 0 & \text{if } i \text{ is odd,} \\ \sum_{U \in \mathscr{U}_i^c} (-1)^{p(U)} 2^{|C(U)|} \sigma'(U) & \text{if } i \text{ is even.} \end{cases}$$

where $\mathscr{U}_i^c$ is the set of basic figures (disjoint unions of elementary figures) on i vertices in G' containing no odd cycles, $C(U)$ is the set of cycles in U, $p(U)$ is the number of components in U and $\sigma'(U) = \prod_{ij \in E(U)} s_{ij}$.

9.3.5 Eigenspaces

Following [232], let $\mathscr{E}(\lambda)$ and $\mathscr{E}(-\lambda)$ denote the eigenspaces corresponding to the eigenvalues $\lambda \neq 0$ and $-\lambda$ of a bipartite signed graph $\dot{G}$. Similarly, let $\mathscr{E}(\mathrm{i}\lambda)$ and $\mathscr{E}(-\mathrm{i}\lambda)$ denote the eigenspaces corresponding to the eigenvalues $\mathrm{i}\lambda$ and $-\mathrm{i}\lambda$ in the spectrum of an associated oriented graph G'. Then the union $\mathscr{E}(\mathrm{i}\lambda) \cup \mathscr{E}(-\mathrm{i}\lambda)$ is spanned (over $\mathbb{C}^n$) by the combined bases of $\mathscr{E}(\lambda)$ and $\mathscr{E}(-\lambda)$.

Indeed, since $-S_{G'}^2 = A_{\dot{G}}^2$, the eigenspace of λ^2 for $A_{\dot{G}}^2$ coincides with the eigenspace of $-\lambda^2$ for $S_{G'}^2$. Consequently, the former eigenspace is spanned by the union of the eigenbases of λ and $-\lambda$ (for $A_{\dot{G}}$), while the latter is spanned by the union of eigenbases of $\mathrm{i}\lambda$ and $-\mathrm{i}\lambda$ (for $S_{G'}$). By the same reasoning, if 0 is an eigenvalue of $\dot{G}$, then the corresponding eigenspaces for $\dot{G}$ and G' coincide.

Furthermore, it is straightforward to observe that if $\mathbf{x}$ is an eigenvector associated with the eigenvalue $\lambda \neq 0$ of a bipartite signed graph $\dot{G}$, then negating the entries corresponding to one colour class of $\dot{G}$ produces an eigenvector associated with $-\lambda$. We use this observation to prove the following result concerning the bipartite case.

Theorem 9.19 ([232]) *Let* $\mathbf{x}_j = \begin{pmatrix} \mathbf{x}_1 \\ \mathbf{x}_2 \end{pmatrix}$ *and* $\mathbf{x}_{-j} = \begin{pmatrix} -\mathbf{x}_1 \\ \mathbf{x}_2 \end{pmatrix}$ *be eigenvectors associated with distinct eigenvalues* λ_j *and* $-\lambda_j$ *of a bipartite signed graph* $\dot{G}$ *with adjacency matrix*

$$A_{\dot{G}} = \begin{pmatrix} O & B^{\intercal} \\ B & O \end{pmatrix}. \tag{9.10}$$

Let, with a consistent vertex labelling, G' *be an associated oriented graph. Then* $\mathbf{x}_j + \mathrm{i}\mathbf{x}_{-j}$ *and* $\mathbf{x}_j - \mathrm{i}\mathbf{x}_{-j}$ *are eigenvectors associated with* $\mathrm{i}\lambda_j$ *and* $-\mathrm{i}\lambda_j$.

Proof If the adjacency matrix of $\dot{G}$ has the form (9.10), then the skew adjacency matrix of G' is $S_{G'} = \begin{pmatrix} O & B^{\intercal} \\ -B & O \end{pmatrix}$. We compute

$$\begin{aligned} S_{G'}(\mathbf{x}_j \pm \mathrm{i}\mathbf{x}_{-j}) &= \begin{pmatrix} B^{\intercal}\mathbf{x}_2 \\ -B\mathbf{x}_1 \end{pmatrix} \pm \mathrm{i} \begin{pmatrix} B^{\intercal}\mathbf{x}_2 \\ B\mathbf{x}_1 \end{pmatrix} = \begin{pmatrix} \lambda_j\mathbf{x}_1 \\ \lambda_j\mathbf{x}_2 \end{pmatrix} \pm \mathrm{i} \begin{pmatrix} \lambda_j\mathbf{x}_1 \\ \lambda_j\mathbf{x}_2 \end{pmatrix} \\ &= \pm\mathrm{i}\lambda_j \left(\begin{pmatrix} \mathbf{x}_1 \\ \mathbf{x}_2 \end{pmatrix} \pm \mathrm{i} \begin{pmatrix} -\mathbf{x}_1 \\ \mathbf{x}_2 \end{pmatrix} \right) \\ &= \pm\mathrm{i}\lambda_j(\mathbf{x}_j \pm \mathrm{i}\mathbf{x}_{-j}), \end{aligned}$$

as desired. □

Theorem 9.19 reveals that, in the bipartite case, the eigenspaces of $\mathrm{i}\lambda_j$ and $-\mathrm{i}\lambda_j$ of G' contain vectors of the form

$$\mathbf{y}_{\mathrm{i}j} = \begin{pmatrix} \mathbf{x}_1 \\ \mathbf{x}_2 \end{pmatrix} + \mathrm{i} \begin{pmatrix} -\mathbf{x}_1 \\ \mathbf{x}_2 \end{pmatrix}, \qquad \mathbf{y}_{-\mathrm{i}j} = \overline{\mathbf{y}}_{\mathrm{i}j} = \begin{pmatrix} \mathbf{x}_1 \\ \mathbf{x}_2 \end{pmatrix} - \mathrm{i} \begin{pmatrix} -\mathbf{x}_1 \\ \mathbf{x}_2 \end{pmatrix},$$

respectively. Conversely, if these eigenvectors are given, then the eigenvectors associated with λ_j and $-\lambda_j$ of $\dot{G}$ are easily computed. We proceed with the non-bipartite case.

Theorem 9.20 ([232]) *Let* G' *be a non-bipartite oriented graph with eigenvectors* $\mathbf{y}$ *and* $\overline{\mathbf{y}}$ *associated with* $\mathrm{i}\lambda_j$ *and* $-\mathrm{i}\lambda_j$*, respectively. Then the pair*

$$\begin{pmatrix} \mathbf{y} \\ \mathbf{y} \end{pmatrix}, \begin{pmatrix} \mathbf{y} \\ -\mathbf{y} \end{pmatrix} \quad \left(\textit{resp.} \begin{pmatrix} \overline{\mathbf{y}} \\ \overline{\mathbf{y}} \end{pmatrix}, \begin{pmatrix} \overline{\mathbf{y}} \\ -\overline{\mathbf{y}} \end{pmatrix} \right)$$

is associated with $\mathrm{i}\lambda_j$ *(resp.* $-\mathrm{i}\lambda_j$*) in the bipartite double* $\mathrm{bd}(G')$*, where*

$$S_{\mathrm{bd}(G')} = \begin{pmatrix} 0 & 1 \\ 1 & 0 \end{pmatrix} \otimes S_{G'}.$$

Let $\dot{H}$ *be a signed graph associated with* $\mathrm{bd}(G')$ *and* k *be the multiplicity of* λ_j *in the spectrum of* $\dot{H}$*. The first (resp. second) pair is spanned by* $\mathbf{x}_{jl} + \mathrm{i}\mathbf{x}_{-jl}$

(resp. $\mathbf{x}_{jl} - \mathrm{i}\mathbf{x}_{-jl}$*) for* $1 \le l \le k$*, where* $\mathbf{x}_{jl}$ *and* $\mathbf{x}_{-jl}$ *span the eigenspaces of* λ_j *and* $-\lambda_j$*, respectively.*

Proof Since $S_{\mathrm{bd}(G')}$ is the tensor product, the eiegnvectors of $\mathrm{i}\lambda_j$ for $\mathrm{bd}(G')$ are $(1,1)^\mathsf{T} \otimes \mathbf{y}$ and $(1,-1)^\mathsf{T} \otimes \mathbf{y}$, and similarly for $-\mathrm{i}\lambda_j$. This establishes the proof of the first part of the statement.

By Theorem 9.19, the eigenvectors corresponding to the eigenvalue $\mathrm{i}\lambda_j$ of $\mathrm{bd}(G')$ are given by

$$\mathbf{x}_{jl} + \mathrm{i}\mathbf{x}_{-jl}, \quad 1 \le l \le k.$$

These vectors are clearly linearly independent since $\mathbf{x}_{jl}$ and $\mathbf{x}_{-jl}$ themselves are linearly independent, so they span the first pair. The second pair is considered analogously. □

9.3.6 Transferring Results

We employ Theorems 9.13 and 9.16 to transfer certain results from the theory of signed graphs to oriented graphs, thereby either establishing new results or recovering several known facts for oriented graphs through their associated signed graphs. To this end, we introduce the following definitions.

An even cycle C'_{2l} is said to be *oriented uniformly* if, when traversing the cycle in either direction (clockwise or counterclockwise), the number of edges oriented in the direction of traversal is odd when l is odd, and even when l is even.

A *canonical orientation* of a connected bipartite graph G is an orientation in which all edges are oriented from one colour class to the other. In the disconnected case, the orientation is canonical if it is canonical on each connected component. It is clear that under such an orientation, every cycle is oriented uniformly.

Example 9.21 Note that a bipartite graph endowed with a canonical orientation corresponds to a homogeneous signed graph, which is necessarily switching equivalent to its underlying unsigned graph and thus shares the same spectrum (cf. Theorem 1.25). Since bipartite oriented graphs are switching equivalent if and only if their associated signed graphs are switching equivalent, we obtain the following characterization, originally conjectured in [68] and subsequently proved in [18]: if the skew eigenvalues of a bipartite oriented graph $G' = (G, \sigma')$ are given by $\pm\mathrm{i}\lambda_1, \pm\mathrm{i}\lambda_2, \ldots, \pm\mathrm{i}\lambda_n$, then the eigenvalues of the underlying graph G are $\pm\lambda_1, \pm\lambda_2, \ldots, \pm\lambda_n$ if and only if G' is switching equivalent to a canonically oriented graph.

The next result is proved in the same spirit. The original proof, which does not rely on signed graphs, is given in the corresponding reference.

Theorem 9.22 ([191]) *A graph G with eigenvalues $\lambda_1, \lambda_2, \ldots, \lambda_n$ is a forest if and only if $\mathrm{i}\lambda_1, \mathrm{i}\lambda_2, \ldots, \mathrm{i}\lambda_n$ are the skew eigenvalues of $G' = (G, \sigma')$, for any orientation σ'.*

An oriented graph is *maximal* for some property if it is not an induced subgraph of some other oriented graph with the same property.

Theorem 9.23 ([219, 234]) *Every maximal connected oriented graph whose skew spectral radius does not exceed 2 is equivalent to one of oriented graphs illustrated in Figure 9.3.*

The proof leverages the result of Theorem 6.4, which, combined with Theorems 9.13 and 9.16, immediately settles the bipartite case. The remaining argument relies on Theorem 9.16(ii) and a detailed analysis of bipartite doubles. In particular, it is shown that the oriented graphs T'_{2k} for odd k can be derived from their bipartite doubles T'_{4k}, and similarly S'_8 is obtained from its bipartite double Q'_4.

In the figure, the bipartite and non-bipartite families are shown separately, as the bipartite families correspond to the associated signed graphs and serve as bipartite doubles of the non-bipartite ones.

9.3.7 Cospectrality

Following the approach in [216], we examine the connections between cospectral signed graphs and cospectral oriented graphs, where the latter term denotes switching non-isomorphic oriented graphs that share the same skew spectrum. The subsequent result follows directly from Theorem 9.15.

Theorem 9.24 *If $S_1, S_2, \ldots, S_k$ are all the sets of cospectral oriented graphs with a bipartite underlying graph G, then there are exactly k sets of cospectral signed graphs with the same underlying graph, say $A_1, A_2, \ldots, A_k$. Moreover, the latter sets can be arranged in such a way that there is a bijection between S_i and A_i for $1 \leq i \leq k$.*

We proceed with some constructions of cospectral bipartite oriented graphs from cospectral non-bipartite signed graphs.

Theorem 9.25 *Let $\dot{G}$ and $\dot{H}$ be cospectral non-bipartite signed graphs, and assume that their bipartite doubles are switching non-isomorphic. Then the oriented graphs G' and H' associated with $\mathrm{bd}(\dot{G})$ and $\mathrm{bd}(\dot{H})$ are cospectral.*

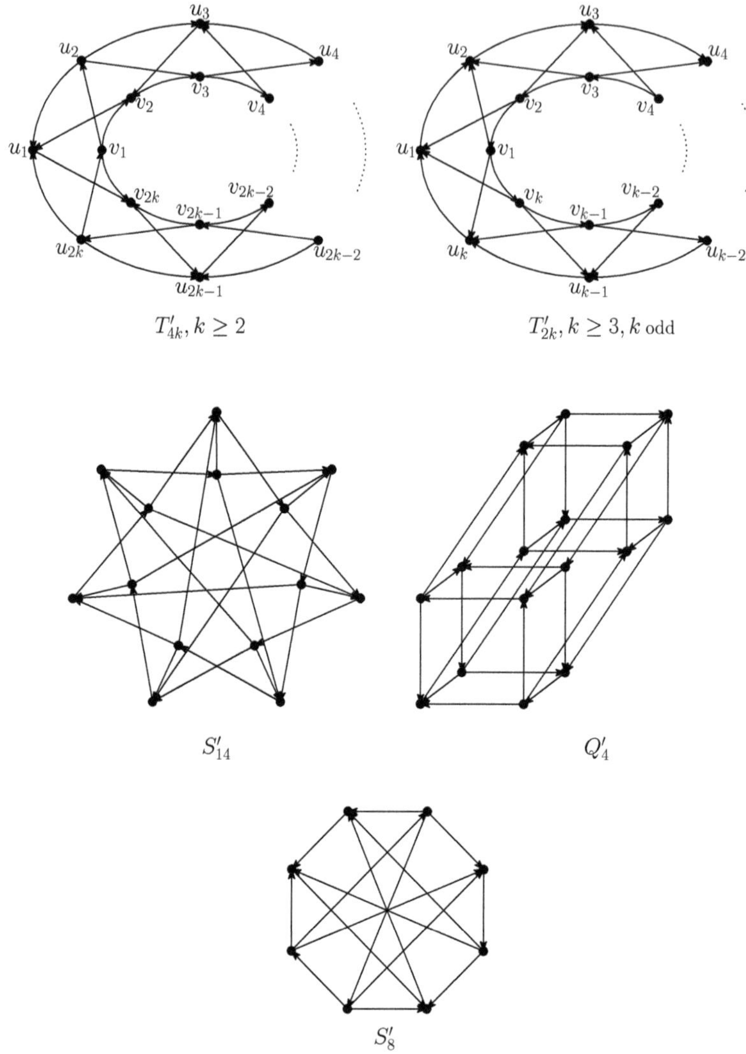

Figure 9.3 Maximal connected oriented graphs with skew spectrum in $[-2\mathrm{i}, 2\mathrm{i}]$. Reproduced from [219].

Proof If the signed graphs $\dot{G}$ and $\dot{H}$ are cospectral, then their bipartite doubles $\mathrm{bd}(\dot{G})$ and $\mathrm{bd}(\dot{H})$ are also cospectral. Hence, by Theorem 9.16, the associated oriented graphs G' and H' share the same skew spectrum. Moreover, since $\mathrm{bd}(\dot{G})$ and $\mathrm{bd}(\dot{H})$ are not switching isomorphic, the same non-isomorphism holds for their associated oriented graphs. □

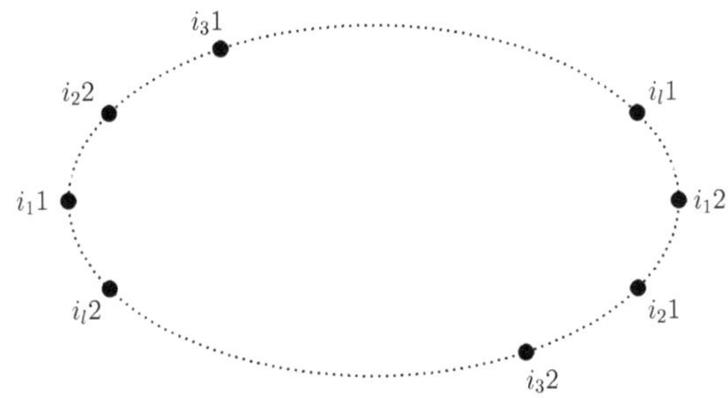

Figure 9.4 A sketch of $\dot{C}_{2l}$ for the proof of Theorem 9.26.

Consider now a simple construction in the domain of oriented graphs. An alternative proof is based on Theorem 9.18.

Theorem 9.26 *Let $G'_1 = (G, \sigma'_1)$ and $G'_2 = (G, \sigma'_2)$ be oriented graphs such that the orientation of every even cycle of G is the same in G'_1 and G'_2. Then G'_1 and G'_2 share the same skew spectrum.*

Proof We begin by proving that an odd cycle C'_l in an oriented graph G' induces a negative cycle $\dot{C}_{2l}$ in the signed graph $\dot{H}$ associated with the bipartite double $\mathrm{bd}(G')$. To this end, let $i_1, i_2, \dots, i_l$ be the vertices of C'_l, indexed in their natural order. Each edge $i_u i_v$ of this oriented cycle corresponds to two edges, $i_u 1 i_v 2$ and $i_u 2 i_v 1$, in the corresponding cycle of $\dot{H}$, as illustrated in Figure 9.4.

The orientation of the edges $i_u 1 i_v 2$ and $i_u 2 i_v 1$ in $\mathrm{bd}(G')$ is fixed by the orientation of $i_u i_v$ in G'; specifically, if $i_u \to i_v$, then $i_u 1 \to i_v 2$ and $i_u 2 \to i_v 1$. Considering the cycle C'_{2l} in $\mathrm{bd}(G')$, it follows that it contains an even number of edges oriented along the route direction, that is, it is not oriented uniformly. Consequently, the corresponding cycle $\dot{C}_{2l}$ in $\dot{H}$ is negative.

Furthermore, by the definition of the bipartite double, every even cycle in G' produces two edge-disjoint copies in $\mathrm{bd}(G')$.

Let $\dot{H}_1$ and $\dot{H}_2$ be the signed graphs associated with $\mathrm{bd}(G'_1)$ and $\mathrm{bd}(G'_2)$, respectively. Since both $\dot{H}_1$ and $\dot{H}_2$ are constructed from the same underlying graph G, they share the same underlying graph $\mathrm{bd}(G)$. By the argument in the previous part of the proof, every cycle of length $2l$ in $\mathrm{bd}(G)$, where l is odd, corresponds to negative cycles of the same length in both $\dot{H}_1$ and $\dot{H}_2$. For even l, the corresponding cycles in $\dot{H}_1$ and $\dot{H}_2$ have identical signatures. Hence, $\dot{H}_1$ and $\dot{H}_2$ possess the same sets of positive and negative cycles, implying that they are switching equivalent. Consequently, $\dot{H}_1$ and $\dot{H}_2$ share the same spectrum, which by Theorem 9.16(ii) entails that G'_1 and G'_2 share the same skew spectrum. □

Here is an immediate consequence concerning oriented graphs without even cycles.

Corollary 9.27 *If G has no even cycles, then all oriented graphs with underlying graph G share the same skew spectrum.*

In fact, this corollary follows as a direct consequence of a stronger result established in [59]: if all oriented graphs with underlying graph G share the same skew spectrum, then G contains no even cycles. From this implication, we immediately obtain the following corollary.

Corollary 9.28 *If a graph G contains an even cycle, then there are oriented graphs $G'_1 = (G, \sigma'_1)$ and $G'_2 = (G, \sigma'_2)$ such that signed graphs associated with their bipartite doubles are not cospectral.*

Proof Indeed, G'_1 and G'_2 can be chosen in such a way that they are not cospectral, which means that the corresponding signed graphs are not cospectral, by Theorem 9.16(ii). □

Problems

9.1 Prove that if A is the adjacency matrix of a signed graph with eigenvalues λ and $-\lambda$, and a codeword $\mathbf{x}$ is a linear combination of k rows of A, then the weight satisfies $w(\mathbf{x}) \equiv k\lambda^2 \pmod 3$.

9.2 For signed graphs $\dot{G}$ and $\dot{H}$, prove that if $A_{\dot{H}} = S^{-1}A_{\dot{G}}S$, then the codes $\mathscr{C}_{\alpha A_{\dot{G}}+\beta I+\gamma J}$ and $\mathscr{C}_{\alpha A_{\dot{H}}+\beta I+\gamma S^{-1}JS}$ are equivalent.

9.3 Prove that if the matrix M from Theorem 9.3 is the adjacency matrix of a signed graph $\dot{G}$, then the dimension of the code $\mathscr{C}_A$ satisfies

$$\dim(\mathscr{C}_A) \leq \min\left\{2(r - |N(i) \cap N(j)| + 1) + \left\lfloor \frac{|N(i) \cap N(j)|}{2} \right\rfloor : i, j \in V(\dot{G})\right\},$$

where $N(i)$ denotes the set of neighbours of vertex i in $\dot{G}$.

9.4 Let A be the $\binom{n}{2} \times \binom{n}{2}$ adjacency matrix of the signed line graph $\mathscr{L}(K_n)$. Prove the following statements:

(i) If $n \equiv 1 \pmod 3$, then $\mathscr{C}_{A-I} = \mathscr{C}_A^{\perp}$ with $|\mathscr{C}_{A-I}| = n-1$ and $|\mathscr{C}_A| = \binom{n-1}{2}$.
(ii) If $n \equiv -1 \pmod 3$, then $\mathscr{C}_{A-I} = \mathscr{C}_{A+I}^{\perp}$ with

$$|\mathscr{C}_{A-I}|, |\mathscr{C}_{A+I}| \in \left\{n-1, \binom{n-1}{2}\right\}.$$

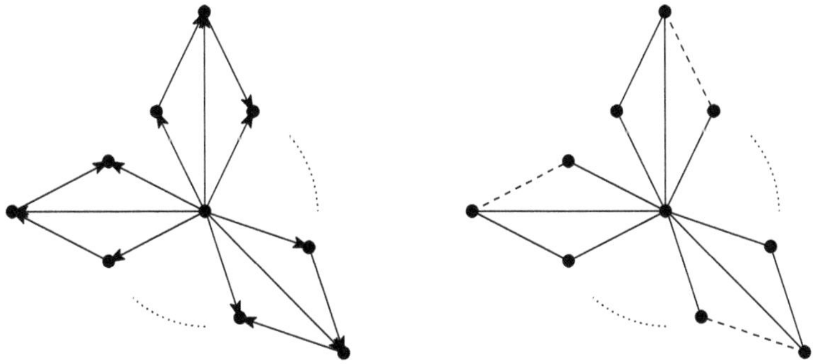

Figure 9.5 An infinite family of oriented graphs and an infinite family of signed graphs. Reproduced from [232], licensed under CC BY 4.0.

(iii) Otherwise, $\mathscr{C}_{A-I}$ is self-orthogonal and $|\mathscr{C}_{A-I}| \leq \min\{n-1, \binom{n-1}{2}\}$.

9.5 ([211]) Following the proof of Theorem 9.1, derive a similar result for strongly regular signed graphs in the class $\mathscr{C}_3$.

9.6 Let A be the Seidel matrix of the Petersen graph. Prove that the code $\mathscr{C}_A$ has dimension 4 and minimum distance 6. Conclude that it is optimal and attains the Griesmer bound mentioned in Subsection 9.1.4.

9.7 Generalise Example 9.10 to compute the quadruple (n, r, k, ψ) and the corresponding angles for the signed line graph $\mathscr{L}(\ddot{C}_n)$.

9.8 Figure 9.5 illustrates a family of oriented graphs G' and a family of signed graphs $\dot{G}$. Show that $-S_{G'}^2 = A_{\dot{G}}^2$. Compute the spectrum of $\dot{G}$ and the skew spectrum of G'.

9.9 Construct families of non-bipartite graphs that serve as underlying graphs of signed and oriented graphs whose spectra are identical up to multiplication by the imaginary unit.

9.10 Construct an example of a signed graph and an oriented graph that do not share the same underlying graph but satisfy the condition stated in the previous problem.

9.11 Let G' and $\dot{G}$ be an oriented bundle and a signed bundle, respectively, sharing the same underlying graph (see the beginning of Subsection 8.3.3 for definitions). Under what conditions does the equality $-S_{G'}^2 = A_{\dot{G}}^2$ hold?

9.12 Characterize oriented graphs with skew spectrum $\left[\sqrt{5}^{n/2}, (-\sqrt{5})^{n/2}\right]$. (Use the corresponding result for signed graphs and the 'transferring' results of Section 9.3.)

9.13 ([236]) Prove that the spectral radius $\rho(G')$ of an oriented unicyclic graph $G' = (G, \sigma')$ of order n and girth g (length of the induced cycle) satisfies

$$\rho(G') \begin{cases} < \zeta(n,g) & \text{if } g \text{ is odd,} \\ \leq \zeta(n,g) & \text{if } g \text{ is even,} \end{cases}$$

where $\zeta(n,g)$ is the largest root of

$$(x^2 - (n-g))\Phi_{P_{g-1}}(x) - 2x\left(\Phi_{P_{g-2}}(x) - 1\right),$$

and Φ_{P_g} is the characteristic polynomial of the adjacency matrix of the path P_g. Show that for even g, equality holds if and only if G is obtained by attaching $n-g$ vertices to a fixed vertex of the cycle C_g and the cycle of G' is oriented uniformly.

10

Generalizations

Chapter 1 began with the definition of a signed graph, to which we now return. Recall that a signed graph is a graph in which each edge is endowed with a sign, whereas an unsigned graph can be viewed as a signed graph possessing an all-positive signature. We emphasize the term 'viewed as' since, strictly speaking, an unsigned graph is not a signed graph. Nevertheless, the adjacency matrix of an unsigned graph arises as a special case of the adjacency matrix of a signed graph, and the same correspondence holds for the Laplacian matrix and the net Laplacian matrix.

Similarly, there exist several types of labelled graph that are closely related to signed graphs, yet are mathematically distinct. Zaslavsky identifies at least four such structures [258]. The first are *group-signed graphs*, in which edges are labelled by elements of a two-element group and these labels multiply around a circle. The second type refers to *sign-coloured graphs*, where edges are labelled from a two-element set acted upon by the sign group, with the operation '$-$' interchanging labels and '$+$' leaving them unchanged. Third, *weighted graphs* assign edge labels from the set $\{1, -1\}$ within the integers or another additive domain; in these graphs, weights behave like numbers rather than signs. Finally, *labelled graphs* are those in which the edge labels carry no inherent algebraic structure or properties.

Although these structures differ from signed graphs, whenever their associated matrices coincide with those of signed graphs, they may be interpreted as such, at least in the sense of sharing the same matrix representation.

In this chapter, we provide a non-exhaustive overview of certain structures that generalize the notion of a signed graph. The term 'generalize' is used informally here to indicate that whenever a matrix associated with a given structure extends or encompasses the corresponding matrix of a signed graph, we set aside formal definitions. Given the wealth of results available for each particular structure, this overview is necessarily brief, focusing primarily on the structures themselves rather than on detailed results or developments.

10.1 Graphs of Hermitian Matrices

For an ordinary graph G of order n, denote by $\mathscr{H}(G)$ the set of all $n \times n$ Hermitian matrices whose graph is G, meaning that if $M = (m_{ij}) \in \mathscr{H}(G)$, then $m_{ij} \neq 0$ for $i \neq j$ holds if and only if ij is an edge of G. The diagonal entries are real, but are otherwise unrestricted.

The inverse eigenvalue problem for graphs concerns determining which spectra can be realized by matrices in the set $\mathscr{H}(G)$. Numerous results focus on graphs G that admit matrices in $\mathscr{H}(G)$ with an extreme number of distinct eigenvalues. For instance, it has been shown in [124] that any tree requires at least as many distinct eigenvalues as the length of its longest path. As observed in [125], trees typically exhibit multiplicity lists corresponding to a large number of distinct eigenvalues. At the other extreme, graphs whose Hermitian matrices have two distinct eigenvalues have been studied in [63], where characterizations, necessary conditions and specific constructions for such graphs have been established.

There is a relaxed approach in which M is required to be real and symmetric with a zero diagonal. If $\mathscr{S}(G)$ denotes the set of all $n \times n$ real symmetric matrices with zero diagonal based on G, and $q(M)$ denotes the number of distinct eigenvalues of $M \in \mathscr{S}(G)$, we set $q(G) = \min\{q(M) \colon M \in \mathscr{S}(G)\}$. The quantity $q(G)$ is determined for all graphs with at most six vertices [40].

A compelling relaxed problem is determining which graphs have $q(G) = 2$. An approach is to determine a lower bound on the number of edges that a graph of order n with $q(G) = 2$ can have. According to [21], if a connected graph G of order n satisfies $q(G) = 2$, then $|E(G)| \geq 2n-4$ if n is even and $|E(G)| \geq 2n-3$ if n is odd. Equality can be achieved for any n, and graphs achieving equality are characterized in the same reference. Another approach, studied in [22], is to consider graphs with a large number of edges: for a fixed n, determine the largest positive integer l such that the removal of any set of l edges from the complete graph K_n results in a graph G with $q(G) = 2$. It is conjectured that if G is a graph on at least three vertices whose complement has at most $n-3$ edges, then $q(G) = 2$.

We note that a classification of regular graphs of degree at most 4 with $q(G) = 2$ is given in [23].

Every matrix in $\mathscr{S}(G)$ can be seen as the adjacency matrix of a weighted graph whose underlying graph is G. In this setting, the entries m_{ij} correspond to edge weights. Notably, the Laplacian matrix, formed by placing the sums of the weights on the main diagonal, and its associated spectrum are frequently studied within this framework. The Laplacian eigenvalues and their corresponding eigenspaces for weighted graphs have been investigated in [38, 41, 42]. Additionally, certain classes of weighted graphs and the spectra of their associated matrices are examined in [17, 86, 88, 156]. For example, it is established in [17] that the entire Laplacian spectrum of a weighted threshold graph can be expressed in terms of a conjugate partition of the associated weighted Ferrers diagram.

When edge weights are non-negative integers, the corresponding matrix can be viewed as the adjacency matrix of a multigraph, that is, a graph allowing multiple parallel edges between vertices. It is worth noting that the Sachs formula, as stated in Theorem 1.9, extends to this context with slight modifications (cf. [73, Theorem 1.3]).

10.2 Gain Graphs

A *gain graph* is a triple $\Phi = (G, \mathfrak{G}, \varphi)$ consisting of the underlying graph $G = (V, E)$, the gain group $\mathfrak{G}$ and the function $\varphi\colon E(G) \to \mathfrak{G}$ (called the *gain function*), such that

$$\varphi(e_{ij}) = \varphi(e_{ji})^{-1}.$$

The notation is consistent with the corresponding literature. This definition requires $E(G)$ to contain two copies of each edge with opposite orientations.

The *adjacency matrix* $A_\Phi = (a_{ij}) \in \mathbb{C}^{n\times n}$ is defined by

$$a_{ij} = \begin{cases} \varphi(e_{ij}) & \text{if } v_i \text{ is adjacent to } v_j, \\ 0 & \text{otherwise.} \end{cases}$$

If v_i is adjacent to v_j, then $a_{ij} = \varphi(e_{ij}) = \varphi(e_{ji})^{-1} = \overline{a}_{ji}$. Therefore, A_Φ is Hermitian and its eigenvalues are real. The *Laplacian matrix* L_Φ is

$$L_\Phi = D_\Phi - A_\Phi,$$

where D_Φ is the diagonal matrix of degrees of vertices in G. Therefore, L_Φ is also Hermitian. Restricting the graph to have specified gains can produce other familiar matrices, as well.

For brevity, we write $\Phi = (G, \varphi)$ for a gain graph if the gain group is clear, and call Φ a $\mathfrak{G}$-gain graph. G is usually assumed to be simple. The set of vertices is $V = \{v_1, v_2, \dots, v_n\}$, and then the edges are denoted by $e_{ij} = v_i v_j$.

To define balance and switching equivalence for gain graphs, we first introduce two additional notions. The gain $\phi(W)$ of a walk is the product of gains of edges contained, that is, $\varphi(W) = \varphi(e_{12})\varphi(e_{23})\cdots\varphi(e_{k-1,k})$. A walk W is *neutral* if $\phi(W) = 1 = 1_{\mathfrak{G}}$.

Taking into account that a cycle is a particular case of a walk, we say that a $\mathfrak{G}$-gain graph is *balanced* if $\phi(C) = 1$ holds for every cycle C, that is, if every cycle is neutral.

A *switching function* is any function $\zeta\colon V \to \mathfrak{G}$. Switching a $\mathfrak{G}$-gain graph $\Phi = (G, \varphi)$ means replacing φ by φ^{ζ}, defined by

$$\varphi^{\zeta}(e_{ij}) = \zeta(v_i)^{-1}\varphi(e_{ij})\zeta(v_j),$$

producing the $\mathfrak{G}$-gain graph $\Phi^{\zeta} = (G, \varphi^{\zeta})$. Accordingly, Φ_1 and Φ_2, based on the same underlying graph, are *switching equivalent* when there exists a switching function ζ, such that Φ_2 coincides with Φ_1^{ζ}. Switching equivalence forms an equivalence relation on gain functions for a fixed underlying graph, and an equivalence class is a switching class of φ.

Theorem 10.1 ([176]) *Let $W = v_1e_{12}v_2e_{23}\cdots v_ke_{k1}v_1$ be a closed walk in a gain graph Φ. If $\zeta\colon V \to \mathfrak{G}$, then $\varphi^{\zeta}(W) = \zeta(v_1)^{-1}\varphi(W)\zeta(v_1)$.*

Proof From

$$\begin{aligned}\varphi^{\zeta}(W) &= \varphi^{\zeta}(e_{12})\varphi^{\zeta}(e_{23})\cdots\varphi^{\zeta}(e_{k1})\\ &= \zeta(v_1)^{-1}\varphi(e_{12})\zeta(v_2)\cdot\zeta(v_2)^{-1}\varphi(e_{23})\cdots\zeta(v_k)^{-1}\varphi(e_{k1})\zeta(v_1)\\ &= \zeta(v_1)^{-1}\varphi(W)\zeta(v_1),\end{aligned}$$

we get the desired result. □

By setting $\mathfrak{G} = \mathbb{T} = \{z \in \mathbb{C}\colon |z| = 1\}$, the multiplicative circle group of all complex numbers with modulus 1, we arrive at $\mathbb{T}$-*gain graphs*. We believe that the reader has detected the facts formalized in the next remark.

Remark 10.2 A natural consequence of studying complex gain graphs is that both signed and ordinary graphs arise as special cases. Specifically, a signed graph can be regarded as a $\mathbb{T}$-gain graph whose gains are restricted to $\{1, -1\}$, while an ordinary graph corresponds to a $\mathbb{T}$-gain graph in which the gain is uniformly 1.

Here is a balance criterion. A *potential function* for φ is a function $\theta\colon V \to \mathbb{T}$ such that, for every $e_{ij} \in E(G)$,

$$\theta(v_i)^{-1}\theta(v_j) = \varphi(e_{ij}).$$

Theorem 10.3 ([259]) *Let $\Phi = (G, \varphi)$ be a $\mathbb{T}$-gain graph. Then the following are equivalent:*

(i) *Φ is balanced.*
(ii) *Φ switches to the $\mathbb{T}$-gain graph with all neutral edges.*
(iii) *φ has a potential function.*

Several fundamental concepts discussed in the previous chapters extend naturally to gain graphs, particularly to $\mathbb{T}$-gain graphs. A spectral analysis of balance in this context is developed in [56, 109], while eigenvalue bounds are established in [177]. For investigations related to NEPSes and other graph products, see [29, 57]. The interplay between eigenvalues and graph regularity or other structural properties is examined in [49, 57]. Studies on line graphs of gain graphs can be found in [26, 176]. The class of $\mathbb{T}$-gain graphs exhibiting two eigenvalues is addressed in [257]. Switching methods analogous to those described in Subsections 7.2.1 and 7.2.2 are explored in [2], while spectral properties and characterizations are further investigated in [4, 155, 177, 256].

Notably, $\mathbb{T}$-gain graphs with gains restricted to $\{1, -1, \mathrm{i}, -\mathrm{i}\}$ or to $\{1, -1, \mathrm{i}\}$ have been studied under various terminologies, as detailed in [20, 128, 129, 130, 169]. We return to this in the following section.

10.3 Mixed Graphs and Extended Mixed Graphs

A *mixed graph* is a structure consisting of a vertex set, a set of unoriented edges and a set of oriented edges. An unoriented edge can be viewed as a bi-oriented edge, that is, an edge with implicit orientations toward both of its endpoints. An alternative formalization involves an underlying graph with edge-specific weights or attributes. A further generalization arises when edges are allowed to be oriented simultaneously from both of their endpoints, a concept introduced in [204]. We refer to such structures as *extended mixed graphs*. In what follows, we denote both mixed and extended mixed graphs by Γ, with no ambiguity arising from context. It is assumed that Γ does not have multiple edges.

Edges that are oriented toward both ends (resp. from both ends) can naturally be interpreted as positive (resp. negative) edges, in accordance with the terminology of signed graph theory. In this light, an extended mixed graph may be regarded as a hybrid structure combining a signed graph (determined by edges declared as positive or negative) and an oriented graph (determined by arcs, i.e., oriented edges).

To indicate the presence of an edge between vertices i and j, we write $i \sim j$, as before. We also adopt the notations i — j, i --- j and $i \rightarrow j$ to designate the existence of a positive edge, a negative edge and an arc from i to j, respectively. These conventions are consistent with those introduced in the previous chapters.

The *adjacency matrix* $A_\Gamma = (a_{ij})$ of an n-vertex extended mixed graph Γ is the $(1,0,-1)$-matrix such that

$$a_{ij} = \begin{cases} 1 & \text{if } i - j, \\ -1 & \text{if } i \dashrightarrow j, \\ -1 & \text{if } i \to j, \\ 1 & \text{if } i \leftarrow j, \\ 0 & \text{otherwise.} \end{cases}$$

It combines the adjacency matrix of a signed graph and the skew adjacency matrix of an oriented graph, defined earlier, together with an observation that if $i \to j$, then the entries a_{ij} of A_Γ and s_{ij} of the skew adjacency matrix have opposite signs. This modification in A_Γ can be viewed as a subtle adjustment that anticipates the forthcoming notions of agreeing and disagreeing edges. In the case of the absence of oriented edges (resp. positive and negative edges), A_Γ is symmetric (resp. skew symmetric).

An alternative formulation assigns complex weights to oriented edges, replacing ± 1 with $\pm \mathrm{i}$. Under this approach, extended mixed graphs are viewed as special cases of $\mathbb{T}$-gain graphs whose gains belong to the set $\{1, -1, \mathrm{i}, -\mathrm{i}\}$; ordinary mixed graphs can be interpreted analogously.

In the remainder of this section, we adhere to the proof of Theorem 2.27 to demonstrate the computation of an upper bound for the largest singular value of an extended mixed graph. In this setting, loops, in the form of positive edges at vertices, are permitted and contribute to the diagonal entries of the corresponding matrix representation.

Two edges ij and ik are said to *agree* at their common vertex i if they are either both oriented towards i or both oriented away from i. Equivalently, agreement occurs when the corresponding matrix entries satisfy $a_{ij}a_{ik} = 1$. Conversely, the edges are said to *disagree* at i if one is oriented towards i and the other away from it, that is, if $a_{ij}a_{ik} = -1$. Thus, every pair of edges incident to a common vertex either agree or disagree at that vertex.

The following list contains the remaining terminology and the corresponding notation:

- d_i The degree of the vertex i (loops are neglected).
- d_i^{s} The 'signed neighbourhood' of the vertex i, that is, the sum of the number of positive edges and the number of negative edges incident with i (loops are neglected).

d_i^{o} The 'oriented neighbourhood' of the vertex i, that is, the number of oriented edges incident with i; equivalently, $d_i^{\mathrm{o}} = d_i - d_i^{\mathrm{s}}$.

m_i The average degree in the neighbourhood of the vertex i, that is, $m_i = (1/d_i)\sum_{j\sim i} d_j$.

l_i The number of loops at the vertex i.

n_i The average number of loops in the 'signed neighbourhood' of the vertex i, that is, $n_i = (1/d_i^{\mathrm{s}})\sum_{j\overset{\mathrm{s}}{\sim}i} l_j$ and $j \overset{\mathrm{s}}{\sim} i$ designates the existence of either a positive or negative edge.

$\Xi(\Gamma)$ The difference between the largest number of loops and the smallest number of loops in an extended mixed graph Γ, that is, $\Xi(\Gamma) = \max\{|l_i - l_j| \colon 1 \leq i, j \leq n\}$.

$T_{i,j}^{+}$ The number of triangles containing the vertices i and j, such that the edges of every triangle agree at each vertex distinct from i.

$T_{i,j}^{-}$ The number of triangles containing the vertices i and j, such that the edges of every triangle disagree at each vertex distinct from i.

Theorem 10.4 ([204]) *Let ϑ be the largest singular value of an extended mixed graph Γ with n vertices and l_i loops at a vertex i $(1 \leq i \leq n)$. Then ϑ does not exceed*

$$\max_{1\leq i\leq n}\left\{\sqrt{l_i^2 + d_i m_i + d_i^{\mathrm{s}}(l_i + n_i) + d_i^{\mathrm{o}}\Xi(\Gamma) - \sum_{j\sim i}\left(T_{i,j}^{+} + T_{i,j}^{-} - |T_{i,j}^{+} - T_{i,j}^{-}|\right)}\right\}.$$

The equality is attained for ordinary regular graphs.

Proof Let ϑ^2 denote the largest eigenvalue of the matrix $A_\Gamma^{\intercal}A_\Gamma$, and let $\mathbf{x} = (x_1, x_2, \ldots, x_n)^{\intercal}$ be a corresponding unit eigenvector. Assume that x_i attains the maximum absolute value among all entries of $\mathbf{x}$. Without loss of generality, we may take $x_i > 0$. Let $A'_\Gamma = (a'_{ij})$ denote the adjacency matrix of Γ with all loops removed, and let B_Γ be the diagonal matrix of loop weights, where the diagonal entry (i,i) is equal to l_i, the weight of the loop at vertex i.

From $\vartheta^2\mathbf{x} = A_\Gamma^{\intercal}A_\Gamma\mathbf{x}$, we find

$$\vartheta^2\mathbf{x} = (B_\Gamma + A'_\Gamma)^{\intercal}(B_\Gamma + A'_\Gamma)\mathbf{x} = (B_\Gamma^2 + B_\Gamma A'_\Gamma + A'^{\intercal}_\Gamma B_\Gamma + A'^{\intercal}_\Gamma A'_\Gamma)\mathbf{x},$$

since $B_\Gamma^{\intercal} = B_\Gamma$. This yields

$$\begin{aligned}\vartheta^2 x_i &= l_i^2 x_i + l_i\sum_{j=1}^{n} a'_{ij}x_j + \sum_{j=1}^{n} a'_{ji}l_j x_j + \sum_{k=1}^{n}\sum_{j=1}^{n} a'_{ji}a'_{jk}x_k \\ &= l_i^2 x_i + l_i\sum_{j\sim i} a'_{ij}x_j + \sum_{j\sim i} a'_{ji}l_j x_j + \sum_{k\sim j}\sum_{j\sim i} a'_{ji}a'_{jk}x_k.\end{aligned} \tag{10.1}$$

In what follows, we write $j \overset{\mathrm{o}}{\sim} i$ to designate the existence of an oriented edge between j and i. By considering the second and the third terms on the right-hand side of (10.1), we find

$$\begin{aligned} l_i\sum_{j\sim i} a'_{ij}x_j + \sum_{j\sim i} a'_{ji}l_jx_j &= \sum_{j\sim i}(l_ia'_{ij}+l_ja'_{ji})x_j \\ &\leq \Big(\sum_{j\overset{\mathrm{s}}{\sim} i}(l_i+l_j) + \sum_{j\overset{\mathrm{o}}{\sim} i}|l_i-l_j|\Big)x_i \\ &\leq \big(d_i^{\mathrm{s}}(l_i+n_i)+d_i^{\mathrm{o}}\Xi(\Gamma)\big)x_i. \end{aligned} \tag{10.2}$$

Consider now the double summation of (10.1):

$$\begin{aligned} \sum_{k\sim j}\sum_{j\sim i} a'_{ji}a'_{jk}x_k = {} & \sum_{a'_{ji}a'_{jk}=1} x_k - \sum_{a'_{ji}a'_{jk}=-1} x_k = \sum_{j:\, a'_{ji}a'_{jk}=1}\ \sum_{k:\, a'_{ki}a'_{kj}=1} x_k \\ & + \sum_{j:\, a'_{ji}a'_{jk}=1}\ \sum_{k:\, a'_{ki}a'_{kj}\neq 1} x_k - \sum_{j:\, a'_{ji}a'_{jk}=-1}\ \sum_{k:\, a'_{ki}a'_{kj}=-1} x_k \\ & - \sum_{j:\, a'_{ji}a'_{jk}=-1}\ \sum_{k:\, a'_{ki}a'_{kj}\neq -1} x_k = \sum_{j\sim i} T^+_{i,j}x_j - \sum_{j\sim i} T^-_{i,j}x_j \\ & + \sum_{j:\, a'_{ji}a'_{jk}=1}\ \sum_{k:\, a'_{ki}a'_{kj}\neq 1} x_k - \sum_{j:\, a'_{ji}a'_{jk}=-1}\ \sum_{k:\, a'_{ki}a'_{kj}\neq -1} x_k \\ \leq {} & \sum_{j\sim i}(T^+_{i,j}-T^-_{i,j})x_j + \sum_{j\sim i, k\sim j:\ [i,j,k]\notin T^+_{i,j}\cup T^-_{i,j}} x_i \\ \leq {} & \sum_{j\sim i}|T^+_{i,j}-T^-_{i,j}|x_i + \sum_{j\sim i}(d_j - T^+_{i,j} - T^-_{i,j})x_i \\ = {} & d_im_ix_i - \sum_{j\sim i}\left(T^+_{i,j}+T^-_{i,j}-|T^+_{i,j}-T^-_{i,j}|\right)x_i, \end{aligned} \tag{10.3}$$

where the latter inequality follows from $(T^+_{i,j}-T^-_{i,j})x_j \leq |T^+_{i,j}-T^-_{i,j}|x_i$.

By inserting (10.2) and (10.3) in (10.1), we obtain

$$\vartheta^2x_i \leq \Big(l_i^2 + d_im_i + d_i^{\mathrm{s}}(l_i+n_i) + d_i^{\mathrm{o}}\Xi(\Gamma) - \sum_{j\sim i}\left(T^+_{i,j}+T^-_{i,j}-|T^+_{i,j}-T^-_{i,j}|\right)\Big)x_i,$$

which, after eliminating x_i and taking the maximum over all vertices, leads to the desired result.

Suppose that Γ is an ordinary regular graph. The inequality (2.9) reduces to $\vartheta \leq \sqrt{d_im_i} = d_i$, since $l_i = 0$ for all i and $T^-_{i,j} = 0$ for all i, j. On the other hand, ϑ coincides with the index of Γ, necessarily equal to d_i. □

The theorem applies to mixed graphs, signed graphs and oriented graphs, as well as ordinary graphs. The corresponding upper bounds on the largest singular value can be derived directly from the statement.

Problems

10.1 What is the minimum number of eigenvalues of a Hermitian matrix whose underlying graph is a path or a cycle?

10.2 It is known that a connected strongly regular unsigned graph has three eigenvalues. Is it possible to associate a Hermitian matrix with such a graph having only two eigenvalues?

10.3 Observe that every oriented graph can be viewed as a $\mathbb{T}_4$-gain graph – that is, a $\mathbb{T}$-gain graph whose gains are restricted to $\{1,-1,\mathrm{i},-\mathrm{i}\}$ – with the additional restriction that all gains lie in $\{\mathrm{i},-\mathrm{i}\}$. However, use the definition of a gain graph to prove that this observation is insufficient to conclude that the class of oriented graphs belongs to the family of gain graphs.

10.4 Extend the star complement technique to the domains of oriented graphs and gain graphs, exploring how its core principles and constructions translate to these more general settings. In particular, adapt the results of Chapter 5 to these two domains, carefully examining which properties and proofs carry over and which require modification. Investigate whether Theorem 4.2 remains valid for oriented and gain graphs, and, if so, provide its exact formulation in these contexts. Additionally, discuss potential applications or limitations of the extended technique.

10.5 Following Section 2.5, extend some of the given spectral bounds to the setting of $\mathfrak{G}$-gain graphs.

10.6 Provide a definition of strongly regular $\mathfrak{G}$-gain graphs.

10.7 ([257]) Classify $\mathbb{T}$-gain graphs with vertex degree at most 4 that have exactly two eigenvalues.

10.8 ([146]) Identify all $\mathbb{T}$-gain graphs with exactly one positive eigenvalue.

10.9 Generalize the switching operations of Subsections 7.2.1 and 7.2.2 to the context of $\mathfrak{G}$-gain graphs.

10.10 Classify mixed and extended mixed graphs with vertex degree at most 4 that have two eigenvalues. Also, classify those with exactly one positive eigenvalue.

10.11 The *mixed line graph* $\mathscr{L}(\Gamma)$ of a mixed graph Γ was defined in [24] as follows: $\mathscr{L}(\Gamma)$ has the edges of Γ as its vertices, and for $e, f \in E(\Gamma)$ the corresponding vertices of $\mathscr{L}(\Gamma)$ satisfy

$$\begin{cases} e - f & \text{if } e \text{ and } f \text{ agree at the common vertex,} \\ e \to f & \text{if } e \text{ and } f \text{ disagree at the common vertex and } f \text{ is oriented from it,} \\ e \nsim f & \text{if } e \text{ and } f \text{ do not have a common vertex.} \end{cases}$$

Prove that the spectrum of $\mathscr{L}(\Gamma)$ is bounded from below by -2.

10.12 For every mixed triangle Γ, construct all mixed graphs Θ satisfying $\mathscr{L}(\Theta) \cong \Gamma$. (Show that the underlying graph of Θ is either a triangle or a four-vertex star, then examine all possibilities.)

10.13 (Krausz theorem for mixed graphs [203]**)** A *canonically oriented complete mixed graph* (COCM) is a complete mixed graph admitting a vertex partition $V_1 \sqcup V_2$ such that every edge between V_1 and V_2 is oriented from V_1 to V_2, and all remaining edges are bi-oriented. One of V_1 or V_2 may be empty.

Prove that a mixed graph is a mixed line graph if and only if its edges can be partitioned into COCMs such that:

(i) each vertex belongs to at most two COCMs;
(ii) two COCMs share at most one common vertex;
(iii) a common vertex of two COCMs is not the head of two arcs belonging to distinct COCMs (where for an arc $i \to j$, vertex j is called the head).

10.14 Generalize the previous two problems to the context of extended mixed graphs.

References

[1] A. Abdollahi, S. Janbaz, M.R. Oboudi, Distance between spectra of graphs, *Linear Algebra Appl.*, **466** (2015), 401–408.

[2] A. Abiad, F. Belardo, A.P. Khramova, A switching method for constructing cospectral gain graphs, *Discrete Math.*, **374** (2024), 113838.

[3] B.D. Acharya, Spectral criterion for cycle balance in networks, *J. Graph Theory*, **4** (1980), 1–11.

[4] V. Adish, M. Khosravi, On some spectral properties of complex unit gain graph, *Filomat*, **39** (2025), 975–981.

[5] L.A. Agong, C. Amarra, J.S. Caughman, A.J. Herman, T.S. Terada, On the girth and diameter of generalized Johnson graphs, *Discrete Math.*, **341** (2018), 138–142.

[6] S. Akbari, F. Belardo, E. Dodongeh, M. A. Nematollahi, Spectral characterizations of signed cycles, *Linear Algebra Appl.*, **553** (2018), 307–327.

[7] S. Akbari, S. Dalvandi, F. Heydari, M. Maghasedi, Signed complete graphs with maximum index, *Discuss. Math. Graph Theory*, **40** (2020), 393–403.

[8] S. Akbari, F.M.C. França, E. Ghasemian, M. Javarsineh, L.S. de Lima, The main eigenvalues of signed graphs, *Linear Algebra Appl.*, **614** (2020), 270–280.

[9] S. Akbari, H.R. Maimani, L. Parsaei Majd, On the spectrum of some signed complete and complete bipartite graphs, *Filomat*, **32** (2018), 5817–5826.

[10] B. Alshamary, Z. Stanić, Computing the determinant of a signed graph, *Open Math.*, **22** (2024), 20230188.

[11] M. Anđelić, T. Koledin, Z. Stanić, Cubic and quartic net-regular strongly regular signed graphs, *Discrete Appl. Math.*, **366** (2025), 216–225.

[12] M. Anđelić, T. Koledin, Z Stanić, Inequalities for Laplacian eigenvalues of signed graphs with given frustration number, *Symmetry*, **13** (2021), 1902.

[13] M. Anđelić, T. Koledin, Z Stanić, On regular signed graphs with three eigenvalues, *Discuss. Math. Graph Theory*, **40** (2020), 405–416.

[14] M. Anđelić, T. Koledin, Z. Stanić, Relationships between net regularity, strong regularity and walk regularity of signed graphs, *Linear Multilinear Algebra*, **73** (2025), 1062–1077.

[15] M. Anđelić, T. Koledin, Z. Stanić, Signed graphs whose all Laplacian eigenvalues are main, *Linear Multilinear Algebra*, **71** (2023), 2409–2425.

[16] M. Anđelić, T. Koledin, Z. Stanić, J. Wang, Signed graphs with integral net Laplacian spectrum, *AKCE Int. J. Graphs Combin.*, **20** (2023), 177–184.

[17] M. Anđelić, Z. Stanić, Laplacian eigenvalues of weighted threshold graphs, *Comput. Appl. Math.*, **44** (2025), 38.

[18] A. Anuradha, R. Balakrishnan, X. Chen, X. Li, H. Lian, W. So, Skew spectra of oriented bipartite graphs, *Electron. J. Combin.*, **20** (2013), #P18.

[19] R.B. Bapat, *Graphs and Matrices*, Springer, London, 2014.

[20] R.B. Bapat, D. Kalita, S. Pati, On weighted directed graphs, *Linear Algebra Appl.*, **436** (2012), 99–111.

[21] W. Barrett, S. Fallat, V. Furst, F. Kenter, S. Nasserasr, B. Rooney, M. Tait, H. van der Holst, Sparsity of graphs that allow two distinct eigenvalues, *Linear Algebra Appl.*, **674** (2023), 374–395.

[22] W. Barrett, S. Fallat, V. Furst, S. Nasserasr, B. Rooney, M. Tait, Graphs with bipartite complement that admit two distinct eigenvalues, https://arxiv.org/abs/2411.12917.

[23] W. Barrett, S. Fallat, V. Furst, S. Nasserasr, B. Rooney, M. Tait, Regular graphs of degree at most four that allow two distinct eigenvalues, *Linear Algebra Appl.*, **679** (2023), 127–164.

[24] L.W. Beineke, J.S. Bagga, *Line Graphs and Line Digraphs*, Springer, Cham, 2021.

[25] F. Belardo, Balancedness and the least eigenvalue of Laplacian of signed graphs, *Linear Algebra Appl.*, **446** (2014), 133–147.

[26] F. Belardo, M. Brunetti, Line graphs of complex unit gain graphs with least eigenvalue -2, *Electron. J. Linear Algebra*, **37** (2021), 14–30.

[27] F. Belardo, M. Brunetti, Limit points for the spectral radii of signed graphs, *Discrete Math.*, **347** (2024), 113745.

[28] F. Belardo, M. Brunetti, M. Cavaleri, A. Donno, Constructing cospectral signed graphs, *Linear Multilinear Algebra*, **69** (2021), 2717–2732.

[29] F. Belardo, M. Brunetti, S. Khan, NEPS of complex unit gain graphs, *Electron. J. Linear Algebra*, **39** (2023), 621–643.

[30] F. Belardo, S.M. Cioabă, J. Koolen, J.F. Wang, Open problems in the spectral theory of signed graphs, *Art Discrete Appl. Math.*, **1** (2018), P2.10.

[31] F. Belardo, E.M. Li Marzi, S.K. Simić, Combinatorial approach for computing the characteristic polynomial of a matrix, *Linear Algebra Appl.*, **433** (2010), 1513–1523.

[32] F. Belardo, E.M. Li Marzi, S.K. Simić, Signed line graphs with least eigenvalue -2: The star complement technique, *Discrete Appl. Math.*, **207** (2016), 29–38.

[33] F. Belardo, P. Petecki, Spectral characterizations of signed lollipop graphs, *Linear Algebra Appl.*, **480** (2015), 144–167.

[34] F. Belardo, S.K. Simić, On the Laplacian coefficients of signed graphs, *Linear Algebra Appl.*, **475** (2015), 94–113.

[35] F. Belardo, Z. Stanić, T. Zaslavsky, Total graph of a signed graph, *Ars Math. Contemp.*, **23** (2023), P1.02.

[36] F. Belardo, Y. Zhou, Signed graphs with extremal least Laplacian eigenvalue, *Linear Algebra Appl.*, **497** (2016), 167–180.

[37] F.K. Bell, P. Rowlinson, On the multiplicities of graph eigenvalues, *Bull. London Math. Soc.*, **35** (2003), 401–408.

[38] A. Berman, M. Farber, A lower bound for the second largest Laplacian eigenvalue of weighted graphs, *Electron. J. Linear Algebra*, **22** (2011), 1179–1184.

[39] A. Bhattacharya, S. Friedland, U.N., Peled, On the first eigenvalue of bipartite graphs, *Electron. J. Combin.*, **15** (2008), R144.

[40] B. Bjorkman, L. Hogben, S. Ponce, C. Reinhart, T. Tranel, Applications of analysis to the determination of the minimum number of distinct eigenvalues of a graph, *Pure Appl. Funct. Anal.*, **3** (2018), 537–563.

[41] M. Bolla, Spectra and structure of weighted graphs, *Electron. Notes Discrete Math.*, **38** (2011), 49–54.

[42] M. Bolla, G. Tusnády, Spectra and optimal partitions of weighted graphs, *Discrete Math.*, **128** (1994), 1–20.

[43] B. Bollobás, V. Nikiforov, Cliques and the spectral radius, *J. Combin. Theory B*, **97** (2007), 859–865.

[44] A.E. Brouwer, A.M. Cohen, A. Neumaier, *Distance-Regular Graphs*, Springer, Berlin, 1989.

[45] A.E. Brouwer, W.H. Haemers, *Spectra of Graphs*, Springer, New York, 2012.

[46] A.E. Brouwer, A. Neumaier, The graphs with spectral radius between 2 and $\sqrt{2+\sqrt{5}}$, *Linear Algebra Appl.*, **114/115** (1989), 273–276.

[47] R. Brualdi, A.J. Hoffman, On the spectral radius of $(0,1)$-matrices. *Linear Algebra Appl.*, **65** (1985), 133–146.

[48] R.A. Brualdi, E.S. Solheid, On the spectral radius of complementary acyclic matrices of zeros and ones, *SIAM J. Algebra Disc. Methods*, **7** (1986), 265–272.

[49] M. Brunetti, Eigenvalues of complex unit gain graphs and gain regularity, *Spec. Matrices*, **12** (2024), 20240005.

[50] M. Brunetti, M. Cavaleri, A. Donno, A lexicographic product for signed graphs, *Australas. J. Combin.*, **74** (2019), 332–343.

[51] M. Brunetti, M. Cavaleri, A. Donno, Erratum to the article 'A lexicographic product for signed graphs', *Australas. J. Combin.*, **75** (2019), 256–258.

[52] M. Brunetti, Z. Stanić, Ordering signed graphs with large index, *Ars Math. Contemp.*, **22** (2022), P4.05.

[53] M. Brunetti, Z. Stanić, Unbalanced signed graphs with extremal spectral radius or index, *Comput. Appl. Math.*, **41** (2022), 118.

[54] P.J. Cameron, J.M. Goethals, J.J. Seidel, E.E. Shult, Line graphs, root systems, and elliptic geometry, *J. Algebra*, **43** (1976), 305–327.

[55] M. Cavaleri, D. D'Angeli, A. Donno, Characterizations of line graphs in signed and gain graphs, *European J. Combin.*, **102** (2022), 103479.

[56] M. Cavaleri, A. Donno, S. Spessato, A group representation approach to balance of gain graphs, *J. Algebraic Combin.*, **54** (2021), 265–293.

[57] M. Cavaleri, A. Donno, S. Spessato, Gain total graphs and their spectra via G-phases and group representations, *Linear Algebra Appl.*, **721** (2025), 484–512.

[58] M. Cavaleri, A. Donno, S. Spessato, On the complement of a signed graph, *Discrete Math.*, **348** (2025), 114433.

[59] M. Cavers, S.M. Cioabă, S. Fallat, D.A. Gregory, W.H. Haemers, S.J. Kirkland, J.J. McDonald, M. Tsatsomeros, Skew-adjacency matrices of graphs, *Linear Algebra Appl.*, **436** (2012), 4512–4529.

[60] A. Cayley, A theorem on trees, *Quart. J. Pure Appl. Math.*, **23** (1889), 376–378.

[61] S. Chaiken, A combinatorial proof of the all minors matrix tree theorem, *SIAM J. Algebraic Discrete Methods*, **3** (1982), 319–329.

[62] M.-Q. Chen, Eigenpairs of adjacency matrices of balanced signed graphs, *Spec. Matrices*, **12** (2024), 20240007.

[63] Z. Chen, M. Grimm, P. McMichael, C. R. Johnson, Undirected graphs of Hermitian matrices that admit only two distinct eigenvalues, *Linear Algebra Appl.*, **458** (2014), 403–428.

[64] F. Chen, X. Yuan, Turán problem for $\mathcal{K}_4^-$-free signed graphs, *Appl. Math. Comput.*, **477** (2024), 128814.

[65] X.-M. Cheng, A.L. Gavrilyuk, G.R.W. Greaves, J.H. Koolen, Biregular graphs with three eigenvalues, *European J. Combin.*, **56** (2016), 57–80.

[66] X.-M. Cheng, G.R.W. Greaves, J.H. Koolen, Graphs with three eigenvalues and second largest eigenvalue at most 1, *J. Combin. Theory B*, **129** (2018), 55–78.

[67] C.J. Colbourn, J.H. Dinitz (Eds.), *Handbook of Combinatorial Designs*, Chapman and Hall/CRC, Boca Raton, 2007.

[68] D. Cui, Y. Hou, On the skew spectra of Cartesian products of graphs, *Electron. J. Combin.*, **20** (2013), P#19.

[69] D.M. Cvetković, Chromatic number and the spectrum of a graph, *Publ. Inst. Math. (Beograd)*, **14**(28) (1972), 25–38.

[70] D. Cvetković, On reconstruction of the characteristic polynomial of a graph, *Discrete Math.*, **212** (2000), 45–52.

[71] D. Cvetković, M. Doob, I. Gutman, On graphs whose spectral radius does not exceed $\sqrt{2+\sqrt{5}}$, *Ars Combin.*, **14** (1982), 225–239.

[72] D. Cvetković, M. Doob, I. Gutman, A. Torgašev, *Recent Results in the Theory of Graph Spectra*, North-Holland, Amsterdam, 1988.

[73] D. Cvetković, M. Doob, H. Sachs, *Spectra of Graphs – Theory and Application*, third edition, Johann Ambrosius Barth Verlag, Heidelberg-Leipzig, 1995.

[74] D. Cvetković, M. Lepović, Seeking counterexamples to the reconstruction conjecture for the characteristic polynomial of graphs and a positive result, *Bull. Acad. Serbe Sci. Arts, Cl. Sci. Math. Natur., Sci. Math.*, **116** (1998), 91–100.

[75] D. Cvetković, P. Rowlinson, S. Simić, *An Introduction to the Theory of Graph Spectra*, Cambridge University Press, Cambridge, 2010.

[76] D. Cvetković, P. Rowlinson, S. Simić, *Eigenspaces of Graphs*, Cambridge University Press, Cambridge, 1997.

[77] D. Cvetković, P. Rowlinson, S. Simić, *Spectral Generalizations of Line Graphs – On Line Graphs with Least Eigenvalue* -2, Cambridge University Press, Cambridge, 2004.

[78] E.R. van Dam, Graphs with few eigenvalues: An interplay between combinatorics and algebra, Doctoral thesis, Tilburg University, 1996.

[79] E.R. van Dam, Nonregular graphs with three eigenvalues, *J. Combin. Theory B*, **73** (1998), 101–118.

[80] E.R. van Dam, Regular graphs with four eigenvalues, *Linear Algebra Appl.*, **226–228** (1995), 139–162.
[81] E.R. van Dam, E. Spence, Small regular graphs with four eigenvalues, *Discrete Math.*, **189** (1998), 233–257.
[82] K.C. Das, An improved upper bound for Laplacian graph eigenvalues, *Linear Algebra Appl.*, **368** (2003), 269–278.
[83] E. Dedò, Sulla ricostruibilità del polinomio caratteristico del commutato di un grafo, *Boll. Unione Mat. Ital.*, **18** (1981), 423–429.
[84] S. Fallat, Y.-Z. Fan, Bipartiteness and the least eigenvalue of signless Laplacian of graphs, *Linear Algebra Appl.*, **436** (2012), 3254–3267.
[85] L. Feng, Q. Li, X.-D. Zhang, Some sharp upper bounds on the spectral radius of graphs. *Taiwanese J. Math.*, **11** (2007), 989–997.
[86] R. Fernandes, Location of the eigenvalues of weighted graphs with a cut edge, *Linear Multilinear Algebra*, **58** (2010), 305–322.
[87] M. Fiedler, A property of eigenvectors of nonnegative symmetric matrices and its application to graph theory, *Czechoslovak Math. J.*, **25** (1975), 619–633.
[88] C.M da Fonesça, Interlacing properties for Hermitian matrices whose graph is a given tree, *SIAM J. Matrix Anal. Appl.*, **27** (2005), 130–141.
[89] S. Friedland, The maximal eigenvalue of 0–1 matrices with prescribed number of ones, *Linear Algebra Appl.*, **69** (1985), 33–69.
[90] K.R. Garren, Bounds for the eigenvalues of a matrix, Dissertations, Theses, and Masters Projects, William & Mary, (1965), 1539624585.
[91] K.A. Germina, S. Hameed, T. Zaslavsky, On products and line graphs of signed graphs, their eigenvalues and energy, *Linear Algebra Appl.*, **435** (2011), 2432–2450.
[92] E. Ghasemian, G.H. Fath-Tabar, On signed graphs with two distinct eigenvalues, *Filomat*, **31** (2017), 6393–6400.
[93] E. Ghorbani, W.H. Haemers, H.R. Maimani, L. Parsaei Majd, On sign-symmetric signed graphs, *Ars Math. Contemp.*, **19** (2020), 83–93.
[94] E. Ghorbani, A. Majidi, Complete signed graphs with largest maximum or smallest minimum eigenvalue, *Discrete Math.*, **347** (2024), 113860.
[95] E. Ghorbani, A. Majidi, Signed graphs with maximal index, *Discrete Math.*, **344** (2021), 112463.
[96] C.D. Godsil, I. Gutman, On the theory the matching polynomial, *J. Graph Theory*, **5** (1981), 137–144.
[97] C.D. Godsil, B. McKay, Constructing cospectral graphs, *Aequationes Math.*, **25** (1982), 257–268.
[98] M.J.E. Golay, Notes on digital coding, *Proc. I.R.E.*, **37** (1949), 657.
[99] G. Greaves, J. Koolen, A. Munemasa, Y. Sano, T. Taniguchi, Edge-signed graphs with smallest eigenvalue greater than -2, *J. Combin. Theory B*, **11** (2015), 90–111.
[100] G.R.W. Greaves, Z. Stanić, Signed $(0,2)$-graphs with few eigenvalues and a symmetric spectrum, *J. Comb. Des.*, **30** (2022), 332–353.
[101] D.L. Greenwell, Reconstructing graphs, *Proc. Amer. Math. Soc.*, **30** (1971), 431–433.
[102] J.H. Griesmer, A bound for error-correcting codes, *IBM J. Res. Dev.*, **4** (1960), 532–542.

[103] J.-M. Guo, A new upper bound for the Laplacian spectral radius of graphs, *Linear Algebra Appl.*, **400** (2005), 61–66.

[104] K. Guo, B. Mohar, Hermitian adjacency matrix of digraphs and mixed graphs, *J. Graph Theory*, **85** (2017), 217–248.

[105] I. Gutman, D.M. Cvetković, The reconstruction problem for characteristic polynomials of graphs, *Univ. Beograd, Publ. Elektrotehn. Fak. Ser. Mat. Fiz.*, **498–451** (1975), 45–48.

[106] W.H. Haemers, L. Parsaei Majd, Spectral symmetry in conference matrices, *Des. Codes Cryptogr.*, **90** (2022), 1983–1990.

[107] W.H. Haemers, H. Topcu, The signed graphs with two eigenvalues unequal to ± 1, *Appl. Math. Comput.*, **463** (2024), 128348.

[108] S. Hameed, Biju K., Z. Stanić, Spectral measures of balance for signed graphs, *Australasian J. Combin.*, **93** (2025), 198–210.

[109] S. Hameed K, K.A. Germina, Balance in gain graphs – A spectral analysis, *Linear Algebra Appl.*, **436** (2012), 1114–1121.

[110] S.K. Hameed, T.V. Shijin, P. Soorya, K.A. Germina, T. Zaslavsky, Signed distance in signed graphs, *Linear Algebra Appl.*, **608** (2021), 236–247.

[111] P.L. Hammer, A.K. Kelmans, Laplacian spectra and spanning trees of threshold graphs, *Discrete Appl. Math.*, **65** (1996), 255–273.

[112] M. Harada, A. Munemasa, On the classification of weighing matrices and self-orthogonal codes, *J. Combin. Des.*, **20** (2012), 45–57.

[113] F. Harary, On the notion of balance of a signed graph, *Michigan Math. J.*, **2** (1953), 143–146.

[114] A.J. Hoffman, On eigenvalues and colorings of graphs. In B. Harris (Ed.), *Graph Theory and Its Applications*, Academic Press, New York, 1970, pp. 79–91.

[115] A.J. Hoffman, On limit points of spectral radii of non-negative symmetric integral matrices. In Y. Alavi et al. (Eds.), *Graph Theory and Applications*, Lecture Notes Math., vol. 303, Springer-Verlag, Berlin, 1972, pp. 165–172.

[116] Y. Hong, A bound on the spectral radius of graphs, *Linear Algebra Appl.*, **108** (1988), 135–139.

[117] Y. Hong, Bounds of eigenvalues of graphs, *Discrete Math.*, **123** (1993), 65–74.

[118] Y. Hong, J.-L. Shu, K. Fang, A sharp upper bound of the spectral radius of graphs. *J. Combin. Theory B*, **81** (2001), 177–183.

[119] Y.P. Hou, Bounds for the least Laplacian eigenvalue of a signed graph, *Acta Math. Sin. – Engl. Series*, **21** (2005), 955–960.

[120] Y. Hou, J. Li, Y. Pan, On the Laplacian eigenvalues of signed graphs, *Linear Multilinear Algebra*, **51** (2003), 21–30.

[121] Y. Hou, Z. Tang, D. Wang, On signed graphs with just two distinct adjacency eigenvalues, *Discrete Math.*, **342** (2019), 111615.

[122] H. Huang, Induced subgraphs of hypercubes and a proof of the Sensitivity Conjecture, *Ann. Math.*, **190** (2019), 949–955.

[123] M.A. Iranmanesh, M. Saheli, Toward a Laplacian spectral determination of signed ∞-graphs, *Filomat*, **32** (2018), 2283–2294.

[124] C.R. Johnson, A. Leal-Duarte, On the minimum number of distinct eigenvalues for a symmetric matrix whose graph is a given tree, *Math. Inequal. Appl.*, **5** (2002), 175–180.

[125] C.R. Johnson, A. Leal-Duarte, On the possible multiplicities of the eigenvalues of a Hermitian matrix whose graph is a tree, *Linear Algebra Appl.*, **348** (2002), 7–21.

[126] I. Jovanović, Z. Stanić, Spectral distances of graphs, *Linear Algebra Appl.*, **436** (2012), 1425–1435.

[127] I. Jovanović, Z. Stanić, Spectral distances of graphs based on their different matrix representations, *Filomat*, **28** (2014), 723–734.

[128] D. Kalita, Spectra of weighted directed graphs, Doctoral thesis, IIT Guwahati, 2012.

[129] D. Kalita, S. Pati, A reciprocal eigenvalue property for unicyclic weighted directed graphs with weights from $\{\pm 1, \pm \mathrm{i}\}$, *Linear Algebra Appl.*, **449** (2014), 417–434.

[130] D. Kalita, S. Pati, On the spectrum of 3-colored digraphs, *Linear Multilinear Algebra*, **60** (2012), 743–756.

[131] M.R. Kannan, S. Pragada, Signed spectral Turán type theorems, *Linear Algebra Appl.*, **663** (2023), 62–79.

[132] P. Kelly, On isometric transformations, Doctoral thesis, University of Wisconsin, 1942.

[133] S. Kirkland, Laplacian integral graphs with maximum degree 3, *Electron. J. Combin.*, **15** (2008), R120.

[134] T. Koledin, Z. Stanić, Connected signed graphs of fixed order, size, and number of negative edges with maximal index, *Linear Multilinar Algebra*, **65** (2017), 2187–2198.

[135] T. Koledin, Z. Stanić, Notes on Johnson and Hamming signed graphs, *Bull. Math. Soc. Sci. Math. Roumanie*, **62** (2019), 183–189.

[136] T. Koledin, Z Stanić, On a class of strongly regular signed graphs, *Publ. Math. Debrecen*, **97** (2020), 353–365.

[137] T. Koledin, Z. Stanić, Regular bipartite graphs with three distinct non-negative eigenvalues, *Linear Algebra Appl.*, **438** (2013), 3336–3349.

[138] D. König, *Theorie der endlichen und unendlichen Graphen*. Mathematik und ihre Anwendungen, Band 16. Akademische Verlagsgesellschaft, Leipzig, 1936. Reprinted by Chelsea, New York, 1950.

[139] P.W.H. Lemmens, J.J. Seidel, Equiangular lines, *J. Algebra*, **24** (1973), 494–512.

[140] M. Lepović, I. Gutman, No starlike trees are cospectral, *Discrete Math.*, **242** (2002), 291–295.

[141] S. Li, J. Wang, Yet more elementary proof of matrix-tree theorem for signed graphs, *Algebra Colloq.*, **30** (2023), 493–502.

[142] D. Li, M. Yan, J. Meng, Minimizing the least Laplacian eigenvalue of signed complete graphs, *Appl. Math. Comput.*, **484** (2025), 129002.

[143] H. Lin, B. Ning, B. Wu, Eigenvalues and triangles in graphs, *Comb. Probab. Comput.*, **30** (2021), 258–270.

[144] J.H. van Lint, *Introduction to Coding Theory*, third edition, Springer, Berlin, 1999.

[145] M. Liu, B. Liu, Some results on the Laplacian spectrum, *Comput. Math. Appl.*, **59** (2010), 3612–3616.

[146] L. Lu, J. Wang, Q. Huang, Complex unit gain graphs with exactly one positive eigenvalue, *Linear Algebra Appl.*, **608** (2021), 270–281.

[147] E. Máčajová, A. Raspaud, M. Škoviera, The chromatic number of a signed graph, *Electron. J. Combin.*, **23** (2016), #P1.14.

[148] S. Mallik, Matrix tree theorem for the net Laplacian matrix of a signed graph, *Linear Multilinear Algebra*, **72** (2024), 1138–1152.

[149] W. Mantel, Problem 28 (solution by H. Gouwentak, W. Mantel, J. Teixeira de Mattes, F. Schuh and W.A. Wythoff), *Wiskundige Opgaven*, **10** (1907), 60–61.

[150] F. Martin, Frustration and isoperimetric inequalities for signed graphs, *Discrete Math.*, **217** (2017), 276–285.

[151] A. Mathew, K.A. Germina, Mycielskian of signed graphs, https://arxiv.org/abs/2302.00946.

[152] B.D. McKay, Small graphs are reconstructible, *Australas. J. Combin.*, **15** (1997), 123–126.

[153] C. McLeman and E. McNicholas, Spectra of coronae, *Linear Algebra Appl.*, **435** (2011), 998–1007.

[154] J. McKee, C. Smyth, Integer symmetric matrices having all their eigenvalues in the interval $[-2, 2]$, *J. Algebra*, **317** (2007), 260–290.

[155] R. Mehatari, M.R. Kannan, A. Samanta, On the adjacency matrix of a complex unit gain graph, *Linear Multilinear Algebra*, **70** (2022), 1798–1813.

[156] L. Medina, O. Rojo, Eigenvalues of certain weighted graphs joined at their roots having cliques at some levels, *Linear Algebra Appl.*, **437** (2012), 878–898.

[157] R. Merris, Laplacian graph eigenvectors, *Linear Algebra Appl.*, **278** (1998), 221–236.

[158] B. Mohar, Eigenvalues, diameter, and mean distance in graphs, *Graphs Combin.*, **7** (1991), 53–64.

[159] T.S. Motzkin, E.G. Straus, Maxima for graphs and a new proof of a theorem of Turán, *Can. J. Math.*, **17** (1965), 533–540.

[160] R. Mulas, Z. Stanić, Star complements for ± 2 in signed graphs, *Spec. Matrices*, **10** (2022), 258–266.

[161] J. Mycielski, Sur le colouriage des graphes, *Colloq. Math.*, **3** (1955), 161–162.

[162] A. Neumaier, Graph representations, two-distance sets, and equiangular lines, *Linear Algebra Appl.*, **114–115** (1989), 141–156.

[163] V. Nikiforov, A spectral condition for odd cycles, *Linear Algebra Appl.*, **428** (2008), 1492–1498.

[164] V. Nikiforov, Some inequalities for the largest eigenvalue of a graph. *Comb. Probab. Comput.*, **11** (2002), 179–189.

[165] V. Nikiforov, The maximum spectral radius of C_4-free graphs of given order and size, *Linear Algebra Appl.*, **430** (2009), 2898–2905.

[166] V. Nikiforov, The spectral radius of graphs without paths and cycles of specified length, *Linear Algebra Appl.*, **432** (2010), 2243–2256.

[167] V. Nikiforov, Walks and the spectral radius of graphs, *Linear Algebra Appl.*, **418** (2006), 257–268.

[168] N. Nisan, M. Szegedy, On the degree of Boolean functions as real polynomials, *Comput. Complexity*, **4** (1992), 462–467.

[169] L. Parsaei-Majd, Z. Stanić, B. Tayfeh-Rezaie, On weight-symmetric 3-coloured digraphs, *Linear Multilinear Algebra*, **71** (2023), 2744–2762.

[170] S. Pirzada, M.R. Ul Rashid, Z. Stanić, Net Laplacian spectrum of products built on the simple corona of a signed graph, *Discrete Math. Algorithms Appl.*, **18** (2026), 2550019.

[171] F. Ramezani, P. Rowlinson, Z. Stanić, More on signed graphs with at most three eigenvalues, *Discuss. Math. Graph Theory*, **42** (2022), 1313-1331.

[172] F. Ramezani, P. Rowlinson, Z. Stanić, On eigenvalue multiplicity in signed graphs, *Discrete Math.*, **343** (2020), 111982.

[173] F. Ramezani, P. Rowlinson, Z. Stanić, Signed graphs with at most three eigenvalues, *Czechoslovak Math. J.*, **72** (2022), 59–77.

[174] F. Ramezani, Z. Stanić, An upper bound for the Laplacian index of a signed graph, *Discrete Math. Lett.*, **5** (2021), 24–28.

[175] F. Ramezani, Z. Stanić, Some upper bounds for the net Laplacian index of a signed graph, *B. Iran Math. Soc.*, **48** (2022), 243–250.

[176] N. Reff, Oriented gain graphs, line graphs and eigenvalues, *Linear Algebra Appl.*, **506** (2016), 316–328.

[177] N. Reff, Spectral properties of complex unit gain graphs, *Linear Algebra Appl.*, **436** (2012), 3165–3176.

[178] P. Rowlinson, On the maximal index of graphs with a prescribed number of edges, *Linear Algebra Appl.*, **110** (1988), 43–53.

[179] P. Rowlinson, The main eigenvalues of a graph: A survey, *Appl. Anal. Discrete Math.*, **1** (2007), 445–471.

[180] P. Rowlinson, Z. Stanić, Signed graphs whose spectrum is bounded by -2, *Appl. Math. Comput.*, **423** (2022), 126991.

[181] P. Rowlinson, Z Stanić, Signed graphs with three eigenvalues: Biregularity and beyond, *Linear Algebra Appl.*, **621** (2021), 272–295.

[182] A. Samanta, M. Rajesh Kannan, Gain distance matrices for complex unit gain graphs, *Discrete Math.*, **345** (2022), 112634.

[183] A.J. Schwenk, Spectral reconstruction problems, *Ann. N.Y. Acad. Sci.*, **328** (1979), 183–189.

[184] A.J. Schwenk, Computing the characteristic polynomial of a graph. In *Graphs and Combinatorics*, Lecture Notes Math., vol. 406, Springer, Berlin, 1974, pp. 153–172

[185] A.J. Schwenk, Exactly thirteen connected cubic graphs have integral spectra. In Y. Alavi, D. Lick (Eds.), *Theory and Applications of Graphs*, Springer, Berlin, 1978, pp. 516–533.

[186] I. Sciriha, Polynomial reconstruction and terminal vertices, *Linear Algebra Appl.*, **356** (2002), 145–156.

[187] I. Sciriha, M.J. Formosa, On polynomial reconstruction of disconnected graphs, *Utilitas Math.*, **64** (2003), 33–44.

[188] I. Sciriha, Z. Stanić, The polynomial reconstruction problem: The first 50 years, *Discrete Math.*, **346** (2023), 113349.

[189] J.J. Seidel, Graphs and two-graphs. In F. Hoffman et al. (Eds.), *Proceedings of the Fifth Southeastern Conference on Combinatorics, Graph Theory, and Computation*, Utilitas Mathematica Pub., Winnipeg, 1974, pp. 125–143.

[190] J.J. Seidel, Strongly regular graphs. In B. Bolobás (Ed.), *Surveys in Combinatorics*, Cambridge University Press, Cambridge, 1979, pp. 157–180.

[191] B. Shader, W. So, Skew spectra of oriented graphs, *Electron. J. Combin.*, **16** (2009), N#32.

[192] T. Shamsher, S. Pirzada, Z. Stanić, Net Laplacian eigenvalues of certain corona-like products of signed graphs, *Bol. Soc. Mat. Mex.*, **30** (2024), 80.

[193] S.K. Simić, A note on reconstructing the characteristic polynomial of a graph. In J. Nešetřil, M. Fiedler (Eds.), *Fourth Czechoslovakian Symposium on Combinatorics, Graphs and Complexity*, North-Holland, Amsterdam, 1992, pp. 315–320.

[194] S.K. Simić, E.M. Li Marzi, F. Belardo, Connected graphs of fixed order and size with maximal index: Structural considerations, *Le Matematiche*, **LIX** (2004), 349–365.

[195] S.K. Simić, Z. Stanić, Polynomial reconstruction of signed graphs, *Linear Algebr Appl.*, **501** (2016), 390–408.

[196] S.K. Simić, Z. Stanić, Polynomial reconstruction of signed graphs whose least eigenvalue is close to -2, *Electron J. Linear Algebra*, **31** (2016), 740–753.

[197] S.K. Simić, Z. Stanić, On the polynomial reconstruction of graphs whose vertex-deleted subgraphs have spectra bounded from below by -2, *Linear Algebra Appl.*, **428** (2008), 1865–1873.

[198] S.K. Simić, Z. Stanić, The polynomial reconstruction of unicyclic graphs is unique, *Linear Multilinear Algebra*, **55** (2007), 35–43.

[199] J.H. Smith, Some properties of the spectrum of a graph. In *Combinatorial Structures and their Applications*, Gordon and Breach, New York, 1970, pp. 403–406.

[200] H.R.K. Spinoza, On some problems in reconstruction, Doctoral thesis, University of Illinois, 2016.

[201] Z. Stanić, A construction of cospectral signed graphs, *Commun. Combin. Optim.*, https://doi.org/10.22049/cco.2024.30034.2284.

[202] Z. Stanić, A decomposition of signed graphs with two eigenvalues, *Filomat*, **34** (2020), 1949–1957.

[203] Z. Stanić, A note on the Krausz theorem and the Whitney theorem for mixed line graphs, *Czechoslovak Math. J.*, **76** (2026), 17–29.

[204] Z. Stanić, An upper bound for the largest singular value of extended mixed graphs, *Examples and Counterexamples*, **7** (2025), 100188.

[205] Z. Stanić, Bounding the largest eigenvalue of signed graphs, *Linear Algebra Appl.*, **573** (2019), 80–89.

[206] Z. Stanić, Connected non-complete signed graphs which have symmetric spectrum but are not sign-symmetric, *Examples and Counterexamples*, **1** (2021), 100007.

[207] Z. Stanić, Eigenspaces for -2 in signed line graphs, *Amer. J. Combin.*, **4** (2025), 42–51.

[208] Z. Stanić, Estimating distance between an eigenvalue of a signed graph and the spectrum of an induced subgraph, *Discrete Appl. Math.*, **340** (2023), 32–40.

[209] Z. Stanić, *Inequalities for Graph Eigenvalues*, Cambridge University Press, Cambridge, 2015.

[210] Z. Stanić, Integral regular net-balanced signed graphs with vertex degree at most four, *Ars Math. Contemp.*, **17** (2019), 103–114.

[211] Z. Stanić, Linear ternary codes of strongly regular signed graphs, *Discrete Math.*, **347** (2024), 113714.

[212] Z. Stanić, Lower bounds for the least Laplacian eigenvalue of unbalanced blocks, *Linear Algebra Appl.*, **584** (2020), 145–152.

[213] Z. Stanić, Main eigenvalues of real symmetric matrices with application to signed graphs, *Czechoslovak Math. J.*, **70** (2020), 1091–1102.

[214] Z. Stanić, Notes on exceptional signed graphs, *Ars Math. Contemp.*, **18** (2020), 105–115.

[215] Z. Stanić, Notes on the polynomial reconstruction of signed graphs, *B. Malays. Math. Sci. Soc.*, **45** (2022), 1301–1314.

[216] Z. Stanić, On cospectral oriented graphs and cospectral signed graphs, *Linear Multilinear Algebra*, **70** (2022), 3689–3701.

[217] Z. Stanić, On strongly regular signed graphs, *Discrete Appl. Math.*, **271** (2019), 184–190.

[218] Z. Stanić, On the spectrum of the net Laplacian matrix of a signed graph, *Bull. Math. Soc. Sci. Math. Roumanie*, **63(111)** (2020), 203–211.

[219] Z. Stanić, Oriented graphs whose skew spectral radius does not exceed 2, *Linear Algebra Appl.*, **603** (2020), 359–367.

[220] Z. Stanić, Perturbations in a signed graph and its index, *Discuss. Math. Graph Theory*, **38** (2018), 841–852.

[221] Z. Stanić, Recognizing signed line graphs with a single root, *Appl. Anal. Discrete Math.*, **19** (2025), 422–434.

[222] Z. Stanić, *Regular Graphs – A Spectral Approach*, De Gruyter, Berlin, 2017.

[223] Z. Stanić, Relations between the skew spectrum of an oriented graph and the spectrum of an associated signed graph, *Linear Algebra Appl.*, **676** (2023), 241–250.

[224] Z. Stanić, Signed graphs of small order, http://www.matf.bg.ac.rs/~zstanic/siggr.htm.

[225] Z. Stanić, Signed graphs with totally disconnected star complements, *Rev. Un. Mat. Argentina*, **62** (2021), 95–104.

[226] Z. Stanić, Signed graphs with two eigenvalues and vertex degree five, *Ars Math. Contemp.*, **22** (2022), P1.10.

[227] Z. Stanić, Some bounds for the largest eigenvalue of a signed graph, *Bull. Math. Soc. Sci. Math. Roumanie*, **62(110)** (2019), 183–189.

[228] Z. Stanić, Some properties of the eigenvalues of the net Laplacian matrix of a signed graph, *Discuss. Math. Graph Theory*, **42** (2022), 893–903.

[229] Z. Stanić, Some reconstructions in spectral graph theory and graphs with integral Q-spectrum, Doctoral thesis, University of Belgrade, 2007.

[230] Z. Stanić, Some relations between the skew spectrum of an oriented graph and the spectrum of certain closely associated signed graphs, *Rev. Un. Mat. Argentina*, **63** (2022), 41–50.

[231] Z. Stanić, Spectral deviations of graphs, *Spec. Matrices*, **13** (2025), 20240030.

[232] Z. Stanić, Spectra of signed graphs and related oriented graphs, *Ars Math. Contemp.*, **24** (2024), P3.10.

[233] Z. Stanić, Spectra of signed graphs with two eigenvalues, *Appl. Math. Comput.*, **364** (2020), 124627.

[234] Z. Stanić, Star complementary characterization of oriented graphs whose spectral radius does not exceed 2, *Notes Number Theory Discrete Math.*, **32** (2026), 120–132.

[235] Z. Stanić, Star complements in signed graphs with two symmetric eigenvalues, *Kuwait J. Sci.*, **49** (2022), 1–8.

[236] Z. Stanić, A. Vijayakumar, On spectral radius of signed graphs without negative even cycles, *Bull. Math. Soc. Sci. Math. Roumanie*, **64(112)** (2021), 89–95.

[237] D. Stevanović, Research problems from the Aveiro Workshop on graph spectra, *Linear Algebra Appl.*, **423** (2007), 172–181.

[238] D. Stevanović, N.M.M. de Abreu, M.A.A. de Freitas, R. Del-Vecchio, Walks and regular integral graphs, *Linear Algebra Appl.*, **423** (2007), 119–135.

[239] P. Turán, On an extremal problem in graph theory (in Hungarian), *Középiskolai Matematikai és Fizikai Lapok*, **48** (1941), 436–452.

[240] S.M. Ulam, *A Collection of Mathematical Problems*, Wiley, New York, 1960.

[241] M.R. Ul Rashid, S. Pirzada, T. Shamsher, Z. Stanić, Spectra of subdivisions of signed graphs, signed R-graphs and related products, *Rev. Un. Mat. Argentina*, https://doi.org/10.33044/revuma.4483.

[242] M.R. Ul Rashid, S. Pirzada, Z. Stanić, Spectra of the Mycielskian of a signed graph and related products, *Discrete Appl. Math.*, **370** (2025), 124–144.

[243] G.R. Vijayakumar, Signed graphs represented by D_∞, *European J. Combin.*, **8** (1987), 103–112.

[244] Y. Wang, Spectral Turán problem for $\mathscr{K}_5^-$-free signed graphs, *Linear Algebra Appl.*, **691** (2024), 96–108.

[245] D.J. Wang, W.K. Dong, Y.P. Hou, D.Q. Li, On signed graphs whose spectral radius does not exceed $\sqrt{2+\sqrt{5}}$, *Discrete Math.*, **346** (2023), 113358.

[246] D. Wang, Y. Hou, Integral signed subcubic graphs, *Linear Algebra Appl.*, **593** (2020), 29–44.

[247] D. Wang, Y. Hou, Laplacian integral signed subcubic graphs, *Electron. J. Linear Algebra*, **37** (2021), 163–176.

[248] D. Wang, Y. Hou, D. Li, Extremal results for C_3^--free signed graphs, *Linear Algebra Appl.*, **681** (2024), 47–65.

[249] Y. Wang, D. Li, H. Lin, The signed graphs with all but at most three eigenvalues equal to -1, *Linear Algebra Appl.*, **664** (2023), 314–323.

[250] Y. Wang, H. Lin, The largest eigenvalue of C_k^--free signed graphs, *Discrete Appl. Math.*, **372** (2025), 164–172.

[251] D. Wang, K. Peng, Y. Hou, On the maximal index of unbalanced signed graphs with a prescribed number of edges, *Discrete Math.*, **349** (2026), 114871.

[252] W. Wang, L. Qiu, Y. Hu, Cospectral graphs, GM-switching and regular rational orthogonal matrices of level p, *Linear Algebra Appl.*, **563** (2019), 154–177.

[253] J. Wang, J. Wang, M. Brunetti, F. Belardo, L. Wang, Developments on the Hoffman program of graphs, *Adv. Appl. Math.*, **169** (2025), 102915.

[254] W. Wang, Z. Yan, Z. Qian, Eigenvalues and chromatic number of a signed graph, *Linear Algebra Appl.*, **619** (2021), 137–145.

[255] H.S. Wilf, Spectral bounds for the clique and independence numbers of graphs, *J. Combin. Theory B*, **40** (1986), 113–117.

[256] P. Wissing, Spectral characterizations of complex unit gain graphs, Doctoral thesis, Tilburg University, 2022.

[257] P. Wissing, E.R. van Dam, Unit gain graphs with two distinct eigenvalues and systems of lines in complex space, *Discrete Math.*, **345** (2022), 112827.

[258] T. Zaslavsky, A mathematical bibliography of signed and gain graphs and allied areas, *Combinatorics of Electron J. Combin. Dynamic Surveys*, (2018), #DS8.

[259] T. Zaslavsky, Biased graphs. I. Bias, balance, and gains, *J. Combin. Theory B*, **47** (1989), 32–52.

[260] T. Zaslavsky, Matrices in the theory of signed simple graphs. In B.D. Acharya, G.O.H. Katona, J. Nešetřil (Eds.), *Advances in Discrete Mathematics and Applications: Mysore, 2008*, Ramanujan Mathematical Society, Mysore, 2010, pp. 207–229.

[261] T. Zaslavsky, Negative (and positive) circles in signed graphs: A problem collection, *AKCE Int. J. Graphs Comb.*, **15** (2018), 31–48.

[262] T. Zaslavsky, Signed graph coloring, *Discrete Math.*, **39** (1982), 215–228.

[263] T. Zaslavsky, Signed graphs, *Discrete Appl. Math.*, **4** (1982), 47–74.

[264] T. Zaslavsky, Six signed Petersen graphs and their automorphisms, *Discrete Math.*, **312** (2012), 1558–1583.

[265] K. Zarankiewicz, Problem P 101, *Colloq. Math.*, **2** (1951), 301.

[266] M. Zhai, B. Wang, Proof of a conjecture on the spectral radius of C_4-free graphs, *Linear Algebra Appl.*, **437** (2012), 1641–1647.

Index

algebra
  Bose–Mesner, 70
angle
  main, 23
arc, *see* edge, oriented
association scheme
  adjacency matrix, 70
  eigenvalue, 70
  Hamming, 70
  intersection numbers, 70
  Johnson, 70
  rectangular, 71
  symmetric d-class, 69
BIBD
  complementary, 66
  incidence graph, 66
block design
  symmetric balanced incomplete (symmetric BIBD), 66
bound
  Anderson–Morley, 43
  Cvetković, *see* bound, Wilf
  Hoffman, 48
  Merris, 43
  relative (for the number of lines), 190
  Schwenk, 43
  Stanley, 43
  Wilf, 48
class
  switching equivalence, 5
  switching isomorphism, 5
clique number
  balanced, 54
code
  dimension, 183
  distance, 184
  dual, 184
  generator, 184
  Golay ternary, 188
  linear q-ary, 183
  optimal, 189
  self-dual, 184
  self-orthogonal, 184
  size, 184
codes
  equivalent, 184
codeword, 184
  distance, 184
  weight, 184
cograph, 28
  ∇-, 155
  signed, 155
conjecture
  Ulam, 169
connectivity
  vertex, 51
corona
  negative, 33
  positive, 33
counterexample pair (for the polynomial reconstruction), 172
criterion
  Acharya's spectral, 19
  Harary's balance, *see* theorem, Harary balance
  Stanić's spectral, 20
  Zaslavsky's spectral, 20
cube
  r-dimensional, 3
cycle
  even and oriented uniformly, 200

negative, 3
positive, 3
deck
polynomial, 170
degree
negative, 2
positive, 2
design
Hadamard, 69
Menon, 69
digon, 112
negative, 112
distance (between lines), 189
dot notation, 2
edge
negative, 1
orientation, 191
oriented, 191
positive, 1
edges
agree, 212
disagree, 212
eigenvalue
main, 23
strict interlacing, 173
equality
Motzkin–Straus, 54
equation
Laplacian eigenvalue, 12
net Laplacian eigenvalue, 14
figure
basic, 9
elementary, 9
formula
Dedò, 13
Sachs, 9
Schwenk-like, 9
frustration
algebraic, 17
index, 4
number, 4
function
gain, 209
potential, 210
switching, 210
graph, 2
13-Paley, 189
chain, 51
extended mixed, 211
friendship, 55
gain, 209
balanced, 210
switching equivalent, 210
$\mathbb{T}$-, 210
matching, 9
mixed, 211
of a Hermitian matrix, 208
order, 1
size, 1
switching
generalized GM-, *see* graph switching, WQH-
GM-, 157
WQH-, 159
threshold, 28
underlying, 1
inequality
Wilf, 54
join
negative, 16
positive, 16
line systems, 116
lines
extremal, 190
matrix
adjacency, 7
adjacency (of a gain graph), 209
adjacency (of an extended mixed graph), 212
coronal, 33
divisor, 26
incidence, 11
Laplacian, 11
Laplacian (of a gain graph), 209
net Laplacian, 13
parity-check, 184
skew adjacency, 191
spectral decomposition, 22
switching, 8
neighbourhood
negative, 2
positive, 2
net degree, 2
orientation
associated, *see* signature, associated
canonical, 200
vertex-edge, 11
oriented graph, 191
bipartite double, 195
maximal, 201

skew eigenvalues, 191
skew spectrum, 191
switching equivalent, 194

partition
equitable, 25
plane
Fano, 69
polynomial
characteristic, 7
Hoffman, 61
Laplacian characteristic, 11
matching, 9
principle
edge, 16
problem
Brualdi–Solheid, 53
polynomial reconstruction
graphs, 169
signed graphs, 171
spectral Turán, 53
Turán, 53
product
Cartesian, 32
tensor, 32
program
Hoffman, 57
pseudoforest, 13

root systems, 116

sign function (signature), 1
signature
associated, 192
signed cube
r-dimensional, 3
signed doubled graph, 113
signed graph, 1
antibalanced, 4
balanced, 4
bipartite, 2
bipartite double, 33
block, 49
bundle, 178
chromatic number, 48
cone over, 25
cubic, 2
determinant, 28
determined by the spectrum, 156
eigenvalue, 7
eigenvector, 7
exceptional, 116
foundation, 119
Hamiltonian, 50
Hamming, 73
homogeneous, 4
index, 7
infinity, 166
inhomogeneous, 4
integral, 150
Johnson, 73
k-colouring, 48
k-cyclic, 2
labelled, 6
Laplacian eigenvalue, 11
Laplacian eigenvector, 11
Laplacian spectrum, 11
lollipop, 166
maximal extension, 99
μ-extension, 99
Mycielskian, 35
negation, 4
net Laplacian eigenvalue, 13
net Laplacian eigenvector, 13
net Laplacian spectrum, 13
net-regular, 2
Petersen, 5
pseudo-potential, 30
quartic, 2
regular, 2
representation, 116
sign-symmetric, 37
simply, 113
simply dumbbell, 125
simply theta, 125
spectrum, 7
SSNSS, 38
strongly regular (SRSG), 64
subcubic, 154
switching
GM-, 159
WQH-, 161
TU-subgraph, 12
unbalanced, 4
very strongly regular, 65
walk-regular, 79
signed graphs
cospectral, 150
NEPS, 32
spectral deviation, 60
spectral distance, 58
switching equivalent, 4
switching isomorphic, 4
signed line graph

spectral, 111
spanning tree
negative, 14
positive, 14
spectral radii
limit point, 57
SRSG
parameters, 64
star complement, 95
line, 119
star partition, 97
star set, 95
line, 118
system
line, 189

technique
star complement, 98
theorem
Cauchy interlacing, 8
Harary balance, 4
Hoffman, 57
Krausz for mixed graphs, 216
Laplacian interlacing, 12
matrix-tree, 13
reconstruction, 96
threshold graph
∇-, 156
signed, 155
tree, *see* signed graph, k-cyclic

vertex
average 2-degree, 43
degree, 2
neighbourhood, 2
pendant, 10
vertices
partners, 172

walk, 2
negative, 3
neutral, 210
positive, 3

For EU product safety concerns, contact us at Calle de José Abascal, 56–1°, 28003 Madrid, Spain or eugpsr@cambridge.org.

www.ingramcontent.com/pod-product-compliance
Ingram Content Group UK Ltd.
Pitfield, Milton Keynes, MK11 3LW, UK
UKHW020200290926
15471UKWH00005B/979
* 9 7 8 1 0 0 9 8 5 3 3 4 7 *